NEC4: A User's Guide

Dedicated to my family of
Claire, Eloise and Callie-anne
all of whom I love more than words can express
(plus Sophie-Daisy, the cat, who I am rather fond of).

Institution of Civil Engineers Publishing

NEC4: A User's Guide

Dr Jon Broome
Managing Consultant
leading edge projects consulting ltd

Published by ICE Publishing, One Great George Street, Westminster, London SW1P 3AA

Full details of ICE Publishing representatives and distributors can be found at: www.icebookshop.com/bookshop_contact.asp

Other titles by ICE Publishing:
NEC3 and NEC4 Compared.
Robert Alan Gerrard. ISBN 978-0-7277-6201-6
Managing Reality, Third edition. Complete set.
Bronwyn Mitchell and Barry Trebes. ISBN 978-0-7277-6195-8
NEC4 Resolving and Avoiding Disputes
Robert Alan Gerrard and Patrick Waterhouse. ISBN 978-0-7277-6404-1

www.icebookshop.com
A catalogue record for this book is available from the British Library

ISBN 978-0-7277-6205-4

ICE Publishing is a division of Thomas Telford Ltd, a wholly-owned subsidiary of the Institution of Civil Engineers (ICE).

Commissioning Editor: Michael Fenton
Development Editor: Melanie Bell
Production Editor: Madhubanti Bhattacharyya
Marketing Specialist: April Nagy

Typeset by Academic + Technical, Bristol
Index created by Madelon Nanninga-Franssen
Printed and bound in Great Britain by Bell and Bain, Glasgow

Contents

Foreword

Dr Jon Broome must qualify as one of the leading experts on the NEC family of contracts. His expertise, experience and knowledge are yet again, on display in this book.

The use of NEC contracts in both the UK and abroad proceeds apace. Across the UK, public sector NEC contracts have become the contracts of choice. I was involved in the launch of NEC4 in 2017 but it is still early days as far as its adoption is concerned. *NEC4: A User's Guide* is, therefore, very timely.

The great value of this *Guide* lies in its readability and accessibility – it reads extremely well – and in its author's appreciation of the problems and issues facing NEC users. But it also goes further. It challenges a seemingly embedded mindset in the construction industry that is still unwilling to embrace the proactive and collaborative management processes underpinning NEC contracts. I often see this myself in the reluctance to ensure that the risk management mechanisms are installed all along the supply chain.

I have been very keen on the NEC producing very short and simple contracts which is small-business-user friendly. This could be used by tradespeople carrying out very minor works or professionals carrying out small assignments. I was, therefore, very pleased to read that the author was also in favour of this proposal.

This *Guide* is replete with tips and practical advice for both new and experienced NEC users. It is worth finding the time to read it through carefully. Without wishing to diminish or overlook any part of this excellent work, I cannot resist the temptation to highlight some aspects of the text which I believe to be particularly helpful to readers.

- It can be very difficult for NEC users to select the NEC contract that best suits their needs; the author provides invaluable help in respect of the issues likely to be raised by users when considering the most appropriate contract.
- There is an abundance of case studies and examples which illustrate the advice given by the author; I found this to be of great help.
- Many sections are prefaced by good practice tips (the author prefers the use of 'good' in preference to 'best' practice).
- Clear and on-going communication between the parties is an essential prerequisite in the successful management of NEC; the author, rightly, gives space to advising on effective communication.
- 'Z' clauses have generated their fair share of controversy, partly because their drafting often collides with the NEC drafting style; the author addresses this with some very helpful hints on drafting 'Z' clauses.
- A real bonus for me was the clarity in the author's consideration of the main payment options and how one goes about selecting the most appropriate one.

A word of caution. As the author has explained, this is not a text for those looking for a clause-by-clause analysis of NEC contracts. On the contrary, this text is about lifting the confidence of users by providing a very practical insight into the 'everyday' issues arising in the management of NEC contracts.

And finally, there was a passage in the book which struck a very strong chord with me:

> *Ultimately, it is not conditions of contract that build things, but organisations which consist of individuals. If the people involved are not of sufficient quality, in terms of technical, managerial and people skills or have incompatible values and attitudes, then the process of delivery will be adversely affected, which in turn will affect the achievement of objectives.*

When organisations are competing for contracts, this homily should be uppermost in their and their customers' minds.

Professor Rudi Klein
President, NEC Users' Group

About the author

Dr Jon Broome BEng PhD FAPM has been involved with the NEC since 1993, first as an academic, then as a contributor to the second and third editions of NEC and, for the past 20 + years, as an independent consultant. His consultancy experience, both in the UK and internationally, has included

- developing innovative contract strategies for billion-pound programmes and projects
- writing additional conditions of contract for such strategies
- reviewing and improving both additional conditions of contract clauses written by lawyers and technical documents written by engineers
- developing bid evaluation criteria and assessing these with clients
- helping contractors develop bid strategies and writing responses
- giving start-up workshops and training for project teams
- ongoing mentoring and support for both individuals and project teams
- dispute resolution services, including acting as an expert witness for differences of opinion running into hundreds of millions of pounds.

Jon is also a trustee director of the Association for Project Management, the chartered body for project professionals, and was chair of its Contracts and Procurement Special Interest Group (SIG) for more than ten years.

Acknowledgements

This book mirrors the transition between NEC3 and NEC4, in that it is an evolution of my book on the third edition of the NEC Engineering and Construction Contract. In contrast, the previous book was an almost complete rewrite – a revolution – of my first book, which was about the second edition of the contract. That original book was the result of 4 years of applied research, so the first people to thank are all the individuals and companies who contributed their time to the initial research many years ago. In particular, I still owe a huge debt to Professor John Perry, for his time and patience during the research, and for his efforts in 'moulding' me and my thought processes. Without him, I would not have had nearly as fortunate a life as I am having now. Ross Hayes, who is still active in the NEC community, also deserves a mention.

Twenty plus years on, the cornerstone of my business is still NEC consultancy and, to a lesser extent, training. So I would like to thank my clients – especially the repeat-order ones – not only for paying me, but also for sharing their good practice and what are politely described as 'areas for improvement' in how they operate the NEC. In addition, they are an excellent test bed for my ideas, both in critiquing them (as in 'good idea, but that won't work in practice') and putting them into practice and giving feedback. Especially worthy of mention is Connect Plus, for whom I am the Frameworks Advisor.

I would also like to thank my internal reviewers for their line-by-line critique of the near final version of the text, who offered their suggestions for improvements – ranging from simple grammatical changes and rewording of explanations for improved clarity, to extra references or additional points I should make – or just plain disagreed with me. They were

- Richard Patterson of Mott MacDonald
- Sarah Schütte of Schutte Consulting Ltd
- Steve Goodwin of GVE Commercial Solutions Ltd.

Many of their comments I took on board, some I did not, but all were gratefully received. I stress that this work is mine alone and does not necessarily represent their views or opinions.

Almost finally, I would like to thank – and apologise to – Michael Fenton, the commissioning editor of ICE Publishing, for his ongoing patience with me as deadline after deadline passed and his polite persistence in pressing me to finish.

Lastly, an enormous thank you to my wife, Claire, for her love, support and patience in putting up with me; and to my two daughters, Eloise and Callie-anne, for whom writing a book is time which could have been spent with them.

Reviews and endorsements

From high level principles to practical advice and guidance, to more detailed points of interpretation and application, this book brings NEC4 to life in a remarkably compelling and engaging way. It inspires and supports better project delivery for the benefit of all parties.

Sofia Athanassiou
Commercial Director and Executive Director,
M25 DBFO Co, Connect Plus, UK

The structure of NEC4: A User's Guide *is a real help to the reader, set out in the order of project process rather than clause by clause. It is not only a resource for dipping in to for a quick answer, but a book that will reward the reader who reads it cover to cover with fresh perspectives, and ultimately a deeper understanding of the subject.*

Marc Palmer
NEC Advisor and Practitioner,
Marc Palmer Project Management, Australia

Reading NEC4: User's Guide *brings NEC to life by sharing the best of what the industry has to offer. You can learn from it whether you are a seasoned 20 year veteran or a complete novice: it's a guide for all.*

So dive in, enjoy – it's very readable – and keep learning.

Mark Giltsoff FCIOB
Project Director, Balfour Beatty Construction Services, UK

I always recommend Jon Broome's books on NEC to anyone who is serious about the contracts. Why? Because Jon writes as a trainer speaks – in plain, sometimes colloquial, always very readable English.

This is the only NEC book I know that you can enjoy reading cover to cover!

Richard Patterson BA MBA CEng FICE
NEC and Procurement Specialist; Lead for NEC, Mott MacDonald

Jon Broome is clearly on a quest to spread good practice and take away the mystery that otherwise people may have about certain contractual aspects. The book very much embellishes the principle of 'every day is a school day' by creating a very useful, easy to read guide that can be read in one go or dipped in to as and when the reader needs to. He is not afraid to point out potential weaknesses in the contract and offer practical advice as to how to interpret and manage these aspects effectively.

This should be everyone's go-to guide for advice whether you are new to NEC contracts or a seasoned experienced professional.

Glenn Hide
Director, GMH Planning, UK

Whilst NEC4 contracts raise people's expectations for how contracts can be used as a tool to do good instead of harm, users often struggle to execute them as effectively or efficiently as the NEC would like. Jon Broome's book offers great insight into how to get the most out of NEC4 in terms of how to put together great contracts and then how to manage them in order to make sure they deliver. It plugs the gap between theory and practice.

Rather than accept mediocrity from your contracts, read this book and let it show you the art of the possible.

Neil Earnshaw
Procurement and NEC Contract Specialist;
Director, NE Consult, UK

The User Guide *outlines an approach to ensuring good management provisions are made within an NEC4 contract, with the added benefit of examples which assist the reader in understanding why it is beneficial or important to include specific provisions.*

Jillian Ferguson
Client Project Manager, Perth Transport Futures Project, Australia

NEC4: A User's Guide *builds on the reasons why* NEC3: A User's Guide *has been and remains so useful. It continues, as per the previous edition, of clearly identifying the various Contract Options that are available within the NEC family and how there are real benefits to be gained from a robust understanding of how to administer across the integrated suite. The book illustrates how the identification of 'stimuli' to good management and the clarification, from the outset, of clients' and contractors' objectives is key.*

Jon Broome refers to several research examples of where contracts failed to achieve their objectives due to participants not clearly understanding what the other party expected from the contract. This mutual understanding, alongside administering the NEC4 contract correctly, creates an environment to support good project management, develops shared objectives and fundamentally supports the relationship of the project teams.

This latest edition is invaluable in not only highlighting how behaviours and approaches support successful project delivery but also provides excellent technical guidance on the life cycle of project, these activities including approaches to tender appraisals, administration of change and dispute resolution.

Gareth Davies BSc (Hons) FRICS
Head of Capital Cost Management, Dwr Cymru Welsh Water, UK

Back in 2012 I thought that Jon Broome's NEC3 guide was the best NEC book I had seen; now this new NEC4 edition is certainly the best.

NEC4: A User's Guide *provides a practical how-to-do-it-right guide for both new and experienced NEC4 users on both sides of the contract, and is a great reference book to keep handy on the desk. It explains clearly throughout what to do, why to do it and how to best do it. It is written in a conversational tone, is engaging and very easy to read, and includes a generous scattering of anecdotes throughout the text which provide some entertaining lessons.*

If I was hiring a project manager new to NEC, I would buy them this book. I would tell them, 'take this book away for the next couple of days, read it all and absorb its advice, and you will know just about everything you need to know to set up and effectively manage an NEC4 contract'.

Warwick Fergusson
First Chair, Australasia NEC Users' Group, New Zealand

This book focuses on the key ingredients needed in the right place and at the right time to make the most of the NEC.

It covers the basic principles through to the more advanced aspects that even the most experienced of NEC users will find useful. The structure serves for both a quick reference guide or reading straight through, with the benefit that earlier chapters underpin the thinking behind later ones.

Samuel Bizzell MRICS MAPM
Head of Commercial, In-House Services and Estates, UK Parliament

Jon Broome brings humour and real-life to the study and practice of NEC contracts. As a lawyer, this book has provided me with a welcome change to the usual reading list. Jon's depth of experience shines through; it is well-researched and draws generously on a dedicated career. I recommend it to my fellow practitioners as a very practical and engaging read.

Sarah Schütte LLB ACIArb
Solicitor-Advocate; Director, Schutte Consulting Limited, UK

NEC4: A User's Guide *is a refreshing 'how to' book, which is written in a way that is easily readable. The book is full of valuable and constructive information on the practical application of NEC4 ECC. What is particularly good is that the book is structured in such a way that you can easily dip into the sections which deal with different stages and aspects of the contract and shares valuable insights into good practice. Definitely a book to add to your NEC body of knowledge.*

Barry Trebes
NEC Consultant and Facilitator, Trebes Consulting, UK; author of several books on NEC

There are a number of NEC books on the market, each claiming various levels of usefulness and practicality. In my opinion none come as close to being a useful guide for the end user as NEC4: A User's Guide. *Jon Broome combines his wide ranging NEC knowledge and extensive practical experience to produce a book that truly earns its title of 'user guide', explaining not only how to put together and manage an NEC4 ECC contract but also delving into the background of the contract and the rationale behind the clauses. This is an essential addition to the bookshelf of NEC users and practitioners.*

Steven Evans BSc(Hons) LLB(Hons) FCIArb FAMINZ(Arb) FFAVE(Master) FRICS FCIOB FCInstCES FICE PRI
NEC Tutor, Australia

ice Publishing

Broome, Jon
ISBN 978-0-7277-6205-4
https://doi.org/10.1680/necaug.62054.001

Introduction and overview

This book is a 'how to get it right' guide on the operation of the NEC4 Engineering and Construction Contract, and not a legalistic 'how to put the blame on the other party' book for when the contract has gone wrong. From feedback, the first and second editions of this book achieved what it originally set out to do: to help potential and actual users to understand how the contract fitted together, and to provide a multitude of practical tips on its effective use in an engaging conversational style. It was designed so that you can read it from cover to cover, read the relevant part when approaching a phase in the procurement/contract life cycle or dip into it when you have a particular issue.

The original book was the result of 4 years of applied research investigating the use of both the first and second editions of the NEC Engineering and Construction Contract, with less than a year's consultancy and training experience added in. A consistent comment from interviewees was that, while the Guidance Notes that accompanied the contract helped understanding, they tended to concentrate on individual clauses and did not provide a sufficient and succinct overview of the procedures, systems, skills and attitudes necessary for the effective operation of the contract. Similar comments are often made about the more legally based training courses and books that were available.

Since 2012, when the second edition of this book was published, the following have changed.

- The whole family of contracts has been updated to the fourth edition. While this included adding some new contracts, the new edition was publicised as 'evolution, not revolution', and the vast majority of the evolution has been for the good. There have since been two updates to NEC4, the last of which came out the week I was about to send the near completed manuscript out to my internal reviewers, at the end of October 2020.
- The NEC family of contracts is being used ever more widely, both across UK sectors and internationally. For instance, in the UK it has even greater penetration in the building sector and is, I believe, now totally dominant in the water sector, pretty much the standard for road projects and increasingly being used in the rail sector. Internationally, it is now the standard contract for government-funded contracts in Hong Kong, and I see write-ups of successful projects in more and more non-UK countries.
- Both project management generally and NEC accepted 'good practice'[1] have not only advanced, but also become more widespread. Having said that, there is still a lot of bad practice out there!

[1] I prefer the term 'good practice' to 'best practice' because: (a) practice is always advancing, (b) what might be best practice in one context might well not be in another, and (c) 'best practice' tends to be extolled by experts but is resource hungry to implement – the average 'do-er' simply does not have the time and resource to do it. Take PRINCE2 or the profession of risk management as examples.

- Personally, I have had nine more years of training and consultancy experience – now 23 years in total since 1998 – during which I have seen both what works and what does not and been able to test out and refine my ideas with my ongoing clients. They tell me what works and what does not.

  In particular, I have been let loose on the contract strategy front, whereby clients have finally got the idea that for their suppliers to behave in the way and achieve the things they want, they need to motivate them to behave in that way and achieve those things, rather than say one thing but motivate suppliers to do another. In addition, I have worked with a number of teams in a variety of ways to upgrade the more technical parts of the contract documents, such as the Works Information (under NEC3) or Scope (under NEC4). **Note:** throughout this book, underlining is used to highlight changes in the fourth edition relative to the third edition.

  Finally, I have increasingly worked with people and organisations outside of my home sectors of heavy engineering and construction. This has included with others in the Association for Project Management's Contracts and Procurement Specific Interest Group, or APM C&P SIG for short. This has resulted in the production of a guide (APM C&P SIG, 2017) and other assets, such as webinars[2] and a white paper (APM C&P SIG, 2020).

The guide is divided into three chapters.

Chapter 1 is titled **Introducing the NEC4 family of contracts and how they fit together**. This chapter aims to provide a succinct overview of the NEC4 family of contracts, with a focus on the Engineering and Construction Contract (ECC). As with all that is written in this book, much of what is said about the operation of the ECC will be relevant to the operation of all the members of the NEC4 family.

Chapter 1 comprises the following main sections.

1.1. **An overview of the NEC4 family of contracts**, which gives a brief history of the development of NEC4 and describes when to use each member of the family.

1.2. **How flexibility is achieved** (the first objective of the original authors of the NEC), which covers the main payment options, secondary options and other aspects of flexibility, with the aim of providing a succinct, but sufficient, overview of what they are and when you, the reader, would choose a particular option for your contract or subcontracts.

1.3. **How clarity and simplicity are achieved** (the second objective of the original authors), which explains the drafting philosophy behind the contract, to help potential users avoid the pitfalls that are often introduced by those who do not understand it. It also aims to enable the reader to put generic legal criticisms in context (i.e. have the counter-arguments).

[2] The best way to find these webinars is to go to YouTube and search for 'apm webinars contracts procurement'.

1.4. **The mechanisms for stimulating good project management and how they fit together** (stimulating good management being the NEC authors' third objective), which describes how, if used properly, the contract should lead to better project outcomes through stimulating good project management.

1.5. **A quick run through the NEC4 Engineering and Construction Contract**, which aims to give the reader a basic familiarity with the NEC4 ECC. While it focuses on the ECC, it does make some reference to other members of the NEC4 family.

Some lawyers and those with a traditional view of the purpose of a contract may wish to rush straight into the minutiae of each clause, but to do so would miss the whole point of the NEC. I think an explanation of 'the big picture' and how the NEC4 fits together – both as a family and as individual contracts – is necessary, as some participants in NEC4 contracts fail to understand how the contract can aid them in achieving their objectives and, as a consequence, fail to make the changes in culture, skills, systems and organisation (addressed in Chapter 2 and, particularly, Chapter 3) necessary in order to achieve those objectives.

Indeed, I would say the biggest cause of failure to get the best out of the use of the NEC is a failure to make the change in culture from that of using a contract to 'bash' the other party for their mistakes, yet hide behind it when it suits you, to one of using it as manual for good project management.[3] Two changes are necessary to get the best out of the NEC4.

- A change in individual mentality and/or corporate culture, which could be characterised as a change from a 'win–lose' mentality to a 'win–win' mentality.[4] This is often related to the next change.
- Acquiring the basic project management competencies, at individual and/or organisational level, to undertake the actions the contract requires. And when I say 'project management', I mean 'management' and not 'administration'. This relates to how project controls are typically operated: that is, once you have collected information, you do something with it to deliver a better result, rather than just collate it into report on where the contract is heading.

A comment I often receive from people on my training courses, who have some experience of the ECC, is 'Now I understand how it fits together and why I'm meant to be doing what I do'. Chapter 1 is not intended to replace the NEC4 Guidance Notes, which give the thinking behind each clause, but rather to complement them. In fact, for NEC4 the Guidance Notes were renamed as 'User Guides', but I shall continue to refer to them as 'Guidance Notes', to avoid confusion with this book.

Chapter 2 is titled **Good practice prior to entering the contract**.[5] One of the conclusions from my original research echoes what has been found in research on other forms of contract: namely, that

[3] Albeit, at times departing from what the contract says, by agreement, when common sense dictates. Having said this, the contract often allows a fair amount of discretion in 'how' you do things.

[4] In simple terms, a successful contract for a provider is one that makes it a profit and results in a happy customer. If the provider makes a profit by delivering the contract *efficiently* to the required quality, the contract will normally be delivered on time, and on a project-based contract any uplift to the contract Prices will be less if the change to scope is delivered efficiently. A happy customer wants its contract delivered on time, to the required quality and with minimal uplift to contract Prices. Where's the conflict? (I know I have simplified it, but hopefully you get the point!)

[5] See note 1 regarding the term 'best practice'.

a contract team can only do as well as the hand it is dealt. Choosing to use a member of the NEC4 family should improve that hand, especially if you know what you are doing. Conversely, if you do not know what you are doing....

Chapter 2 covers, in three main sections, the things that organisations can do to further improve their hand prior to award of the contract.

2.1. **Introduction** covers the principal issues relating to preparing and tendering an NEC contract.

2.2. **Preparing the contract** – which is done by the *Client*[6] (not the *Employer*, as the *Employer* does not now exist in any of the NEC4 contracts) – covers completing Contract Data part one (a sort of 'contract appendices' in old language), preparing the Scope (for the time being, approximately equivalent to the 'technical specification and drawings' in old language), preparing the Site Information (the conditions the *Contractor* should find when it arrives on Site), and various other issues that need to be thought through prior to putting documents out for tender.

2.3. **Preparing the tender** covers the different or new considerations for Contractors when tendering for an ECC contract, including: accuracy, the programme and the activity schedule, understanding the different Schedules of Cost Components (SCCs), and completing Contract Data part two, including what sort of programme should be submitted and what entries are required for the Early Warning Register.

2.4. **Evaluating the tender** identifies how Clients should go about evaluating returned bids from Contractors for NEC contracts, which might require some changes to the procedures used for traditional conditions of contract.

A finding of my original research, done years ago, was that, on successfully run contracts, the necessary changes in procedures, systems, skills and attitudes had started to be addressed in the pre-contract phases, and were substantially in place within the first month of being on site. This finding fits with much research on project management in general. In my consultancy experience, I have found that where contract teams have recovered from a poor start, it has almost always been done by going back and doing what they should have done in the first months, while at the same time managing a contract in full flow. For some teams, this has only been partially successful.

Chapter 3 is titled **Good practice following award and the crucial first month(s)**. This chapter takes potential users through what is necessary for the effective and efficient operation of the ECC,

[6] I had to scratch my head a fair few times in order to come up with a consistent use of the word 'client' – whether it should have an initial capital or not, and whether it should be italic or not. This is the protocol that I settled on.

- 'client' (no capital, no italics) is used for any client across the engineering, construction or any other industry, regardless of whether they are using NEC. I also use it when referring to my own company's clients.
- 'Client' (capital C, not in italics) is used for general comments about NEC Clients and is therefore typically in the plural.
- '*Client*' (capital C and in italics) is used in the specific context of an NEC contract, either before the contract is entered into or once it has been entered into. It is therefore generally in the single.

I have used a similar convention for contractors, Contractors and the *Contractor* and, when telling stories about the NEC3 edition, for Employers and the *Employer*.

based on observed good practice over the years. It does this in an approximately sequential order, through the following sections.

3.1. **Introduction** covers the principal issues that need to be addressed.

3.2. **Training and common understanding** covers what needs to be done to ensure that all participants, from whichever organisation they are drawn, are on the same page about what the contract says, its implications and how they are going to implement it.

3.3. **Communications: principles and systems** identifies eight principles for effective and efficient communication on NEC contracts which, when implemented, can save participants significant time through amicable management and administration of the contract. The section then looks at the use of cloud-based NEC-specific 'software as a service' information technology to aid this.

3.4. **Operating the payment clauses** first discusses the different payment processes (with and without the application of the two Construction Acts), then focuses on what needs to be in place for the efficient operation of the cost-based options C, D and E.

3.5. **The Accepted Programme** is a long section that first explores what good practice in programming seems to be, then gives a thorough explanation of programming terms in the ECC and their implications – for instance, ownership of different types of float. It then describes what a programme submitted for acceptance could look like and the specifics of getting it accepted promptly. The Accepted Programme has been described as the 'beating heart of the NEC', and programming is at the absolute core of good project management.

3.6. **Early warnings process and application** briefly reviews the fundamentals of risk and issues management and where it typically falls down. It then gives a framework for practical risk and issues management, both in terms of the lead into an NEC contract and its management during the contract. Whereas most of the chapter covers managing out downside risks (threats) and issues, the last section covers how to manage in opportunity, be it time and costs savings, or value enhancements to achieve a better end product for the same money.

3.7. **Agreeing compensation events quickly, amicably and in accordance with the contract** is drawn largely from my initial research, which found that, on some contracts, teams were easily able to agree compensation quotations within the timescales of the contract, in some cases quite substantial ones. This allowed participants to focus on managing and minimising the consequences, resulting in less additional time and cost, as well as much faster settlement of the final account. On the other hand, quite a few contract teams struggled to do this, and many still do today. This section gives contract participants logical frameworks/processes to follow, and practical tips for agreeing both multiple small compensation events and single large ones.

3.8. **Resolving issues and disputes** is the section that I do not want you to be reading, because it generally means communications and management of the contract are breaking down somewhere. The focus throughout the rest of this book, and that of the NEC family of contracts, is to pre-empt this situation. Nevertheless, this section presents the reader with some options for resolving a disagreement before it become a dispute, and briefly discusses tactics for winning a dispute if it does go to adjudication.

Chapter 4 presents a summary of the book and gives a few last thoughts on NEC4.

Perhaps the biggest difference between this book and its previous editions is that I frequently suggest specific enhancements to the ECC (in the form of option Z clauses), mostly to improve project management.

Finally, while this book is published by Thomas Telford Ltd, the publishers of the NEC family of contracts, this guide is my work, and the views and comments expressed are mine alone. As such, they do not necessarily represent the views of my reviewers. Neither do they necessarily represent the views of past members of the NEC Steering Group, ICE Publishing or, indeed, the Institution of Civil Engineers. It was a condition of mine in updating and rewriting this book that my independence remained, and I would like to thank these people and organisations for holding true to that promise.

REFERENCES

APM C&P SIG (2017) *APM Guide to Contracts and Procurement: For Project, Programme and Portfolio Managers*. Association for Project Management, Princes Risborough, UK.

APM C&P SIG (2020) *Procuring for Agile Projects*. Opinion Piece. Association for Project Management, Princes Risborough, UK. Available at: https://www.apm.org.uk/media/42324/procuring-for-agile-projects-white-paper.pdf (accessed March 2020).

ice Publishing

Broome, Jon
ISBN 978-0-7277-6205-4
https://doi.org/10.1680/necaug.62054.007

Chapter 1
Introducing the NEC4 family of contracts and how they fit together

The aim of this chapter is to provide a succinct overview of the NEC4 family of contracts, with a focus on the Engineering and Construction Contract (ECC). The 'how they fit together' in the chapter title applies both to the family of contracts and to the contents of each individual contract. As with all that is written in this book, much of what is said about the operation of the ECC will be relevant to the operation of the other members of the NEC4 family.

This chapter will be especially valuable to new users of NEC4 contracts, or those who just want to know what all the fuss is about. However, it will also be useful to those who

- have had some experience of NEC4 contracts, but have concentrated their efforts in a few limited areas and want to know how the whole fits together to be greater than the sum of its parts (a quality that gives the NEC4 a distinct advantage over other families of contracts, which have copied some of the features of NEC – while often denigrating it – but in piecemeal, stand-alone way)
- have not had an explanation of what NEC4 is trying to achieve (as well as explaining the 'what' and the 'how' at a high level, this chapter explains the 'why', which most adults, in order to change, need to understand)
- want an update on the key changes from NEC3 to NEC4 (I have highlighted these changes – be they new contracts, secondary options or definitions, or changes to a procedure – by underlining words, like this, so readers who feel that the previous bullets are not relevant to them easily identify the information they are looking for).

1.1. An overview of the NEC4 family of contracts
1.1.1 History and development
The history of the NEC ECC can be summarised as follows.

September 1985 The newly formed Legal Affairs Committee of the Institution of Civil Engineers (ICE) led 'a fundamental review of alternative contract strategies for civil engineering design and construction with the objective of identifying the needs for good practice'. From this review came a strong recommendation that it was time to look afresh at conditions of contract. While there were ten reasons to support this view, the most pertinent were

- the proliferation of standard forms of construction contract across the industry (the majority of projects are now multi-disciplinary, yet most contract forms are single discipline in concept, e.g. the JCT (Joint Contracts Tribunal) family for building, the ICC (Infrastructure Conditions of Contract, formerly ICE) suite for civil engineering and the Institution of Engineering and Technology's MF series for manufacturing)
- the fact that the origins of most existing forms of contract dated back to the 1870s, when modern principles of project management were unknown

- the perceived high incidence of disputes and the waste of resources involved in resolving them
- a perception that clients wanted greater certainty that their projects were to be delivered according to their time, cost and performance objectives.

1986 The ICE initiated the development of a new form of contract, with three main aims. These were stated positively as, compared with more traditional forms, greater

- flexibility of use (for different contract strategies, engineering and construction disciplines and in different countries across the world)
- clarity and simplicity
- stimulus to good management.

In reality, the last of these was stimulus to good *project* management: you can confirm this by looking both at the procedures of the contract (as I do in Section 1.4) and at the backgrounds of the two co-instigators.[1] I write on basis of this, rather than with reference to general management.

January 1991 A consultative version of the then New Engineering Contract was published. Written responses were sought and received from a wide range of organisations and individuals, and oral comment was obtained from seminars and conferences. In addition, five employer organisations tested the consultative version on real projects, with feedback being given to the Panel developing the contract.

March 1993 The first edition of the New Engineering Contract (NEC) was launched and a number of companies, at varying rates, increased the use of it on their projects. A 'back-to-back' Subcontract was also published.

1994 The first editions of the Professional Services Contract (PSC) and Adjudicator's Contract were published. These contracts had the same key objectives as the construction contract and were designed to be compatible with it. The PSC was originally designed to be used for design, project or construction management services, and site supervision, either when using the NEC as the construction contract or as a stand-alone contract. An adjudicator is a person brought in to rapidly and inexpensively (compared with arbitration or litigation) resolve a contractual dispute between the parties to a contract. Use of adjudication is embedded throughout the NEC family of contracts.

July 1994 Sir Michael Latham produced his report *Constructing the Team*, which stated that the NEC fulfilled most of his principles for best practice, in a modern form of construction contract that was substantially ahead of other forms. He did, however, include a number of recommendations for its further improvement.

November 1995 The second edition of the NEC – the NEC ECC – was published. The main changes from the first edition were

---

[1] For instance, Dr Martin Barnes began his career looking at scheduling when computers first started to enable PERT and critical path programming to be done electronically. He also co-founded and is a former Chair and President of what is now the Association for Project Management.

- the implementation of all of Sir Michael Latham's recommendations
- small refinements prompted by further comments and feedback from live contracts
- extensive changes to the insurance and adjudication procedures
- re-naming the contract the NEC Engineering and Construction Contract (ECC).

April and June 1998 The second editions of the PSC and Adjudicator's Contract were published. Publication was delayed until the details of the Housing Grants, Construction and Regeneration Act 1996 were finalised (which came into force on 1 May 1998). This Act means that any construction contract has to have certain provisions in it to do with payment and dispute resolution, e.g. have adjudication clauses that comply with the Act. An addendum to the ECC was also published, to change and add relevant clauses to the contract so that it would comply with the Act.

1999 The Engineering and Construction Short Contract was published. It was designed with the same principles in mind as the other members of the NEC family and intended for use on simple or low-risk works.

2001 The Engineering and Construction Short Subcontract was published, for use as a subcontract to either the Short Contract or the full ECC, but for low-risk, low-complexity subcontracts.

March 2005 The whole family was updated, and the name 'New Engineering Contract' was dropped and replaced by the new prefix 'NEC3'. Two new contracts were added to the family

- the Term Service Contract (TSC), which was widely taken up within the housing sector (roughly £1 billion worth of work was let under it within 9 months of its launch)
- the Framework Contract, which employs a contractor, consultant or supplier over a period of time, with individual packages of work, once sufficiently defined, being let under the relevant form of contract for works, services or goods.

September 2008 The Term Service Short Contract, for low-complexity, more reactive type services, was added to the family.

January 2010 The Supply Contract, for supply of goods specifically designed for a project, and the Supply Short Contract, for more standard goods, were launched. Note that both are for the delivery only of goods, and not for installation.

At this point, the publishers of the NEC claimed that the family of contracts covered the procurement of all goods, services and works. While it did indeed cover more than any other family of contracts that I was aware of, I personally thought there were still a few gaps and, it would seem – given that further contracts were subsequently added to the family, both before and when NEC4 came out – so did the NEC Panel.

April 2013 The whole NEC3 family was updated, but in a minor way: a bit like an iPhone being upgraded to an iPhone S. A Professional Services Short Contract was also added, for use on simpler consultancy assignments.

June 2017 The fourth edition of NEC was published, with the whole family of contracts being updated at the same time. All fourth edition contracts are prefixed with 'NEC4', and I adopt this convention from here on. The transition from NEC3 to NEC4 was described as 'evolution, not revolution', both in bringing the contract up to date with developments since 2005 and in pushing good practice. More specifically, the key objectives of this evolution were to (and I quote from the preface to the NEC4 contracts)

- inspire increased use of NEC in new markets and sectors (e.g. the term '*Employer*', which is only used in the construction industry, was changed to '*Client*'[2])
- support new approaches to procurement which improve contract management (e.g. new members were added to the family, such as the Design Build and Operate Contract, and new secondary options introduced, including for early contractor involvement (ECI) and information modelling – building information management/BIM in a construction context)
- provide greater stimulus to good project management (e.g. the 'Risk Register' was renamed as the 'Early Warning Register').

The Adjudicator's Contract was replaced by the Dispute Resolution Service Contract. In addition, the accompanying Guidance Notes were renamed 'User Guides', which causes confusion with this publication! As stated previously, I will continue to refer to them as 'Guidance Notes' to avoid confusion with this book.

June 2018 A consultation version of the NEC4 Alliance Contract, which was published with the other NEC4 contracts, was withdrawn and replaced by a fully authorised version.

January 2019 Some slight amendments were made to various members of the family, most of which were inconsequential. The exception to this was a change to ECC clause 63.5, where the original contract had a non-sensical last paragraph that promoted poor project management. The revised clause is quite good, so make sure you use the January 2019 amendments!

October 2020 (The week I sent this out to my internal reviewers!) Further amendments were made to various contracts, mostly corrections to grammar and minor tweaks in wording. The main revisions to the ECC were the deletion and replacement of the clauses in option X22 (for early *Contractor* involvement) and Y(UK)1 (for the Project Bank Account), and also some tweaking of option Y(UK)2 (the payment clauses for UK construction contracts) to avoid any potential conflict with the wording of the Construction Acts in the light of recent case law.

1.1.2 When to use each member of the NEC4 family

By its nature, this is a high-level overview of which contract to use in which circumstances. My intention is to enable users to focus in rapidly on the most appropriate contract for their circumstances.

Before we start, it is worthwhile briefly mentioning that, with the exception of the Framework Contract and the Dispute Resolution Service Contract, all members of the NEC4 family have the same basic structure of nine core sections, each having a similar order of clauses and, where possible, using similar or the same terminology. This means that whichever member of the family you are using

- training and education needs are greatly reduced, as once a person or team is familiar with one member of the family, they automatically have some familiarity with the others (as a trainer myself, this is bad for business!)

[2] As a result of this change, when I use the term *Employer*, I am referring to historical contracts let under the first, second or third editions. The term *Client* is used for all NEC4 contracts. The words 'client' and 'clients' (no capital initial or italics) are used to refer to general clients and likewise for 'contractor' and 'contractors'.

- you can use substantially the same administrative systems, albeit tweaked for timescales, terminology and clause numbering, because the contract processes are pretty much the same. Consequently, users are familiar with *how to* operate the contract.

This uniformity across contracts is obviously an advantage to organisations that have a portfolio of project types using different members of the NEC family. This is true for clients, but is especially so for contractors, who will have a multitude of subcontractors of various types and sizes. A contractor will therefore be much more back-to-back in terms of its rights and obligations (e.g. how they are paid, risk allocation, contract procedures etc.). This compares with, for example, the case of a multi-disciplinary project, where the main contract is let on the 'best fit' sector-specific form, but subcontracts are let on a multitude of different forms – such as the ICC form for the civil engineering element, the JCT for the building element, the MF contract for the heavy engineering supply element etc. – all of which have different terminologies and procedures, and where risk allocation is incompatible with the main contract. This creates significant risk.

The advantages to both the client and the contractor do not stop there either. The NEC4 family goes much further than traditional forms in allowing for the *management* of risk, as opposed to just its allocation and 'letting it happen and then picking up the pieces' through the administration of the contract. Consequently, the ability of all participants to proactively influence a project's outcome is enhanced by the back-to-back nature of the management provisions.

1.1.2.1 NEC4 Engineering and Construction Contract (ECC or Black Book)

The Black Book is both the original member of the family and the 'big beast' of the family, in terms of the complexity and risk profile of site-based works that can be let under it. It allows for any extent of *Contractor* design, including functional and performance specifications.

It has a *Project Manager*, who manages the contract on the behalf of the *Client* (as previously stated, the term '*Employer*' has disappeared from all NEC4 contracts) and does most of the contract administration. The exception is for the quality management and testing aspects, including identification of Defects, which is done by the *Supervisor*, also on behalf of the *Client*. As with most members of the NEC family, the ECC contains adjudication clauses. In concept, adjudication is a quicker, less expensive and less procedural form of dispute resolution than arbitration or litigation. However, it is an interim form of dispute resolution, so the Parties can still go on to arbitration or litigation. More will be said on this in the discussion on the Dispute Resolution Service Contract.

The ECC has six main payment options (A to F), which reflect different degrees of risk put on the *Contractor*, but equally give the *Client*, via the *Project Manager*, different degrees of control over the *works*: the more risk put on the *Contractor*, the less control the *Client* has, despite its desire to have the best of both worlds! Option A is effectively a fixed-price contract (maximum risk on the *Contractor*); option B transfers the risk of quantity variations to the *Client*; options E and F are cost-plus contracts (maximum risk on the *Client*, but also maximum flexibility), with options C and D being the target-cost options (essentially cost reimbursable in nature, but with a pain/gain formula for sharing risk). The ECC has numerous 'pick and mix' secondary options for the *Client* to choose from in order to fine tune the contract strategy to their particular project without having to delete and amend clauses. More will be said on what the options are and when to select them in Section 1.2.

NEC4 now has 'Scope', which (for those familiar with NEC3) replaces the term 'Works Information' in the Engineering and Construction Contracts, 'Service Information' in the Term Service Contracts and 'Goods Information' in the Supply Contract. The Scope specifies and describes the *works*, as well as stating any constraints on how the *Contractor* is to provide them (e.g. working hours, noise levels, ISO/British Standards to follow). It also has Site Information, which describes the Site and its surroundings (i.e. what the *Contractor* can expect to find when arriving on Site to undertake the *works*).

Consequently, when using the ECC in its unamended form, you need a dedicated Site for where the work will be done and on which will end result (e.g. a building) will be delivered. However, it does not take much imagination to use the ECC, with some amendments, to procure a large, complex, heavy engineering project that will be constructed in one place and then transported elsewhere (e.g. a ship or oil rig).

1.1.2.2 NEC4 Engineering and Construction Subcontract (ECS)

The ECS is the 'back-to-back' version of the Black Book (ECC). As such, it is suitable for reasonably complex subcontracts, typically when there will be a number of trades included in the package (and hence subsubcontracting), as well as subcontract design by specialists who will also do physical work. A typical example would be for the mechanical and electrical subcontract on a building project, where the *Subcontractor* will both be carrying out design based on performance specifications and subsubcontracting a lot of the installation.

The main differences from the Black Book are

- it loses main payment option F (management contract), as it would be unusual to subcontract a management contract
- in terms of actions allocated to a Party or role, the characters move down a contractual level, so the *Client*, *Project Manager* and *Supervisor* become the *Contractor*, the *Contractor* becomes the *Subcontractor*, and Subcontractors become Subsubcontractors
- the timescales for actions change, so that a communication can go along the contractual line and back, with all parties being compliant with their contracts
- slight changes to the liabilities and insurance provisions, to avoid the necessity for 'double' insurance (by the *Contractor* and *Subcontractor*) (in reality, as companies take out annual insurance, these are normally irrelevant).

1.1.2.3 NEC4 Engineering and Construction Short Contract (ECSC or Short Contract)

The Short Contract is for use in similar circumstances to the Black Book, but where the *works* are less complex and lower risk. Note that I do not say 'low value': this is for reasons illustrated by the following examples.

- London Underground, before it was Private Finance 'Initiatived' and back again, trialled the consultative version on three contracts. It apparently worked extremely well on a £2.5 million contract with two trades and repetitive work, but not so well on a £0.5 million contract with five trades and non-repetitive work. In the latter case, the Short Contract's management procedures were insufficient to handle the complexity of the work, and the participants apparently reverted to operating the contract more like the Black Book.

- One of my previous clients in Africa had an 800 hectare site – or 800 football fields – which it wished to put a security fence around. This was a large contract, but relatively simple and linear in what had to be done. Here, the use of the Black Book would have been overkill in terms of the management effort both Parties would have to have put in to be contractually compliant, given the nature of the *works*.

To make the contract simpler and shorter than the Black Book, the NEC Panel increased the emphasis on the clarity and simplicity objective, and reduced the emphasis on the other two objectives, stimulus to good management and flexibility, although it still does have features of both. For example

- while it does not have any payment options, you could still let a contract on a fixed-price basis, a bill-of-quantities basis or a time-charge (daily rate × time used) basis, or a combination of all three
- it has no *Project Manager* or *Supervisor*, with a reduced number of actions falling on the *Client*.

A commonly-asked question is: 'When should I use the Black Book and when the Short Contract? What's the dividing line?' There is no clear-cut line, but I would say the Short Contract should be chosen when there is minimal 'fit for purpose' design (by which I mean the work let under the contract will not impact on how well the asset performs, e.g. painting), and only a few different types of work going on at any time (i.e. where there is little subcontracting between trades, so you do not need the management procedures to manage the complexity and interdependencies that arise when there are many concurrent work types).

My personal view is that the Short Contract is underused, with the Black Book often being used when the Short Contract would do. I think this is partly due to a lack of education and knowledge, and partly because there is a gap in the contract family (see below).

1.1.2.4 A missing contract?

Based on my experience and observations, I believe there is a need for some form of in-between or intermediate contract, between the Black Book and Short Contract. On many occasions, there are provisions in the Black Book that Clients want to use but which are not found in the Short Contract.

The most commonly needed aspects are for simple *Contractor* design and sectional Completion. Let me expand.

- For *Contractor* design, the only references to this in the ECSC are in clause 20.2, which states that 'The *Contractor* does not start work which the *Contractor* has designed until the *Client* has accepted that the design complies with the Scope', and clause 14.1, which states that this acceptance does not transfer liability to the *Client*. There is nothing about incorporating any pre-contract design proposals by the *Contractor* into the contract, the priority of pre-contract design proposals relative to the *Client's* Scope, reasonable skill and care liability, insurance, retention of documents, etc., all of which are relevant when the *Contractor* is carrying out design work.

- For sectional Completion, you can specify in the ECSC (as constraints in the Scope) dates for when the *Contractor* has to complete particular areas, but there is no remedy in the contract for late Completion of an area.

The view of the NEC Steering Group and Users' Group seems to be that if you need these provisions, you should use the full Black Book. However, this would mean you will have all the other provisions to comply with as well. So if, as a domestic client, I – or more precisely my wife – wants an attic extension that is to be designed and built by a contractor, both the contractor and I would have to comply with the full programming and quality management provisions of the ECC, including assessing compensation events according to the full provisions of Section 6 of the contract and completing the Short Schedule of Cost Components. The vast majority of small builders and domestic clients, even if they hired professionals to project manage on their behalf, would be in breach of the contract, both due to a lack of understanding and/or competence regarding the ECC and because many of the ECC's provisions are absolutely irrelevant to a contract of this size and complexity. And if they were to comply, I would have to foot the bill for all this unnecessary management overhead.

So what's the solution? Ideally, the publishers of the NEC would produce an 'Intermediate Form', but I may have a go in the meantime!

1.1.2.5 NEC4 Engineering and Construction Short Subcontract (ECSS or Short Subcontract)

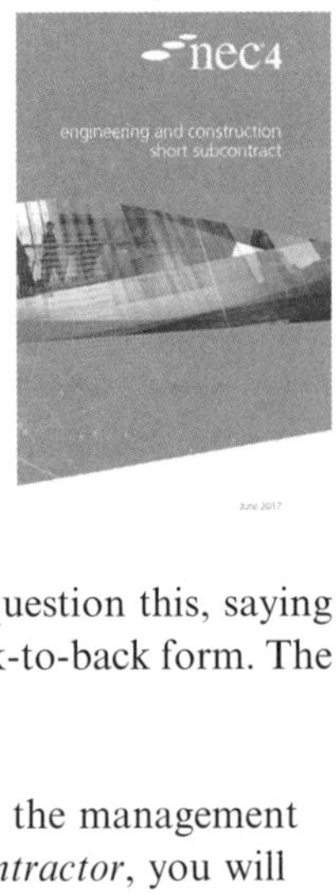

The Short Subcontract is the subcontract version of the Engineering and Construction Short Contract. It can be used as a subcontract to both the Short Contract and the Black Book, although main contractors operating at the main contract level under the Black Book may wish to tweak some of the timescales.

Again the question often arises, but this time from Contractors, 'When should I use the full Subcontract (ECS) (or full back-to-back conditions) and when should I use the Short Subcontract if my main contract is under the Black Book?' The answer is pretty much the same as for the choice between the Black Book and the Short Contract. Lawyers, however, might question this, saying that, in order to protect yourself fully in a legal sense, you need to use the back-to-back form. The counter-arguments to this are as follows.

- If the *Subcontractor* is going to operate the ECS as written, even though the management procedures are not needed given the nature of the work, then, as the *Contractor*, you will be paying for unnecessary management overhead.
- If the *Subcontractor* (because it does not have the management capability) and *your own people* (because they do not have the time and it is not value-adding work) are not going to operate the ECS as written, then why let work on a contract that you know both Parties are going to be in breach of?

Over more than 20 years as a contracts consultant, I have found that the most frequently occurring legal risk is that people do not operate the contract as written. Using the full ECS, when the ECSS will do, positively, but inadvertently, encourages breaches of contract.

1.1.2.6 Another missing contract or two?

My personal view is that the NEC family could do with a 'Really Short Contract', for very simple work involving only one or two trades. For instance, when my wife and I bought our 1930s house, we let three contracts for improvement work, each of a value around £10 000. The Short Contract was suitable. However, once we had settled in, we wanted a single workman in to do 4 days' worth of painting and simple carpentry. The Short Contract is a bit over the top for work like this. Likewise, a 'Really Short Subcontract' might be useful for a simple trade subcontractor.

1.1.2.7 NEC4 Professional Service Contract (PSC)

The PSC is, unsurprisingly, for the provision of a professional service, regardless of whether it is construction based. For at least one *Client*, this includes its legal services, which was a bit of a shock to the lawyers, as the PSC includes requirements for them to programme out their work and communicate in set timescales and, as it is written in plain English, they had trouble understanding it!

What Clients like about the contract is that, in the past, professional service providers – the *Consultant* in the contract – have hidden behind their 'professionalism' or, in reality, lack of. The project management procedures within the PSC put greater rigour on the professional service provider to do what they say they will do by the stated times, and places a requirement on them to project manage their work in order to do this. Consequently, it is suitable for large consultancy assignments where there are multiple deliverables, perhaps being delivered by multiple sub-teams and hence with multiple interdependencies, both in technical decisions and timings. The quid pro quo for this is that the *Client* needs to define more rigorously in the Scope document what it is they want the *Consultant* to deliver, and must keep to its commitment, otherwise compensation events happen (i.e. it is more evident who is not performing).

Like the Black Book, the PSC has a number of main payment options, as well as numerous secondary options. The three main payment options are: a fixed-priced contract, for when the services are well defined; a time-charge-based contract, for when they are not; and a target-cost contract, where the consultant is reimbursed on a time-charge basis, but with a pain/gain formula on top, which implies there are some unknowns or risks in how the desired output (as expressed in the Scope) is to be achieved.

A change from the third edition has been the dropping of main option G (term contract). Under this option, the contract ran for a defined period of time, during which Tasks could be called off as necessary. Work would be reimbursed either on a lump-sum priced basis (if predefined) or time-charge basis (if not).

An argument for doing away with this option was that the provisions of the contract were inadequate for more complex Tasks. However, many long-term consultancy relationships, and hence many PSC contracts, are based on 'business-as-usual' type activities, such as regular monitoring of an asset, which, depending on what is found, can lead to a series of Tasks. For example, if the contract is to carry out annual inspections of a bridge, Tasks could be to: conduct a more detailed/intrusive inspection; prepare an options paper to identify and evaluate options for remediation or repair; prepare detailed design for the remediation or repair; run a competition to select the contractor to do the physical work; and then manage and administrate the construction contract.

Without having the option to instruct Tasks, new contracts would need to be signed for each of these activities, which is administratively burdensome for everybody. For this reason, for a number of Clients I have cut, pasted and tweaked the NEC3 PSC option G term contract clauses into the NEC4 PSC, while at the same time strengthening the management procedures from the NEC3 contract.

Another change from the third edition PSC is the addition of the *Service Manager* role. In the NEC3 PSC, the *Employer* undertook all of the actions, although the *Employer* could specify an '*Employer's* agent' through the use of secondary option X10 and an entry in the Contract Data. I view the greater emphasis on a single person taking responsibility for the management and administration of the contract, as opposed to potentially many individuals in a Client organisation, as a positive change.

1.1.2.8 NEC4 Professional Service Subcontract (PSS)

The PSS is a new contract for NEC4. The most common use of the PSS would be as a subcontract to the ECC on a design and build contract, where the design consultant may be responsible for multiple deliverables that are to be prepared by a number of interdependent teams and which have to be delivered to a timescale driven by the *Contractor's* programme. Programming is therefore very important. Having a standard document for use in such situations, rather than requiring Contractors to amend the full PSC (and each doing it differently) is, in my view, a good thing.

Additionally, and fairly obviously, the PSS can also be used as a subcontract to the full NEC4 PSC.

1.1.2.9 NEC4 Professional Service Short Contract (PSSC)

The NEC4 Professional Services Short Contract (PSSC) is an update of the NEC3 version which was published in April 2013. NEC3 was jointly developed with the Association for Project Management's Contracts & Procurement SIG, led by myself and other members of the SIG.

Like most short contracts in the NEC family, it was designed to be used for lower risk, less-complex assignments, for which the full contract would be excessive. As such, it promotes the clarity and simplicity objective at the expense of flexibility and stimulus to good management.

In the professional services context, a typical assignment might involve only one person or a small team and relatively few deliverables. The *Consultant* can be paid on a time-charge basis (plus identified expenses) or activity-schedule basis (i.e. on completion of defined deliverables). Alternatively, the contract can be run as a term service contract, whereby tasks can be instructed as compensation events (unlike under the full NEC4 PSC). Tasks that are defined before the contract commences are paid on the basis of a tendered schedule of rates – the 'Price List' in NEC language. Other tasks are paid on the basis of tendered *people rates*, whereby

- if the task could be defined and priced up before being instructed, then the *Consultant* is paid on the basis of the estimated time multiplied by the tendered *people rates*
- if the task is too vaguely defined or subject to too much uncertainty, then the *Consultant* is paid on actual time spent multiplied by the tendered *people rates*.

Like most short contracts, I think it is underused. The following are two examples of how it could be used more often.

- A *Client's* managing director said to me, half-jokingly, 'If I was to employ a *Project Manager* for an ECC contract on the full Professional Service Contract, I would then need to employ another project manager as the *Service Manager* on the full Professional Service Contract to manage that contract. And then another one to manage them, and so on.'
  To avoid this prospect, where there is a small or medium-sized professional team, use the PSSC, but specify in the Scope that the team are to carry out the actions of the *Project Manager* and/or *Supervisor* stated in the relevant ECC clauses, but with stated exceptions, and list the monthly reporting/liaising requirements and any circumstances in which they are to consult with the *Client*.
- I try to get my repeat-order clients to use the PSSC as a 'zero hours' term service contract for my training and/or consultancy services. As all the terms and conditions are already agreed, this means a client can just phone me up, have a discussion about what they want, we agree and summarise the task in writing as an instruction (contractually this is a compensation event) and then I do it. If that leads onto another task, the process is repeated. This makes it very easy for both of us. However, one large Client does not have a template for the PSSC, so instead their legal department insists we use the full PSC. Consequently, both of us are in breach of contract!

1.1.2.10 Another missing contract?

Arguably, if you are going to have a full Professional Service Subcontract, you should also have a Professional Service Short Subcontract – a PSSS – because many more consultants will be carrying out smaller value, less-complex assignments as subcontractors. This is regardless of whether the main contract being used is the full PSC, the full ECC or the Short Contract (ECSC).

1.1.2.11 NEC4 Dispute Resolution Service Contract (DRSC)

Historically, the Adjudicator's Contract was the contract you did not want to be using, because to do so would have meant you were in dispute. For NEC4, the NEC3 Adjudicator's Contract was replaced by the DRSC. The new contract contained a range of general refinements and improvements, where the previous contract had been found wanting,[3] and the change of name was intended to reflect the contract's wider application. This is because, in addition to adjudication, the full ECC (but not the other full

[3] For a full review of the changes, you could either

- watch a couple of webinars by myself, which can either be found via my website (https://www.jonbroome.com/resources/videos) or on the Association for Project Management's YouTube channel – search 'Jon Broome NEC4 good bad and ugly'
- read *NEC3 to NEC4 – evolution, not revolution; some fixes and some good ideas!* (Patterson, 2017).

contracts) potentially allows for the use of a Dispute Avoidance Board (DAB) (see Section 1.2.3.3 for more on what a DAB could be).

As stated previously, in concept, adjudication is a quicker, less expensive and less procedural form of dispute resolution than arbitration or litigation. For construction and civil engineering contracts (except where the client is a domestic home-owner), use of adjudication is pretty much mandatory in the UK as an interim form of dispute resolution. The relevant legislation, if you want to look it up, is in the Housing Grants, Construction and Regeneration Act 1996 and the Local Democracy, Economic Development and Construction Act 2009 (which amended the 1996 Act). The 'interim' bit means the parties can still go onto arbitration or litigation, but the adjudicator's decision is enforceable in the meantime.

Consequently, most of the other members of the NEC4 family have adjudication procedures within the contract clauses, covering how the Parties will conduct a dispute if one occurs. This includes the provisions for appointing the *Adjudicator*, which, if and when you do so, would be under the DRSC. The *Adjudicator's* appointment is a three-way contract between the Parties (as parties to the dispute) and the appointed *Adjudicator* (see Section 1.2.3.2 for more information on adjudication).

1.1.2.12 NEC4 Framework Contract (FC)

The Framework Contract was introduced in 2005, when the whole family was updated to NEC3. It is used to set up a framework, which is an arrangement where a contractor, consultant or supplier – all of whom would be termed the *Supplier* in the Framework Contract – is appointed to deliver works, services or goods of a similar type (under separate contracts) for a set period of time. For situations where the European Union (EU) procurement rules apply (and, if you are the UK, they will continue to apply until the Public Contracts Regulations 2015 are changed), the default period is 4 years unless there is a good reason for it to be otherwise. In practice, many frameworks seem to be set up for more than 4 years, so it does not seem hard to come up with a 'good reason'.

When the *Client* wants to let an order, a Work Order (using the terminology of the contract) is let under one of the other forms of contract in the NEC4 family, with much of the non-Work Order-specific Contract Data and specifics having already been agreed at the signing of the Framework Contract. Alternatively, for some simple consultancy, the *Client* can let a Time Charge Order (where the *Supplier* is paid per unit of time expended), which would normally be let under the PSSC.

When, as a client, you have a repeated requirement for works, services or goods of a similar type, a framework offers you the following advantages.

- You only need to go through a time-consuming and expensive initial selection procedure once, at the start of the framework. Once the framework is up and running, it is therefore much easier and quicker to get an individual contract in place, bringing time and cost savings to all concerned.
- There is the potential for bulk purchase discounts, in that the contractor, consultant or supplier is avoiding costs associated with repeated bidding and/or setting up and continuity of use of resources (e.g. avoiding demobilising, doing nothing, then remobilising).

- Continuous improvement, in terms of what is delivered, by when and for what cost, should be a feature. What *should* lead to this is, among other things, a well-defined measurement system (with some rewards for top performance), as well as continuity of workload, and hence people.

However, notice the proviso in the last point above: a well-defined measurement system implies that a client has to accurately describe what they want to be achieved in order that the supplier's performance can be measured against it. Continuity of workload means a client has to supply this, otherwise contractor or consultant resources will be put elsewhere (i.e. it also does not absolve the client from being a good client generally). Poor quality statements of what is required, not delivering on your obligations or not managing your risks etc. will undermine successful delivery. In other words, the 'continuous improvement' bit applies to clients as well!

If subject to EU procurement rules, you must have either one or three or more supplier organisations for a category of work type within a framework (but not two, for some reason). What this lends itself to is reduced but managed competition, and ideally will lead to the avoidance of the excesses of full-on capitalism as well as the complacency of a monopoly situation.

In the Framework Contract, the Contract Data references a number of schedules, in which the detail of how the framework will operate is fleshed out. These comprise

- the *framework scope*, which describes the technical scope of work to be carried out under the contract
- the *selection procedure*, which describes how the *Supplier* is selected for an individual Work Order (e.g. details of how any mini-competitions are to be held or criteria for direct award)
- the *quotation procedure*, which specifies how the contract Prices for a Work Order are built up
- the *quotation information*, which is pricing information provided by the *Supplier* before being appointed to the framework and which is used to price up a Work Order in accordance with the quotation procedure
- the *framework information*, which is any other information needed for the effective operation of the framework over its duration.

1.1.2.13 NEC4 Term Service Contract (TSC)

The TSC is designed for the ongoing servicing of assets, particularly in a planned or condition-based way. It is typically used for

- 'hard' maintenance services, such as highways maintenance and in the housing sector
- 'soft' facilities management, such as for reception, security, catering and cleaning services, as well as ground maintenance (according to my contacts, this is not uncommon).

Another area in which the TSC could be applied is in outsourcing arrangements, such as IT or cloud services, although a few tweaks would need to be made to the clauses to allow for a performance- or outcome-based specification.

The TSC has a *Service Manager* (whose role is equivalent to the *Project Manager* and the *Supervisor* in the ECC) and includes a Scope document (which used to be Service Information under NEC3), with work being done within the Affected Property (a sort of equivalent to the Site in the ECC).

The contract has three main payment options: a fixed-price contract, a target-cost pain/gain share option and a cost-reimbursable option. It also has a range of as secondary options.

A difference between the NEC3 and NEC4 versions is that the Task Order clauses, which allow for improvement projects/upgrades, have ceased to be a secondary X option and have been incorporated into the core clauses. In an IT context, Task Orders could be used for system upgrades. From what I know about software development,[4] the TSC could also be used under the agile project management methodology, again with a few tweaks (e.g. a base service is covered by the 'business-as-usual' clauses and improvement projects are carried out as Tasks).

1.1.2.14 NEC4 Term Service Subcontract (TSS)

The TSS is a new contract for the NEC fourth edition. It would be used for subcontracts that have similar characteristics to the main Term Service Contract. I think this will be a rarely encountered situation.

1.1.2.15 NEC4 Term Service Short Contract (TSSC)

Like other short contracts, the TSSC is intended to be used for less complex work. It therefore promotes the clarity and simplicity objective, at the expense of flexibility and stimulus to good management.

The *Contractor* is instructed to do work by the *Client*, which could be work that has been planned by the *Client* and/or unplanned work, such as repairs. Given the term services context, it is likely there will be more reactive work. As an example, I used to be a director of a small housing association for young homeless people. Because of our customer group (troubled teenagers), we had a high amount of unexpected wear and tear on our assets. As a result, we frequently required reactive repairs, which had to be done within set timescales to satisfy one of our key funder requirements. The Term Service Short Contract is ideal for this sort of service.

In contrast, where the *Client* wants the *Contractor* to plan the service delivery, albeit within set constraints, and to manage the resource, as well as potentially to carry out improvement projects that do not justify a separate contract, the full TSC should be used as it contains the full 'stimulus to good management' clauses.

[4] Actually, what I know about software development is not a massive amount. However, along with people who do know about it, I did contribute to an APM white paper on agile procurement (see APM C&P SIG, 2020).

1.1.2.16 Another missing contract?

It is common for there to be a main contract – under the full TSC – in which the *Contractor* not only does some physical work, but also coordinates the work of quite a few specialists (who are needed to do bits and bobs). The specialists could be needed on a planned basis (e.g. annual servicing of lifts) or in a reactive way (e.g. when a lift has stopped working and needs to be fixed as soon as possible). If there are only a handful of planned activities per year, using the full TSC, with all its management provisions, would be overkill. This is why I am sceptical about how often the full Term Service Subcontract would be used. An NEC4 Term Service Short Subcontract would, in my view, be far more sensible for use in these circumstances and would therefore have greater use.

1.1.2.17 NEC4 Supply Contract (SC)

This contract is for the supply of bespoke goods. Inherent within that is the following.

- The goods are wholly or partly designed and manufactured specifically for the contract. The default is that the *Supplier* designs the goods.
- The goods may have to come from overseas, so the contract allows for the purchase of goods on an international basis.
- In order to supply the goods, other associated services may be needed (e.g. training).
- There is a degree of complexity and risk associated with the purchase, so the monitoring and management procedures to deal with this are defined within the contract clauses.

The Supply Contract can be used by a main contractor as a subcontract for the purchase of goods (e.g. turbines, generators) for inclusion in the main contract works, let under the ECC. On a project for which I facilitated a number of workshops, the *Contractor* used the Supply Contract to purchase some precast concrete slabs – some of which were decorative as well as functional – for storm and wave protection on a seafront. The *Supplier* really liked the contract. Similarly, an article in the *NEC Users' Group Newsletter* enthusiastically detailed how the contract was used for the purchase of tunnel boring machines (TBMs) for the Thames Tideway project.[5]

This enthusiasm for the contract seems to be common among those who have used it for the purchase of major bespoke items. Why would this be? Well, the purchase of these major items is not like buying a premanufactured item, where the seller hands over the product and the buyer hands over the money. Taking the TBMs as an example: the TBMs would almost certainly have unique elements, designed specifically for the circumstances they would encounter; the *Supplier* would want staged payments prior to delivery, while the TBMs were being designed and manufactured; the TBMs would have to be shipped from Europe, where they were manufactured, and assembled on site with, at a minimum, input from the manufacturer; people would need some training on the operation of the TBMs, which would include supply of operation manuals etc.; there would be ongoing maintenance issues; and, once the tunnelling had been completed, the *Supplier* might buy back the TBMs to recondition for another client. Having a contract that

[5] At the time of writing, the easiest way to find this article is to google 'NEC Supply Contract tunnel boring machines'.

covers much of this, and which has a strong focus on project management, can only be beneficial, and will save a lot of time in arguing over the terms for a bespoke contract.

The Supply Contract is between and *Purchaser* (not a *Client*, as per other members of the NEC4 family) and a *Supplier*. It requires a *Supply Manager* to manage and administer the contract, and there is both a Scope document, which describes the goods and services to be provided, and Supply Requirements (which is actually a sub-document to the Scope), which specifies how the goods and services are to be supplied. However, it only has one payment option, which is lump sum fixed price, although the sum is adjusted by compensation events.

But be aware that, in its unamended form, this contract is not a supply *and install* contract, it is a supply *only* contract – that is, it presumes the goods are delivered to the edge of the Working Areas (if using an ECC contract) and that no physical work is done by the *Supplier* outside of its own factory. I think this is a shame as it would not take much to include allowances for installation. There are circumstances where 90% of the value is added in the factory but, for warranties to be valid, the goods need to be installed by the manufacturer.

An example from the UK where the contract was amended to allow for installation was for the supply and installation of a milking system. The system was to be installed in a dairy built under an Engineering and Construction Short Contract (ECSC) by a building contractor. To convert the Supply Contract into a supply and install contract, all that was needed was to add (with tweaks) the insurance provisions for working on site from the ECSC and, because it was in the UK on a construction site, to alter the payment and adjudication provisions.[6] These were cut, pasted and tweaked from the ECSC. An alternative to doing this would have been to use the Supply Contract followed by the ECSC and to have connected the two together using an overarching agreement, which would have required a lot of deletions to avoid repetition of similar, but not identical, terms. Another alternative would have been to use the ECSC, but it would need to have been amended to cover the assessment of compensation events in the factory and to include warranties. Both of the alternatives would have been far more complicated than amending the Supply Contract.

1.1.2.18 NEC4 Supply Short Contract (SSC)

This contract is for the supply of standard manufactured goods and commodities, which can be purchased internationally. Consequently, it does not need complex management procedures, although it does contractually obligate the *Supplier* to tell the *Purchaser* (as there is no *Supply Manager*) about any delays, matters that affect quality, etc.

One of its useful features is that it allows for the call-off of Batch Orders, so that the Prices for goods can be agreed prior to entering the contract and orders for goods called off when needed as the contract progresses.

As per the full Supply Contract, be aware that this is not a supply *and install* contract, but it does not take much to convert it into one. In fact, it would involve pretty much the same amendments

[6] It would now also have needed additional provisions to comply with the Construction (Design and Management) Regulations 2015; these are common to all NEC UK-based construction contracts.

as for the full Supply Contract (i.e. insurance provisions for working on site and, if in the UK, alterations to payment and adjudication clauses, all of which can be cut, pasted and tweaked from the ECSC, plus provisions to comply with the Construction (Design and Management) (CDM) Regulations 2015). A real-life example was the supply and installation of seating at an airport in New Zealand. Another example of where a supply and install contract would be better than an ECSC would be for the supply of windows: although they are manufactured – which is where most of the value is added – specifically for a particular project, they are essentially commoditised items, which you want the manufacturer to install so that the warranty (which is typically for 6+ years) is not invalidated.

Both the Supply Contract and Supply Short Contract should reduce the need for the 'battle of forms'. This occurs where a purchaser and potential supplier attach copies of their preferred terms with every communication prior to formally entering the contract (and often afterwards), in order to try and have the deal done on terms that are most advantageous to themselves. What this results in is (a) game playing before the contract is entered into, as each party tries to get it on their terms, (b) uncertainty over whose terms apply and, therefore, (c) argument. By having contractually fair terms, hopefully both parties will accept them.

1.1.2.19 NEC4 Design Build and Operate Contract (DBO)

The DBO Contract is new to the NEC4 family. As such, it integrates three, arguably two, contracts into one. The 'arguably two' is actually quite a strong argument as the full ECC contains provisions for *Contractor* design. Previously, if you wanted to prepare a contract for a design, build and operate project, you would have had to combine the full ECC and the TSC using an overarching agreement that tied them together and take out all the duplication.[7] The NEC Steering Group and its drafters have now done this for you.

The obvious use of this contract is when a Client wants a Contractor not only to design and build an asset, but also to operate it for a period of time. The *Client* would typically specify the operational requirements as an output-based specification: either performance-based (i.e. based on measurable numbers) or functionally-based (i.e. is descriptive) or a combination. The *Contractor* will therefore think a lot harder about whole life costs than if it just has to design and build the asset and correct any Defects over the next year or two. This contracting strategy appears to be more common in the European continent, with the 'operation' phase being around 5 years. The implication of this is that the *Client* must be quite sure that there will be little change during the operational phase.

Note that this is not a design, build, *finance* and operate (DBFO) contract. DBFO contracts originated in the process sector, where the output could be specified in hard measures (e.g. kilowatts of power) and where payments to the contractor only started once the operational phase commenced and were based on quantity delivered. They became popular in the UK as Private Finance Initiative (PFI) contracts, before being rebranded as Public–Private Partnerships (PPPs), but have since fallen from grace. In truth, the prime driver for their use was to get the 'finance'

[7] An alternative, and far cleaner, option is to have the two contracts entered into on the same date, but to link the *starting date* for the Term Service Contract to the date of Completion of the construction contract.

bit off the public sector balance sheet, so they were massively overused, but they still have their place as a contracting strategy.

So why am I telling you this? For two reasons.

- If, as the *Client,* you wanted to use the DBO Contract for a relatively simple DBFO contract, you would only have to add a few clauses that make clear you will only make payments once the asset is operational. The *Contractor* would then have to arrange finance, but, as far as you're concerned, that's the *Contractor's* business. So apart from making sure that finance has been put in place through a financially strong organisation or consortium, you could be very hands-off in terms of the 'finance' bit.
- If your DBFO contract has more complex financing needs, the DBO Contract is still appropriate. The provisions for the 'DBO' bit have, already been written, and in a widely accepted industry standard form, so the finance professionals just need to do their bit, in whatever form they want. This can then be incorporated either as option Z clauses or a separate contract.

While PFI/PPP contracts are definitely untrendy in the UK at the time of writing, the use of output- and outcome-based contracts is definitely on the rise generally. Consequently, it will be interesting to see how often the NEC4 DBO Contract is used.[8]

1.1.2.20 NEC4 Alliance Contract (ALC)

This is another new contract for NEC4. It was first published in a consultative form when the rest of the NEC family was updated to the fourth edition. The authorised version came out in June 2018, a year later.

But, first of all, what is an alliance? The number one characteristic of an alliance is that the incentivisation arrangements for the main players are closely aligned to the success of the overall project, rather than to the performance of their individual contracts.

So what does this mean in practice? It means that in the ALC

1. a budget is set for the project, which includes both the *Client's* forecast costs and those of the key supply chain players
2. this budget includes allowances for risks and opportunities
3. the key players, including the *Client*, sign up to this budget because they can jointly influence costs within it, including the risks and opportunities
4. costs of the key players are generally paid on an open-book or 'at cost' basis
5. the budget can be changed for far fewer reasons ('compensation events' in NEC language) than under a conventional contract because the budget includes a sufficient allowance for these risks (i.e. the budget is pretty much fixed)
6. the difference between the budget and the key players' final costs (including the *Client's*) is split between the key players in predefined portions.

[8] If you are interested in this application, you should read Patterson and Trebes (2015a), although it was written before the NEC4 DBO Contract came out. This paper was the winner of the Parkman medal in 2015 and can be downloaded for free.

Figure 1.1 Scope of alliancing arrangements

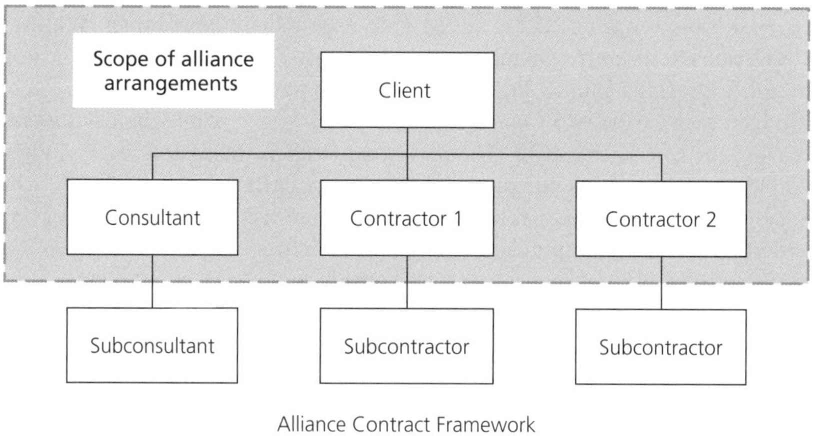

Because of points 5 and 6 above, the members of the alliance (the 'Partners') are motivated to work together to manage in opportunity and manage out downside threat, rather than (if point 5 did not apply) try to pin the blame on each other and the liability on the *Client*. For this arrangement to be successful, there must not only be enough risk and opportunity within the budget – which implies that you enter a contract earlier than you would do conventionally – but also interdependency between the Partners. Although different Partners may be able to manage individual risks effectively, the interconnecting nature of the risks means it is far more efficient and effective to manage them collectively.

However, you need more than an incentive arrangement to make this work: there needs to be a degree of overarching governance, some integration of organisation and systems, and some alignment of cultures. I have just summarised in a sentence what is actually a large commitment needed to make an alliance project successful!

There are three basic types of alliance.

- The first is the 'traditional alliance', whereby the client signs individual contracts with each of the key supply chain partners, shown in Figure 1.1 as a consultant and two contractors. These contracts are normally, but not always, cost-reimbursable in nature and there is an overarching alliancing agreement, jointly signed by all key players, which ties the fortunes of the parties together for the success of the project. The contracts between the parties are shown in Figure 1.1 as black lines, while the alliancing agreement is in grey. The principal measure of success is normally out-turn costs relative to budget, i.e. if out-turn costs are less than the budget, the parties share in the saving, but the opposite occurs if there is a cost overrun. There can, or course, be other measures of success, such as time to completion. Although members of the NEC4 family of contracts could be used as the individual contracts between the *Client* and the *Contractor*, *Consultant* or *Suppliers*, there is no overarching alliancing agreement in the NEC family that covers this arrangement.
- The second is what I call either the 'simple alliance' (because it is the simplest to put in place) or an 'X12 alliance' (because, if using the NEC4 family, it can be put it into place through secondary option X12). Here the *Client* signs individual contracts with each of the

key supply chain partners as per a normal project. However, rather than have an overarching contract that ties the fortunes of all key players into the success or otherwise of the overall project, option X12 is added onto each individual contract. Within the option X12 clauses, incentive payments are set against Key Performance Indicators. For example (referring again to Figure 1.1), of every pound that the project comes in under budget, each of the two *Contractors* gets a 25% share, while the *Consultant* gets 10%, leaving the *Client* with 40%. The reason why this is the easiest form of alliance to put into place is that there is no contract between the Partners in the alliance, which means
  - there are no cross-liabilities, i.e. there is no downside risk if another Partner underperforms or mucks up in some way, and therefore
  - the X12 incentive arrangements only cover upside performance.

  So what contractor or consultant is going to refuse the opportunity to be paid more with no downside? On the other hand, clients quite justifiably have some reservations about paying out for the upside on the business case but not getting anything back for the downside. The counter-argument is that the *Client* does not have any more downside than if there was no X12 arrangement, while increasing the likelihood of upside. I say a lot more about the operation of option X12 in Section 1.2.3.4, but my view is that simple alliances are underused as a means of creating commercial alignment.
- The last is the 'pure alliance', which is what the NEC4 Alliancing Contract creates. Here, all the key players sign the same contract at the same time, so they are all in it together, for better or worse. And in that sense, it is the most integrated and aligned of the three alliancing arrangements, with the same arrangement and liabilities (except for quantum) applying to all the members of the Alliance. Normally, for the participants to get comfortable with this arrangement, there is an 'engagement' period, when the Scope, the Budget and how the Alliance members are going to work together – think systems and organisation – are worked up and the Partners get used to each other, before the full-on 'marriage' takes place (although this does normally happen under the traditional alliance as well). However, there are three main downsides to this alliancing arrangement compared with the previous two.
  - Every member has to sign the same contract at the same time, which means that all the technical and legal issues will have to have been agreed. This can be tricky. Indeed, I heard of one alliance contract under the PPC2000 form – a rival UK contract (which arguably is what prompted the development of the NEC4 Alliancing Contract) – where the contract was never actually signed by all the parties because their lawyers could not agree on particular details. Yet economic necessity meant that delivery started. So what version of the 'draft' contract was the alliance actually let under?
  - A cost-plus arrangement might not be feasible for some members of the alliance. For instance, on a real oil project, one of the principal suppliers was the manufacturer of the pipes for a pipeline to connect the oil field to the shore. It was essential that these pipes were delivered according to a timeline that suited the project, i.e. if the project was going well and was 6 months in advance, as it turned out to be, they did not want to wait 6 months for the pipes to be delivered and hence the oil to start to flow. However, the vast majority of this manufacturer's costs were incurred in its factory mixed in with all the other costs that it was incurring on other contracts. Therefore, to account costs as either being alliance costs or non-alliance costs would have been administratively burdensome and subjective. Under both of the traditional and simple/X12 alliance arrangements, you can have 'fit for purpose' commercial, and hence contractual, arrangements that match the circumstances of each individual contracting party.

– Compared with the traditional alliance and, especially, the simple/X12 alliance, if one Partner is not performing, it can be fiendishly complicated to get them out as termination not only excludes them from being part of the alliance arrangements, but also means their contributions are terminated as well. As a *Client* using the Alliance Contract, you would therefore need to find another *Contractor*, *Supplier* or *Consultant* to provide the works, goods or services.

For these reasons, I generally prefer the traditional and simple alliance arrangements to the pure alliance. I would therefore have preferred the NEC Board to have commissioned an agreement for a traditional alliance. However, as always with contracting strategy, it's horses for courses, so I would not say never to use the Alliance Contract.

1.1.3 Another benefit of having an integrated family

Earlier in this chapter I described three advantages of using an integrated family of contracts, namely: training needs are reduced, systems for contract management are similar across the family and there is more back-to-back legal protection, particularly for a contractor when subcontracting.

From my perspective, as someone who drafts contracts, one of the really useful things that comes out of having an integrated family of documents is that, if you want to enhance one form of contract for a slightly different use, you can simply lift clauses from another member of the family to do so.

To give an example, using the perimeter fencing project that I previously mentioned, the *Employer* (under NEC3) wanted the work to enclose the 800 hectare site done as soon as possible. We simply lifted the 'option X6: Bonus for early Completion' clauses from the full ECC and added them into the Short Contract as additional clauses.

Being able to do this reduces the need for extensive legal reviews and the drafting of specialist clauses, which not only costs clients money, but also increases the likelihood of ambiguity and inconsistency, and hence disputes.

My summary of this section is very brief to avoid repetition, given that this can be a very involved subject – I could write a whole book on it! Figure 1.2 overleaf shows how the different members of the NEC4 family could be used on a common engineering or construction project.

1.2. How flexibility is achieved

1.2.1 Introduction

This section follows naturally on from the previous one, in that having selected the best-fit member of the NEC4 family for the contract package, you now need to select the best-fit main and secondary options plus other aspects of flexibility. The objective of flexibility correlates with the topic heading of 'procurement', of which the three main elements, in a project and programme environment, are

- how a project or programme is broken down into contract packages, and hence what work is included in each contract (e.g. whether detailed design should be included in the construction contract or let as a separate contract)
- how each supplier, contractor or consultant is selected for each contract (e.g. what weighting is given to price factors, the track record of company and/or people, and project-specific proposals)

Figure 1.2 NEC4 project contract structure for a typical engineering or construction project

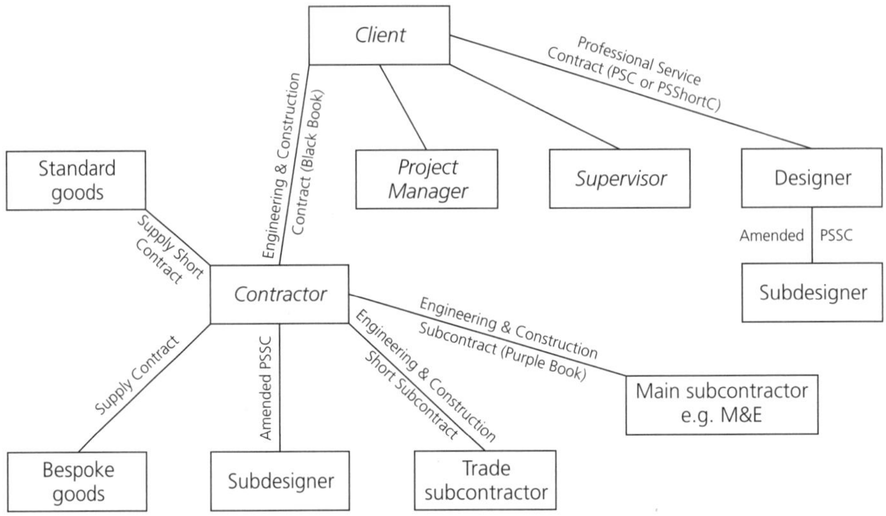

- the contracting strategy, which at a high level is principally about risk allocation in the contract. This impacts on how suppliers, contractors or consultants are rewarded for success or failure, which in turn impacts on how they – and the client – are motivated to behave during the contract. This is what we will consider in this section.

Before proceeding, let us look at why is it important to consider flexibility in the procurement strategy.

- If you take the extreme examples for a single project, a 'spot-on' procurement strategy can save a client around 30% in cost compared with orthodox approaches, and perhaps give a 20% saving in time, which means that the asset will earn a return much sooner. Savings of 10% in cost, compared with established benchmarks in one-off arrangements, are not uncommon.
- If you take the best examples from frameworks then, over a series of projects and years, costs can come down by 50% and durations by 60%.

The aim of this section is to give you an understanding of contract strategy and to outline how the NEC4 family, with particular focus on the ECC, accommodates different contracting strategies. The guidance on appropriateness of different contract strategies given in the NEC document titled *Establishing a Procurement and Contract Strategy* (which was formerly part of the NEC3 Guidance Notes, but is now rebranded as a User Guide) is expanded on considerably (however, the treatment is not nearly as detailed as in my second book (Broome, 2002), which was largely on contract strategy, and in particular those strategies best suited to partnering or collaborative arrangements). I also cover, in terms of what they are and when to use them

- the different main payment options
- the secondary options
- other aspects of flexibility.

It should be noted that the development of the 'right' procurement and contract strategy for a particular contract is as much an art as it is a science. As well as providing clarity about what the project objectives are and their relative importance (e.g. time driven or least-cost driven), there is also an evaluation of the risks and constraints associated with operating on the project, and the financial, managerial and technical ability of the different potential parties to take those risks. Consequently, there is rarely, if ever, a perfect procurement strategy, merely an optimal or best-fit one. If the procurement strategy is inappropriate, the contract that is set up will have a higher potential for failure. This was observed on a few of the contracts included in both my original research sample and my subsequent postdoctoral research. However competent the participants were, and regardless of the goodwill between them and the effectiveness of the mechanisms of the NEC/ECC in stimulating good management, the contracts were effectively set up for trouble.[9] This has subsequently been illustrated spectacularly on some projects[10] – fortunately none that I was involved in at the procurement stage!

1.2.2 Use of the different main payment options

1.2.2.1 Overview

The ECC has six main options, one of which must be chosen. These are as follows.

- Option A: Priced contract with activity schedule
- Option B: Priced contract with bill of quantities
- Option C: Target contract with activity schedule
- Option D: Target contract with bill of quantities
- Option E: Cost reimbursable contract
- Option F: Management contract

These options govern how the *Contractor* is paid and are therefore the primary means of allocating risk under the contract.

Table 1.1 shows the main payment options across the members of the NEC4 family. Whatever member is chosen and whatever main option is chosen, most of the procedures and systems necessary to administer and manage the contract will be the same, as approximately 85% of the text is core and independent of the main option.

The reason why the short contracts have less bold ticks in Table 1.1 is that they do not have options that you select, as there are in the full contracts. However, within the pricing mechanism documents (the 'Price List'), you can alter how the *Contractor* is paid.

[9] Given the implications that the choice of procurement strategy has for projects and programmes, you would think that there would be more literature and courses available on it, both generally in project management and specific to the NEC. However, there are remarkably few, which is reflected in some of the poor decisions made by educated and experienced people, and in a comment I received after giving a course to a top ten consultant firm. At the end, one of the senior participants stated something like 'In the past I've recommended contract strategies to clients, but not really known why. Now I'll know why I am recommending a particular course of action.'

[10] For instance, it is arguable that the collapse of the Heathrow Express tunnelling project in the late 1990s, when a big hole opened up in the middle of Heathrow Airport, was partly due to too much risk having been put on the contractor when an innovative tunnelling method was being used for the prevailing ground conditions. Losing money and behind on time, and without the ability to reclaim any of it through the contract, corners were cut, ultimately costing the contractor, the client and their insurers an awful lot more in both time and money. I can give other examples that are almost as spectacular.

Table 1.1 Contracting strategy at main payment option level

Option	ECC	ECS	PSC/ PSS	TSC/ TSS	DBO	Supply Contract	Short contracts (not Supply)
A: Priced contract with activity schedule	✓	✓	✓	✓	✓	✓	✓
B: Priced contract with bill of quantities	✓	✓					✓
C: Target contract with activity schedule/price list	✓	✓	✓	✓			
D: Target contract with bill of quantities	✓	✓					
E: Cost reimbursable/time-based contract	✓	✓	✓	✓			✓
F: Management contract	✓						

Also, note that under option E, payment is not cost reimbursable under the short contracts, but on a price per unit of input time, which would need to be agreed or tendered prior to entering the contract.

1.2.2.2 Options A and B: priced contracts with activity schedule or bill of quantities

Priced contracts should be used when the client can at the outset provide the contracting party with a complete description of what is required, so that a contractor can price it with a degree of certainty. This does not necessarily mean a complete design, but could be a full and unambiguous statement of what is wanted (e.g. a performance specification or scope design, and a statement of the purpose of the asset, with an adequate description of the existing starting state, which in the ECC should be described in the Site Information document). Provided that the client clearly specifies what is wanted, then the majority of risk is contractually transferred to the contractor and, through the selection procedure, the client can be confident that the contractor can deliver this work scope and manage the risks. The exceptions to this complete risk transfer are those risks listed as compensation events in clause 60.1 and, for option B (priced contract with bill of quantities), changes in quantities and mistakes in the bill of quantities.

Activity schedules are an attempt to move away from the problems associated with the use of bills of quantities. When the NEC originally came out, the normal path was for Employers to run a few projects under option B, before moving over to option A (priced contract with activity schedule). In concept, an activity schedule is similar to a series of bars on a bar or Gantt chart. The difference is that each bar/activity has a price attached to it and the contractor is paid for each *completed* activity at the assessment date following its completion. The 'completed' bit is the key difference between activity schedules and lump sums under more traditional contracts, where, for the latter, the contractor is paid according to the proportion of work under that lump sum that has been completed at each assessment date. Under an activity schedule, to be paid the contractor must have completed the work in that activity – the activity schedule is, therefore, closely linked to the programme. Under the ECC, because the *Contractor* prepares the construction programme, so the *Contractor* would normally also prepare the Activity Schedule, from which the expected cash flow requirements can be derived. A simple example of an activity schedule is shown in Table 1.2.

The original research found that the theoretical advantages that the use of activity schedules should give largely materialised in practice and, in my consultancy experience, I have come across nothing which shows these generalisations do not still hold true. Note, however, that I do say

Table 1.2 Example activity schedule for a stormwater relief storage tunnel

REF	Activity	Price: £
A001	Mobilisation	5 000
A002	Site clearance and establish at shaft B worksite	10 000
A003	Sink shaft B	63 000
A004	Set up pipe jack equipment in shaft B	6 500
A005	Pipe jack shaft B to A	47 500
A006	Site clearance and establish at shaft A worksite	3 000
A007	Sink shaft A	51 000
A008	Site clearance and establish at shaft C worksite	4 000
A009	Sink shaft C	48 000
A010	Set up pipe jack equipment for drive B to C	3 500
A011	Pipe jack shaft B to C	102 000
A012	Establish worksite in the Dell	2 500
A013	Construct outfall to river – excavation and concrete works	7 500
A014	Take delivery of *Employer*-supplied pipes to the Dell	150
A015	Construct 600 diameter drain in Dell – open cut portion	6 000
A016	Construct heading for drain to shaft A	17 500
A017	Gas diversion by Others – liaison	50
A018	Take delivery of *Employer*-supplied pipes at shaft C worksite	200
A019	Construct 600 diameter drain between shaft C and existing sewer	39 000
A020	Remove pipe jack equipment form shaft B	1 000
A021	Air tests	1 500
A022	Benching, ladders, miscellaneous internal works to shafts	10 500
A023	Connect to existing sewer	7 500
A024	Demobilise	4 000
	Total	£440 900

'generalisations'. These advantages (and some of their potential disadvantages) are, briefly, as follows.

- Any significant level of *Contractor* design is more easily accommodated, as design itself can become an activity. Standard methods of measurement for bills of quantities normally assume that the work is fully designed, which is rare in practice. If design is not finalised at tender, then any bill of quantities which might be prepared is, strictly speaking, not in accordance with the method of measurement: at best, it is not complete, and at worst it can provide an opportunity for 'contractual games'. Also, if the *Contractor* is responsible for design, yet is paid for the quantity of work done, then its motivation is to design for as much quantity as possible, which is why bills of quantities are rarely used on contractor-designed contracts. In contrast, lump sums can be used to pay for both the design and the contractor-designed physical elements.
- Contractors have to plan the job before they can prepare their Activity Schedule. Because they start from a blank sheet, rather than being given a prepared bill of quantities, Contractors are forced to prepare a more thorough tender, giving greater confidence in their tendered total of the Prices. This, however, is more time-consuming, and increases the

person-hours needed to prepare a tender. Some rules of thumb taken from a number of Contractors are that
- it takes 10% more time to prepare a tender if given the quantities, as the Contractor will check the main quantities regardless, although focusing more on the programme
- on civil engineering contracts, it takes 20–30% more time if the Contractor has to take off quantities, although not necessarily in accordance with and to the detail in the standard method of measurement
- on building contracts, it takes at least 50% more time if done in accordance with the method of measurement.

Consequently, responsible Clients have adopted a number of measures to reduce the costs incurred by Contractors when preparing tenders (see Section 2.2.4.3).

- As payment is linked to completion of an activity or group of activities, the cash flow requirements for both Parties are more visible.
- In order to receive payment as planned, the *Contractor* has to complete activities by the forthcoming assessment date. Consequently, it has to programme realistically, and is motivated to keep to that programme during construction. It is in the *Contractor's* interest keep an eye on the programme, and to speed up work on any a particular path of activities on a monthly basis, rather than try to recover time towards the end of the contract period in one big rush (or as one *Contractor* project manager put it, 'it's a series of mini-panics at the end of each month, rather than one big panic at the end of the contract'. This is good project management!) This has the advantage that the overall programme keeps its shape, giving greater certainty and avoiding the need to re-sequence, break links etc., throughout the contract, which in turn may mean re-jigging the Activity Schedule so that it continues to correlate with the Accepted Programme.
  However, in reality (and although not in accordance with the contract) many *Project Managers* understand the importance of cash flow to the *Contractor* and certify for payment an amount equivalent to the percentage completion of activities if (a) the *Contractor* is substantially on programme, (b) the activity will be finished by the time the *Project Manager* actually certifies the payment, (c) the *Contractor* is showing some give and take to the *Project Manager* and/or (d) the activity's completion has been delayed by a compensation event. However, if not done carefully, this can undermine one of the principal advantages of option A compared with traditional lump-sum contracts and the bill of quantities option (option B; i.e. the monthly cash flow incentive to keep to programme). The danger can be that if this is done enough times then it can be argued that the conditions of contract have been changed through the *Project Manager's* repeated actions, i.e. estoppel by convention. Consequently, when the *Project Manager* then wants to operate the contract as written, for whatever reason, there is a potential for not just for an argument, but one with a legal angle.
- Contractors are not paid for changes in quantity of the permanent work, unless an instruction changing the original Scope is issued by the *Project Manager*. This transfers the risk of quantity variation to the *Contractor*. However, if quantities are too uncertain, the *Client* will pay a premium for this risk transfer, which may not represent good value for money.
- The risk transfer to the *Contractor* is increased further because it is Contractors who normally prepare the activity schedules at tender. Consequently, if there is an error, omission or ambiguity in the Activity Schedule, it is the *Contractor* who normally suffers, not the *Client*. For bills of quantities, which are normally prepared by or on behalf of the *Client*, it is the other way around.
- The assessment of the effect of a compensation event is easier and fairer with an activity schedule than with a bill of quantities. This is because any change in resources or

methods associated with an activity can be compared with those stated in the Accepted Programme before the compensation event occurred. However, with activity schedules the process is more rigorous than when using bill rates, and therefore takes longer during the contract. For this reason, both Parties may be happy to use a straight bill rate for the assessment of a simple compensation event, where additional cost is closely correlated with the change in quantity multiplied by the relevant rate. Indeed, the contract allows for this under core clause 63.2, and the relevant rate may be one in a subcontract plus an amount for the *Contractor* (e.g. the *fee percentage*). However, using straight bill rates ignores the delay and disruption costs associated with any change, which are often assessed as a claim after construction has finished, and which research has found, on average, are approximately twice the direct costs (Revay, 1992). It makes sense, therefore, to use the programme as the primary means of evaluating change, as this correlates with how costs are actually incurred! (See Section 3.7 for much more detail on how to agree compensation events.)

- Assessment of the amounts due to the *Contractor* is easier with an activity schedule and involves many fewer person-hours than with a bill of quantities. This and the preceding point may well account for a general feeling among participants on completed projects that the total administrative and management input from commencement of a contract (the Contract Date) to settlement of final account is less under a priced contract with an activity schedule (option A) than with a conventional contract form, and slightly more with a priced contract with a bill of quantities (option B). However, as changes are agreed as the contract progresses, this does not necessarily mean fewer person-hours are needed during the construction phase! In some cases, where there are few compensation events, it is much less. On the other hand, where there have been more compensation events than either Party probably anticipated, more resource has been needed during the contract, but they have avoided a long tail to settle the final account at the end.

So, if these advantages of activity schedules are generally true, why use bills of quantities? I have more sympathy with the use of bills of quantities when contractors' costs are more quantity related and/or when there is greater likelihood of quantity variation that the tendering contractors cannot realistically price. For the former case, in building contracts, many of the trade subcontractors price on, and are happy to be paid on, a bill of quantities basis, and so, in turn, main contractors' costs are quantity proportional. It is worth noting that option B does allow for lump sums, so fixed and time-related costs can be taken out of the bill rates to give method-related items in the Bill of Quantities.

In the case of there being a greater likelihood of quantity variation that the tendering Contractors cannot realistically price, provided the *Contractor* has built up the lumps sums and bill rates in such a way that reflect its costs, then the Bill of Quantities would be an effective way of reimbursing the *Contractor* for changes in quantity in a way which matches its cost base. As an example of this, one of my regular clients is a water company which regularly lets contracts to 'plug' holes in dams. The costs of setting up and maintaining the site establishment are fixed, as are time-related costs per week on site, and so these can be paid for as lump sums. The cost of actually plugging the holes is then paid for on a bill of quantities basis.

However, I do have serious concerns with the option B clauses as written in the third edition of the ECC and which (apart from replacing Works Information with Scope) have not been changed in NEC4. The concerns are outlined in Table 1.3.

Table 1.3 Engineering and Construction Contract, option B and D clauses

Clause No.	The clause says:	Interpretation/commentary
60.4	A difference between the final total quantity of work done and the quantity stated for an item in the Bill of Quantities is a compensation event if	Fine with this. Expressed another way, 'A change in quantities is a compensation event if'.
	■ the difference does not result from a change to the Scope	No instruction is issued. It just turns out that a greater or lesser quantity of work is needed than was stated in the original bill of quantities document.
	■ the difference causes the Defined Cost per unit of quantity to change and	As the *Contractor*, I would argue, if it suits me, that this is always true, as my costs always have a fixed and time-related element. Therefore, if costs are plotted against quantity of work done, the cost line will never go through the origin, so any change in quantity will always cause my cost per unit of quantity to change. It will generally suit the *Contractor* to argue this when quantities go down, but not necessarily when they go up.
	■ the rate in the Bill of Quantities for the item multiplied by the final total quantity of work done is more than 0.5% of the total of the Prices at the Contract Date.	On a £1 million contract, 0.5% is £5000. But, note that it does not say that the *change* in quantity multiplied by rate is greater than this, but the *final* quantity multiplied by rate. If the second bullet is always true, this means that any change, however small, in quantity for items above the 0.5% threshold is a compensation event.
	If the Defined Cost per unit of quantity is reduced, the affected rate is reduced.	No problem in itself, but in implication it conflicts with the opening sentences of 60.4 and 63.15, which are more definitive (see below).
63.15 opening sentence	Assessments for changed Prices for compensation events are in the form of changes to the Bill of Quantities.	Along with the opening sentence of 60.4, this makes it clear that it is the *changed* quantities that are assessed as a compensation event, i.e. the original rates for the original quantity are not reassessed. Consequently, you potentially have one rate for the original quantity and another for the changed quantity.
63.15 final sentence	For the whole or a part of a compensation event for work already done, the change is a new lump sum item.	As we are looking at the difference between the quantity stated in the original Bill of Quantities and the final quantity, this is how the assessment will be reflected in the Bill of Quantities. But as it is the change in quantities that is the compensation event, the original rate for the original quantity still stands, with a lump sum for the changed quantities. This lump sum could be a negative amount if the quantities have gone down.

Consequently

- for rates multiplied by a final quantity that is below the 0.5% threshold (which you will only know on completion of the item, as it is the 'final quantity'), option B is a re-measurement contract.
- for rates multiplied by final quantity above the 0.5% threshold
  - the original quantity is paid for at the original rate multiplied by the quantity
  - for the changed quantities, whether more or less, the change in what the *Contractor* is to be paid is assessed, in accordance with clause 63.1, as the changes in the Defined Cost and the Fee. Thinking it through, there may be arguments about which part of the item coverage is the changed quantity, i.e. if a quantity has increased, the *Contractor* would argue for the most expensive part of the item to be assessed as a compensation event, while the *Project Manager* may well argue the opposite, and vice versa if there are reduced quantities.

Furthermore, if the work has already been done, then, in accordance with the last sentence of clause 63.15 (see Table 1.3), the adjustment is made in the form of a lump sum. If less work has been done than stated in the original Bill of Quantities, then you would have to multiply the original rate by the original quantity, then subtract a lump sum to adjust for the quantity not done.

If forecasting, then, in accordance with the first bullet of clause 63.15, you might end up with two rates for the same item: one for the original quantity and one for the changed quantity. What is more, if the changed quantity is a reduction, this implies you have a rate for work *not* done, which you then subtract from the work done!

Lastly, if you read clause 60.5 (in conjunction with clause 63.5), any upward change in a quantity, no matter how small, that is on the critical path entitles the *Contractor* to a compensation event, which can put back the contractual Completion Date and, in doing so, entitle the *Contractor* to time-related costs.

Confused? If you want to bottom it out, read it again slowly with the contract to hand. Alternatively, take my word for it that it is confusing and heinously hard to operate. Consequently, when I have recommended the use of bills of quantities under the ECC, regrettably I have also recommended some redrafting of these clauses. Perhaps this is also one of the reasons why option A has proved far more popular in use than option B.

Note: Under option A (priced contract with price list) of the NEC4 Term Service Contract, the *Contractor* is paid both for *completed* lump sums and, if specified as a rated item, the quantity of work done multiplied by the rate. So, when setting up the Price List, the *Client* can choose to retain the risk of quantity variations for certain items.

1.2.2.3 Option E: cost-reimbursable contract

I cover option E before options C and D (the target contracts) because it is useful to explain how a cost-reimbursable contract works first, and then to explain the differences.

Under Option E of the ECC, the *Contractor* is reimbursed its Defined Cost plus a Fee, to give the Price for Work Done to Date (i.e. in simplified terms, what the *Contractor* is owed at any point in

Figure 1.3 How the Contractor is reimbursed under a cost-reimbursable with fee percentage contract

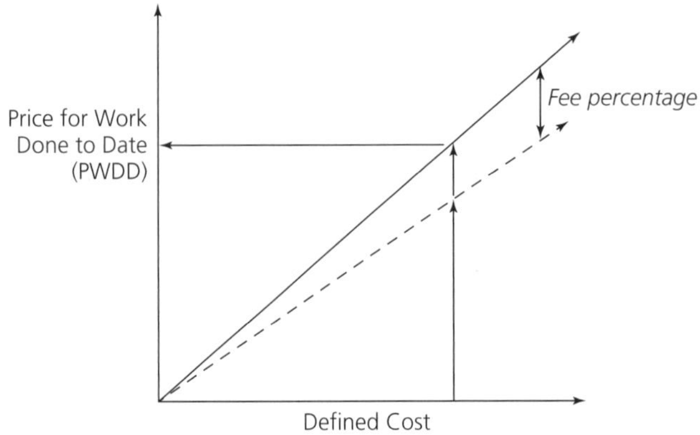

time during the contract). (To be more precise, the *Contractor* is reimbursed those costs which it has incurred and is expected to incur in the period up to the next assessment date.) This is illustrated in Figure 1.3.

An important change across the whole of the NEC4 family, compared with NEC3, is that there is now only one *fee percentage*, which applies to both the Defined Cost incurred directly by the *Contractor* and the Defined Cost incurred by its Subcontractors. This contrasts to the NEC3 provisions, where there was a *direct fee percentage* and a *subcontracted fee percentage*.

For a cost to be included in the Defined Cost, it has to be identified in the Schedule of Cost Components, which, under NEC4, includes for subcontracted work. Under NEC3, the Schedule of Cost Components only covered the *Contractor's* direct, not subcontracted, costs (with Subcontractor costs being part of the definition of Defined Cost). The Fee is therefore calculated by applying the *fee percentage* – as tendered by the *Contractor* – to the Defined Cost. The *fee percentage* effectively covers the *Contractor's* off-site/head-office overheads and the cost of insurances and profit, and includes an allowance for Disallowed Cost, together with anything else not listed in the Schedule of Cost Components. This commercial arrangement gives little incentive for the *Contractor* to minimise costs once on site (unless there is an overarching business arrangement, where the incentive is repeat-order business, e.g. a framework agreement). This particular contract strategy should be used either or both where

- time or quality are overriding priorities for the *Client*
- the scope is not sufficiently defined at the outset or the contract is subject to a high level of uncertainty (e.g. in planning interfaces or risks encountered, such that the *Contractor* is unable to price the *works* with any degree of accuracy, as is often the case when a project is time driven).

A potential advantage of cost-reimbursable contracts is that the open-book accounting procedures act as a catalyst to openness in other aspects of contract administration and management. However, for this to happen, the *Contractor* has to be able to do open-book accounting i.e. demonstrate that a particular cost was incurred on the *Client's* contract and not somewhere else in its business.

1.2.2.4 Options C and D: target-cost contracts

Target-cost contracts are a development of cost-reimbursable contracts and were traditionally used in less extreme circumstances, such as where the requirements are sufficiently developed that a contractor can put an approximate price to the work.[11] The requirements and/or solution would still require some development, with the key point being that *both* Parties could influence the outcome through some joint management of risk – whether threat or opportunity – within the target. Once the contract is underway, the *Contractor* is reimbursed Defined Cost plus a Fee in the same way as in the cost-reimbursable option (option E). However, any cost overrun or underrun relative to the agreed target (or the 'Prices' in ECC terminology) is split between the *Contractor* and the *Client* in pre-agreed proportions at the end of the contract. This is illustrated in Figure 1.4, which shows a cost underrun in part (a) and the more general case in part (b).

Target-cost contracts therefore very effectively align the motivations of the Parties to decrease costs. In contrast, on a lump sum (option A) ECC contract, the *Client* and its team have no direct interest in the *Contractor's* actual costs. Target-cost contracts have proved to be the most common form of contracting arrangement for partnering/collaborative relationships, both because of the commercial alignment and because of the transparency and openness engendered by open-book accounting. Their use may be appropriate

- when the extent of the *works*, in terms of detail, is not fully defined at the outset (e.g. on design and build)
- when a reasonable degree of flexibility with regard to design or programming changes is anticipated
- where both Parties can contribute to decreasing the Defined Cost, such as through the use of value engineering and risk management techniques
- for sharing risk – if risks were to be allocated to the *Contractor* in the traditional way (e.g. in a priced contract), Contractors would include a high risk premium in their tender prices, or not bid at all; the *Client* may not wish to pay for risk that does not eventuate, but does wish to motivate the *Contractor* to manage it
- for any combination of the above, including being the means for aligning motivations in partnering/collaborative relationships.

To emphasise what was said above, sitting behind all these scenarios is a reality that both Parties to the contract can jointly contribute to managing out the negative risks or threats and managing in the opportunities that are within the Prices. If only the *Contractor* has the ability to contribute, go for option A. If only the *Client* can contribute, go for option E.

The caveat to all of this is do not use a cost-based contract where the *Contractor* (or Subcontractor, from a Contractor's perspective) does not have the financial sophistication to do open-book

---

[11] It is important to emphasise that, in an ideal world, the Scope will be a complete statement of what is wanted, expressed to a level of detail whereby the *Client* is certain that it is going to get the asset it wants. If, subsequently, work needs to be added or the requirements refined, it will be treated as a change to the Scope, which is a compensation event, and so likely to increase the target Prices and maybe delay Completion. In reality, this is not always achievable, but you should be able to get close. If you cannot get close, then use a cost-reimbursable contract instead.
Obviously, the less well the Scope is defined, the more risk there is that a contractor's single point estimate will turn out to be wrong. If it is downside risk, then (for any given share profile), the contractor will add more contingency into its estimate to arrive at the target Prices. If a client does not wish to pay this premium, then a way to reduce this risk contingency is to alter the share profile.

Figure 1.4 How the pain/gain share works in a target-cost contract

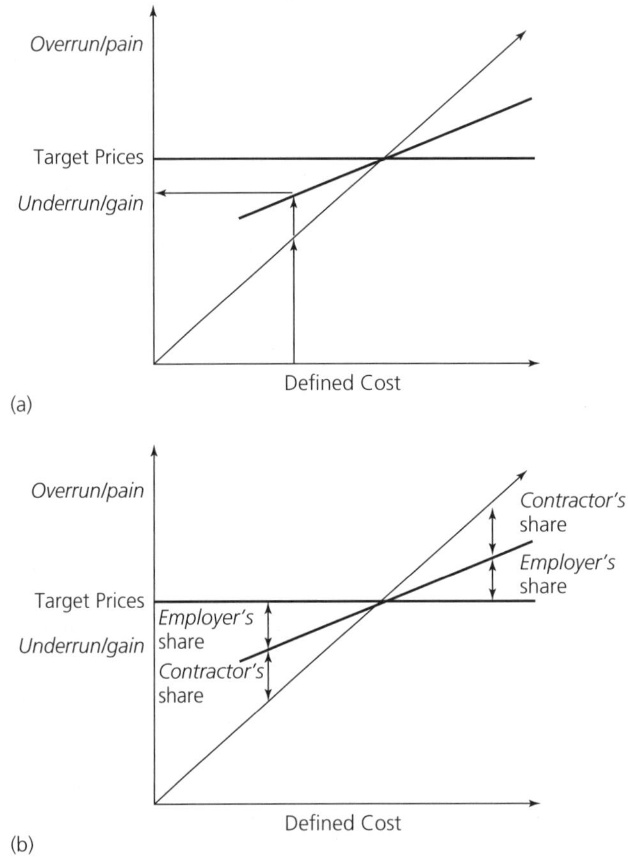

accounting in real time. By this I mean

- they cannot develop costs from first principles, but instead price on the basis of the market rate
- they do their accounts on a 2-monthly plus basis, as opposed to a monthly basis
- they are unable to allocate costs to an individual contract.

If this is the case with the *Contractor* or Subcontractor, then not only will the reasons for using the contracting strategy be undermined, but you will end up with an administrative nightmare trying to ascertain Defined Cost.

When doing my postdoctoral research, which was partly on target cost-contracts, one of my first phone calls was to a Client who I wanted to interview about a contract. They told me not to bother! Why? Because almost 3 years after the contract had been completed, the Parties were still arguing over which of the *Contractor's* costs had been incurred on this contract and which were actually from other contracts that the *Contractor* had at the same time.

Greater client involvement is essential in this contractual arrangement, and unfamiliar administrative procedures, which include open-book accounting, may lead to higher administrative costs. Contracts of this type are, therefore, less likely to be as suitable for low-value contracts, although as their use has become more common (together with enabling software), the threshold for their use has become lower.

Under option C (target contract with activity schedule), the target Prices are negotiated, tendered or built up on an open-book basis, and presented in much the same way as under an option A contact. Having said this, because the *Contractor* is reimbursed Defined Cost plus a Fee as the contract progresses, the Activity Schedule does not need to be as detailed, as payment is not dependent upon it.[12]

Under option D (target contract with bill of quantities), the *Client* has chosen to take the risk of quantity variations and errors in the Bill of Quantities, so the initial target Prices are presented in the same way as for a traditional bill of quantities but, like option C, the *Contractor* is reimbursed Defined Cost plus a Fee as the contract progresses. This has a number of consequences.

- The target will be a running total as the contract progresses, because it depends on the quantities of work done multiplied by tendered rates, and the proportions of lump sum operations completed. That is, during the contract, the Total of the Prices (which becomes the fulcrum around which the pain/gain share operates on Completion) progressively increases as work is completed. So the tendered total of the Prices (the amount entered into Contract Data part two by the *Contractor* as a headline sum) has no contractual significance. It is the tendered lump sums and rates that matter.
- This means that quantity surveyors and the like are kept extremely busy, both counting Defined Cost for payment purposes and keeping track of quantities of work done, to ensure that the Total of the Prices, around which the pain/gain share operates, is up to date.
- Furthermore, changes in quantities are addressed using the same obtuse contractual rules as apply in option B, which means that the quantity surveyors are kept even busier!

In the same way that priced contracts need a method of adjusting the prices if an event for which the client is liable occurs (i.e. a compensation event), so target-cost contracts need a mechanism for adjusting the target prices. In the ECC, the mechanism for evaluating the change in Prices is identical whichever main option is chosen, be it a priced contract, cost-reimbursable or target-cost option. However, it should be noted that with the target-cost contracts, like the priced contracts, there is a temptation for the *Contractor* to maximise the financial assessment of compensation events, so as to maximise the difference between the final target Prices and final Defined Cost, thus maximising its share of the savings. In this respect, these contracts do not align the motivations of participants (although the transparency of open-book accounting means it is more difficult to get away with unreasonable upward adjustments). If the extent and value of the *works* are changed too much and too often, so that the changes are assessed retrospectively, the participants have effectively reverted to using a cost-reimbursable contract.

[12] As a side comment, the requirement to show 'how each activity on the Activity Schedule relates to the operations on each programme' in clause 31.4 of the NEC3 ECC has disappeared from option C of the NEC4 ECC. I think this is a backward step as it undermines the monitoring of progress against value delivered. I recommend putting this requirement back into an option C contract as an option Z clause.

1.2.2.5 Management-based contracts – option F (management contract) or construction management under the PSC

Management-based contracts are generally suitable when

- there is a need to coordinate a considerable number of works contractors and suppliers
- the client does not have sufficient staff and/or expertise to manage the procurement of the asset
- the timescale of the project is tight, necessitating an early start of construction. (The key point is that the extent of the project is not fully developed, and so a price for the full *works* cannot be accurately determined at tender, but the price for the initial works packages can be. As the scope is developed and construction progresses, successive works contracts are let, and the interfaces between these packages are managed.)

Management contracting, where works contracts are let between the management contractor and works subcontractors, is enacted in the ECC by using Option F. Under this option, it is expected that design, construction and installation will be subcontracted, leaving the *Contractor* to concentrate on the management of the work. The *Contractor* is paid Defined Cost plus a Fee, with the Fee being calculated by applying the tendered *fee percentage* to the *Contractor's* Defined Cost. But what is included in the Defined Cost? Under Option F, Defined Cost is what the *Contractor* pays to its Subcontractors in accordance with their subcontracts, plus any other *prices* tendered by the *Contractor* for small amounts of physical work that the *Contractor* does itself. The Fee, therefore, needs to cover all of the *Contractor's* internal costs, both within the Working Areas and elsewhere.

The management contracting approach peaked in popularity in the 1980s, being largely superseded by the construction management approach, where all the works contracts are let directly between the client and the works contractors. The works contracts are then managed and administrated on behalf of the client by a construction management company, which, unencumbered by many of the commercial drivers embedded in a management contractor's contract, can more freely act in a professional role. The two approaches are illustrated in Figure 1.5.

Figure 1.5 Contractual and communication paths for the two management contract commercial arrangements

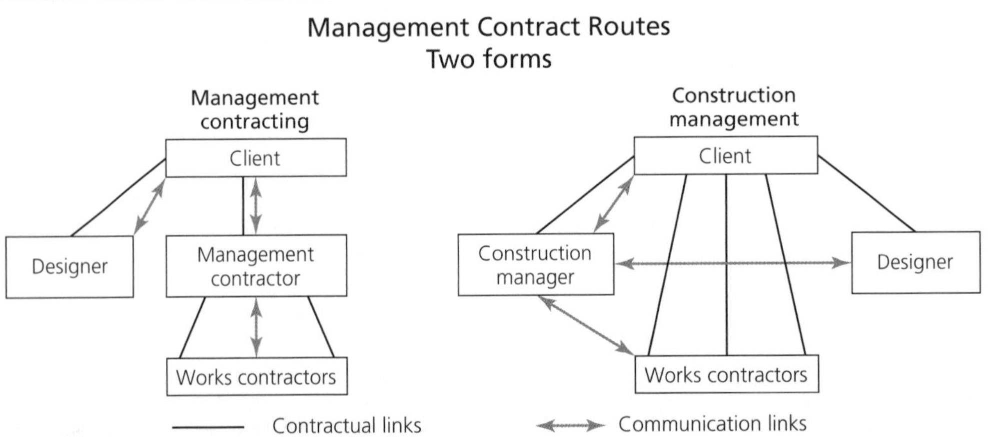

The rise in popularity of construction management occurred for a number of reasons.

- Works contractors, under the construction management approach, are less inhibited in making their design contribution, and are more committed to the client's objectives, both in terms of human interactions and relationships and because the individual contractual arrangements can be tailored to the client's objectives as opposed to the management contractor's.
- Under the management contracting approach, clients tended to assign more and more risk to the management contractor, which it could neither manage nor price. While the out-turn price was not a contractor's risk – as the contractor would be paid what it had paid out, plus a fee – achieving the completion date was, especially if it had large damages attached. The contractor would thus be tempted to both pass the risk down to the works contractors and, when problems occurred, adopt a defensive stance to minimise its potential liabilities. This undermines the objective of achieving less adversarial behaviour. Conversely, the role of the construction manager is unambiguously professional.
- It is clearer that the client retains risks under the construction management approach, especially the risk of default by a works contractor.
- The direct contracts between a client and the works contractors can encourage early payment, which enhances the likelihood of good site relationships and higher productivity, leading to lower bids and potential long-term relationships.
- If the relationship between the client and the construction manager breaks down during a contract, the client can terminate the contract and retain the works contractors. Under the management contract, the original works contracts would have to be novated or assigned to the new management contractor.

The above points explain why the construction management approach became accepted as a better alternative to management contracting in many, but not all, circumstances, including for use with the ECC. However, the more defined the *works* are, the more the management contracting approach comes into play. Equally, strong leadership by the client, in terms of good project management discipline, is needed, otherwise the project can run away.[13]

To use NEC4 for construction management, Clients should let the construction management and design-only contracts on the PSC, and the works packages should be let on the ECC or the Short Contract.

The term 'delivery partner' has been bandied around a lot recently, particularly in the wake of the London 2012 Summer Olympics. From what I can gather, a delivery partner is a heavily incentivised construction manager.

1.2.3 Secondary options

1.2.3.1 Overview

This section covers all the secondary options of the NEC4 family, briefly explaining what they are, when to use them and, where they have a similar function to other secondary options, their advantages and disadvantages. Where a secondary option is in the ECC, it will be explained using

[13] Probably the most high-profile case of this is the Scottish Parliament building in Edinburgh, which was let under the construction management approach, but not the NEC. Its costs rose from £40 million to £400 million, due, in large part, to there being poor leadership, an inexperienced client and an inappropriate contract strategy.

ECC terminology, but the words and effect will predominantly be the same whichever member of the family is chosen.

Table 1.4 gives a summary of all the secondary options and which members of the NEC4 family have them.

Table 1.4 The secondary options within the NEC4 family

Secondary option	ECC	ECS	TSC/ TSS	PSC/ PSSC	DBO Contract	Supply Contract	Alliance Contract
W1: Adjudication (if not in UK construction industry)	✓	✓	✓	✓	✓	✓	
W2: Adjudication (if in UK construction industry)	✓	✓	✓	✓	✓		
W3: Dispute Avoidance Board	✓						
X1: Price adjustment for inflation	✓	✓	✓	✓		✓	
X2: Changes in the law	✓	✓	✓	✓		✓	✓
X3: Multiple currencies	✓	✓	✓	✓	✓	✓	
X4: Ultimate holding company guarantee	✓	✓	✓	✓	✓	✓	✓
X5: Sectional Completion	✓	✓		✓			
X6: Bonus for early Completion	✓	✓		✓			
X7: Delay damages	✓	✓		✓		✓	
X8: Undertakings to *Client* or Others	✓	✓	✓	✓	✓		
X9: Transfer of rights	✓	✓		✓	✓		✓
X10: Information modelling	✓	✓	✓	✓	✓		✓
X11: Termination by the *Client*	✓	✓	✓	✓			
X12: Multiparty collaboration	✓	✓	✓	✓		✓	
X13: Performance bond	✓	✓	✓	✓	✓	✓	
X14: Advanced payment to the *Contractor*	✓	✓			✓	✓	
X15: The *Contractor's* design	✓	✓					
X16: Retention	✓	✓				✓	✓
X17: Low performance damages	✓	✓	✓			✓	
X18: Limitation of liability	✓	✓	✓	✓		✓	✓
X19: Termination by either Party			✓				
X20: Key Performance Indicators	✓	✓	✓	✓			
X21: Whole life cost	✓	✓	✓				
X22: Early *Contractor* involvement	✓						
X23: Extending the Service Period			✓		✓		
X24: The *accounting periods*			✓				
X25: Supplier warranties						✓	
X26: Programme of work							✓
Y(UK)1: Project Bank Account	✓	✓	✓	✓	✓	✓	✓
Y(UK)2: HGCR Act (payment)	✓	✓	✓	✓	✓	✓	✓
Y(UK)3: Rights of third parties	✓	✓	✓	✓	✓	✓	✓
Z: *Additional conditions of contract*	✓	✓	✓	✓	✓	✓	✓

Note: The Engineering and Construction Short Contract, the Engineering and Construction Short Subcontract, the Professional Service Short Contract and the Supply Short Contract effectively have the delay damages embedded in the core clauses of the contracts.

1.2.3.2 Dispute resolution options W1 and W2 for adjudication

Options W1 and W2 are the NEC's dispute resolution options, but in both cases we are talking about adjudication. To explain what adjudication is and the key differences between the two options, I find it easiest to give a give a quick history of adjudication.

- The first edition New Engineering Contract was the first construction contract to include an adjudication procedure. Essentially, adjudication is a less procedural form of dispute resolution compared with arbitration and litigation. As the process has to be completed over a much shorter timescale – from the time of referral, the *Adjudicator* has 4 weeks to make the contractually enforceable decision – it is considerably quicker and less expensive than arbitration or litigation. Adjudication has been called 'quick and dirty justice', but a losing party is able to take the dispute on to either arbitration or litigation, depending on what form of *tribunal* is specified in the Contract Data.
- Sir Michael Latham, in his 1994 report *Constructing the Team*, liked the idea of adjudication and recommended that it became mandatory in all UK construction contracts, with an adjudication's decision being enforceable in law unless and until overturned by arbitration or litigation.
- The Housing Grants, Construction and Regeneration Act 1996, which came into force in in 1998, made Latham's recommendation law in England and Wales. There are equivalent Acts for Northern Ireland and Scotland, in which the relevant clauses are the same. However, at that time, the NEC adjudication clauses did not comply with the Act because they included a 4-week time bar from the time of the alleged incorrect or missing decision. Under the Act, you have to able to go to adjudication 'at any time'.
- This ultimately led to the introduction of the option W1 and W2 clauses, with W2 being the option that has to be used if the contracting party is carrying out 'construction operations' as defined in the original Act. This would apply to most civil engineering or building operations in the UK – provided they are not for a domestic residence, with the client being the resident.
- The original Act was enhanced in 2009 by provisions in the Local Democracy, Economic Development and Construction Act 2009 – the two Acts together are often referred to as the Construction Acts – which closed out a number of loopholes that parties were using to avoid justice. Amendments to option W2 were incorporated into the NEC3 family, first by a published addendum and then by inclusion in the April 2013 update.

The NEC4 dispute resolution clauses, whether in option W1 or option W2, are substantially the same as the April 2013 NEC3 ones, except for a few clauses that allow for senior management resolution. This is a stage of dispute resolution prior to going to adjudication, whereby *Senior Representative* from the two Parties, who are detached from the day-to-day, possibly month-to-month, operation of the contract, try to come to an acceptable commercial resolution. Under W1, this senior management resolution stage is mandatory under the contract, but under W2 it is not, because under the Construction Acts you have to be able to go adjudication 'at any time'. But why would you not try to resolve a difference this way first!

So, to summarise where we are now, you have to specify option W2 if your contract is in the UK and is considered to be for 'construction operations' (i.e. most civil engineering and building contracts, excluding domestic building projects where the *Client* is the resident).

But there are other sectors where construction works are carried out but which are not considered 'construction operations'. For instance, works for the oil and chemical industries and the nuclear sector. However, I have seen and given training on contracts for these sectors in which option W2, rather than W1, has been specified. The reasons for this are probably twofold.

- A pure 'construction' subcontract on these projects would probably be covered by the Construction Acts. Consequently, in case there is a back-to-back dispute, it makes sense to have the same procedure in the main contract and the subcontract.
- Option W1 puts a time bar on when a Party can notify a dispute after the disputed action or inaction. However, the unhappy Party would then go straight to the *tribunal* as defined in the contract, either litigation or arbitration, both of which are more expensive and time-consuming than adjudication. It makes sense that a dispute can always be referred to adjudication first, and not to time bar a Party from doing so. Similar reasoning could be applied to international contracts.

A final comment on adjudication is that, at the time of writing (early 2021), statutory adjudication exists in some form in Australia (depending on the state), New Zealand, Singapore, the Isle of Man, Malaysia and Ireland, with Hong Kong likely to follow at some point soon. So, if you are thinking of contracting 'construction operations' in these countries, you need to make sure that whichever option you select complies with the relevant legislation and, if necessary, modify it accordingly.

1.2.3.3 Dispute avoidance option W3 for a Dispute Avoidance Board

Option W3 is new to NEC4 and is only an option under the ECC. But what is a Dispute Avoidance Board (or DAB for short)? In simple terms, it is small group of 'elder statesman' type people who are brought in at the first signs of a dispute and whose task is to try and nip the potential dispute in the bud while at the same time getting to its root causes. This can include addressing management and personality issues, not just any legal issues over which the Parties are having a fight: in my experience, the legal issue is the surface problem, but dig deeper and there are almost always more fundamental issues. As such, the DAB can only make recommendations, but it must try to bring the Parties along with it so that its recommendations are likely to be implemented.

Under NEC4, a DAB consists of one or three members. If three members, one is appointed by each Party, with the third being agreed between them. Depending on its brief, the DAB can regularly visit the Site to provide proactive advice and guidance, to help prevent disagreements from becoming potential disputes in the first place (i.e. by helping to address a more fundamental issue). However, the DAB could be seen as an expensive overhead, particularly if the contract is relatively small, so it would only be worthwhile using option W3 on larger contracts, which justify the additional overhead costs.

Alternatively, the DAB could be one person who periodically visits the Site, or even only when asked, in which case it potentially offers much better value than adjudication, provided the DAB's recommendations are decent and are acted upon. (As a sort of NEC/collaboration/partnering coach, I have performed this sort of role and can testify to how effective it can be in resolving issues early and positively.) Although, if the DAB clauses are specified but are used to employ a single person to provide an ongoing 'project coach' role, then the Dispute Avoidance Board title becomes a bit misleading.

There are three important things to note about the DAB clauses.

- As written, the wording is not compatible with UK legislation with respect to the right to go to adjudication. This is a shame because through a little bit of tweaking of the clauses, the DAB could be part of a structured issue-resolution ladder, at a level before adjudication. The NEC Board seems to have recognised this and has since brought out Practice Note 5, which provides an amendment for option W2 that replaces the *Senior Representatives* provisions with the DAB ones (NEC, 2019). Given that the October 2020 revision came out after the Practice Note, it would have been good if the contract had included the changes as an option W4.
- The DAB only provides recommendations to the Parties, not enforceable decisions. By comparison, an adjudicator's decision is enforceable by court order. This can, though, be a good thing, because addressing the root causes of a potential dispute will push the Parties to take ownership of the problem, rather than simply blame each other.
- A Party cannot refer a dispute to the *tribunal* (typically either arbitration or litigation) until after it has referred the dispute to the DAB and receive the DAB's recommendations.

1.2.3.4 The X options

Before I start explaining the X clauses, I should mention that you will see far more italicised words in this subsection than in the preceding ones. These words are also italicised in the ECC. In clause 11.1, the ECC states that 'terms identified in the Contract Data are italicised', which means there is a contract-specific entry for them (see sections 1.2.4.4 and 1.5.1 for more explanation.

OPTION X1: PRICE ADJUSTMENT FOR INFLATION

This option is typically used for longer term contracts, with a duration longer than 18–24 months, and transfers the risk of inflation largely over to the *Client* which, in the Contract Data, specifies the indices to be used and the proportions in which the Prices are adjusted by these indices. The risk (and opportunity) for the *Contractor* is that these proportions do not match the proportions in which it will actually incur the costs, or that the indices do not reflect the rate of inflation actually encountered. In its unamended form, the Prices are adjusted not only each time one of the referenced indices is updated, but also after every assessment interval. The option is often amended so that the Prices are adjusted on an annual basis, normally on the anniversary of the Contract Date (i.e. the date on which the Parties entered into the contract).

For framework arrangements, the duration of the framework is usually a minimum of 4 years, but each contract within the framework might well run for less than 1 year. Consequently, an equivalent clause is normally written into the Framework Contract, located in the option Z (*additional conditions of contract*) clauses, to allow for inflation when pricing each individual contract package.

OPTION X2: CHANGES IN THE LAW

This option makes changes in the law of the country in which the Site is located a compensation event, and it is one of the few compensation events that can reduce the Prices. As a result, it is often used on longer term contracts, where the *Contractor* would not be able to foresee these changes. In the UK, changes in the law can be by an Act of Parliament (where parliamentarians vote on the issue) or by a change to a statutory instrument (as authorised by a Secretary of State, or by a delegated body such as the Health and Safety Executive). Clients have come to realise how wide ranging this can be: for instance, the budget is an annual Act of Parliament, so if a tax that affects the *Contractor's* cost is introduced or increased, then the *Client* becomes liable. Consequently, this option is often amended to exclude or include particular types of change in the law, not just all changes in the law.

From the *Contractor's* perspective, if the option is not amended in the contract and you know that a new law is going to be enacted sometime during the contract period, you are likely to price the work accordingly (a real-life example from some years ago was an increase in the standard of protection required for working at height). However, if you are already doing the work that meets the new legal requirement, while there would be a compensation event at the time of change in the law, there would be no additional compensation as there would be no additional costs over those you are incurring already.

A recent example of this coming into play was caused by the COVID-19 pandemic. The UK government quickly enacted the Coronavirus Act 2020, which in itself did not cause any changes in the law directly relevant to construction. However, it gave a Secretary of State power to do various things through regulations that sat beneath the Act. It was under these regulations that particular types of premises were shut down etc. (See Appendix 1 for a copy of an article I wrote on this.)

OPTION X3: MULTIPLE CURRENCIES

This option is used with main options A and B, the priced contracts, if there is significant risk of currency fluctuation as, without it, this risk sits entirely with the *Contractor*. Having specified that the option applies, the *Client* specifies what items it will pay for in which currency and the maximum amounts that can be paid in each currency, as well as which indices will be used for currency conversion calculations. The option therefore transfers the risk of currency fluctuation onto the *Client*, but only for the items identified and only up to the specified limits.

Under the cost-based main options C, D, E and F, the *Client* states in Contract Data part one which *exchange rates* will be used for any currency conversion under core clauses 50.7 or 50.8. However, in the example given in the NEC4 Guidance Notes, the *exchange rates* used are tied to a reference date stated in the Contract Data. These *exchange rates* may be significantly different to the actual exchange rates encountered months or years after the reference date. An alternative would be to tie the *exchange rates* (in italics) into the exchange rates of your referenced source on either the date of the relevant transaction or the next assessment date. Whichever you do, to me this seems a bit odd as the principle is that you are paying actual cost, which would be the currency conversion the *Contractor* actually gets, plus any relevant charges!

So, under option E and F, it is the *Contractor* that takes the risk of currency fluctuation from the date(s) specified, which could be positive or negative. The same applies under options C and D, albeit the risk is shared as part of the pain/gain share around the target Prices. And this could be quite a significant risk, especially if the *exchange rates* are those referenced at the Contract Date.

OPTION X4: ULTIMATE HOLDING COMPANY GUARANTEE

This option has been revised for NEC4. The NEC3 option referred just to 'parent company guarantee'. The objective of specifying the guarantee is to give the *Client* financial recourse in the event of a default by the *Contractor*, by making it easier to claw back money. On large projects that have gone wrong, this prevents the temptation for a parent company to shut down a contractor subsidiary – which does happen – as the parent company will still have a liability. The substitution, in NEC4, of the word 'ultimate holding' for 'parent' is presumably so that, if there is a chain of company ownership, it is the ultimate owner that is promising the guarantee.

However, under NEC4, if that ultimate holding company is located abroad, say in Bermuda (as big family-owned companies may be), then by the time you get a legally enforceable direction for that ultimate parent company to pay up, the money and assets may well have been shifted into another company, perhaps in Guernsey. So there is a commercial decision to be made here, regarding the likelihood of getting money from a parent company based in the country of business versus getting it from the ultimate holding company. In contrast, as I understand, in the UK it is possible to get an injunction effectively to stop money being moved out of a UK parent company to elsewhere. For this reason, UK Clients often want the ultimate UK parent company to be the guarantor.

It is useful to note that publicly traded companies dislike ultimate holding/parent company guarantees because the full value of a guarantee has to appear on their balance sheet, regardless of the likelihood of it being called in. This makes them appear even more indebted than they normally are, which deters investors and future clients, who like strong balance sheets.

For a *Client* wanting financial recourse in the case of default, another alternative is to specify option X13 (performance bond), which is effectively insurance in case of a default. This, in turn, has its pros and cons, which are discussed below.

OPTION X5: SECTIONAL COMPLETION

This option allows the *works* to be divided up into sections for which different *completion date*s can be specified, and for which the next two secondary options can be used. Of note is that, while it uses construction terminology for a geographical area (i.e. a *section*), it could, with a little tweak of terminology, apply to a completed state. For instance, in the process industry, 'sections' could be used for: installation and commissioning; optimisation (during which the *Client* may well be getting some benefit); continuous running (even more benefit); and full Completion of the whole of the *works,* including supply of all certificates and documentation.

OPTION X6: BONUS FOR EARLY COMPLETION

Bonuses, in my opinion, are massively underused in the UK. Substantial research in the USA, where positive incentives are used much more widely to encourage collaborative relationships (whereas in the UK we have used target-cost contracts), strongly indicates that bonuses help to deliver superior performance. This is because people from across the different organisations tend to focus jointly on achievement of bonus measures, despite the difficulties (i.e. the client wants the better-than-needed level of performance and the contractor wants the bonus, so there is alignment). In contrast, damages have typically been shown to stimulate negative performance, as people devote their efforts to blaming others for the issue in order to avoid liability, rather than trying to resolve the issue.[14]

[14] If you want to read more about how to go about developing an incentive plan, you can find an article titled *Incentivisation Under the NEC* on my website (at www.jonbroome.com/resources/articles). It was published in the *NEC Users' Group Newsletter* in 2010, but is still relevant today.

Option X6 applies a bonus for each day that Completion is achieved ahead of the contractual Completion Date. Obviously, if the *Client* gains no benefit from earlier Completion, then it undermines the reasons for specifying this option. Note also that the Completion Date will tend to be extended due to compensation events, so if, as a *Client*, a set date matters to you – such as the opening of an athletics event – you might wish to detach the incentive from the contractual Completion Date and instead anchor it to the date that matters to you.

If the option is used in conjunction with option X5 (sectional Completion), the *Contractor* will get a bonus for early Completion of each *section*.

OPTION X7: DELAY DAMAGES

This option is the equivalent to the 'liquidated damages' clauses found in most traditional contracts – if selected, delay damages will apply if Completion by the *Contractor* is later than the contractual Completion Date. If option X5 is also used, delay damages will also apply for late Completion of *sections*. There are two main things to note about damages.

- In most of the world (except, I believe, the USA and the Gulf states), the stated amount for damages used to have to be a genuine estimate, made at the time of entering the contract, of the likely loss to the client, or it could be less. However, if this amount exceeded the loss the client was actually likely to suffer, then it could be challenged by the contracting party, who would then become liable for damages at large. Consequently, as a client, you needed to have some justifiable rationale for setting the rate for damages at a particular financial level. However, the UK's Supreme Court has recently expanded what losses can be included[15] (such as from reputational damage, through being reflected in share price) and made it clear that it is very unwilling to intervene in a commercial agreement unless the damages are genuinely exorbitant or unconscionable against the greatest likely loss.
- Because of the negative behaviour damages often engender, it is claimed that they are not appropriate for partnering/collaborative relationships. This is a simplification, because contractors are generally extremely wary of entering a contract without specified damages, and they cap their liability for delay. Instead, damages are specified, but are set at or close to zero.

A couple of other things to note are as follows.

- In the Contract Data entry for option X7 (whether used on its own or with option X5) there is no place to put in a financial limit on the total delay damages payable. This means damages are essentially 'unlimited', and so contractors are likely either to price their tenders accordingly or not to bid for the work. Consequently, it is not unusual for Clients to insert an extra entry for an overall cap on the delay damages that can be applied.
- The October 2020 amendments also tweaked these clauses across the whole NEC4 family, to clarify that any delay damages will cease to apply following termination. After termination, any further delay costs are considered as part of the general costs/damages due as a result of the termination of the supplier's obligation to provides the goods, works or services.

[15] See *Cavendish Square Holdings BV* v. *Talal El Makdessi* and *ParkingEye Limited* v. *Beavis* [2015] UKSC 67. In its opinion, the Supreme Court recast the true test of whether something is a penalty as 'whether the impugned provision is a secondary obligation which imposes a detriment on the contract-breaker out of all proportion to any legitimate interest of the innocent party in the enforcement of the primary obligation'.

OPTION X8: UNDERTAKINGS TO THE *CLIENT* OR OTHERS

In NEC3, this option was called 'collateral warranty agreements' and, for some reason, was only included in the Professional Services Contract, which I could never really understand. It is now available in other forms, with the name changed to avoid using a very UK-centric term.

As an 'undertaking to the *Client* or Others' will normally take the form of a collateral warranty, I will describe that first. So, what is a warranty? As a simple example, consider the situation when you buy something from a shop, say a watch. When you buy the watch, you enter a legal contract with the shop and therefore have some rights, as described (in the UK) in the Consumer Rights Act 2015. However, when you open the packaging, you will also find a warranty from the watch-maker. This provides you with a direct legal relationship with the watchmaker and extra rights if something goes wrong with the watch, usually in addition to your rights with the shop that sold the watch to you (e.g. the watch is guaranteed for a longer period). While the lawyers might argue over the exact wording, this is a simple form of collateral warranty, although it is not named as such.

Collateral warranties are needed because of the legal doctrine of privity of contract, which generally prevents third parties relying on other parties' contracts. A collateral warranty enables a third party to claim its losses directly from the person who caused the loss; without one, the third party may be unable to make an effective claim. On a construction or engineering project, a collateral warranty is a contract under which a professional consultant, a building contractor or a subcontractor warrants to a third party that they have complied with their professional appointment, building contract or subcontract.

So, let us take an example of a developer who enters into an agreement with a purchaser to sell a building once it is completed. The developer also enters into a contract with a main contractor to construct the building, who in turn subcontracts the design and installation of the mechanical and electrical (M&E) services, who in turn buys an air-conditioning system from a manufacturer. After a couple of years' use by the building purchaser, the air-conditioning system needs to be replaced due to a design defect inherent in the manufactured product. Without a collateral warranty, the purchaser would try to claim from the developer, who in turn would try to claim from the main contractor, who in turn would try to claim from the M&E subcontractor, who would in turn try to claim from the manufacturer. It is far from certain that such a claim would be successful, even if all the relevant parties were still in existence to respond to the claims, and it would be very expensive. If, however, the building purchaser had received a collateral warranty from the manufacturer of the air-conditioning system, they could make a claim directly against them. This example illustrates why it makes sense that this option is now included in other members of the NEC4 family, although I do not know why it is not included in the Alliance Contract, for those in the supply chain who are not part of the Alliance.

In terms of the PSC, option X8 would be used to oblige professional consultants to provide collateral warranties to any relevant third parties, such as funders, purchasers or tenants of the development.

So why the change from 'collateral warranty' to 'undertakings to the *Client* or Others'? Apart from expanding the wording (from one short sentence to five clauses) to deal with things more comprehensively, I presume it is to expand what can be warranted, while avoiding the use of construction-oriented terminology.

Lastly, in the UK, a potential alternative to requiring collateral warranties/undertakings to the *Client* and Others is to use option Y(UK)3 (see Section 1.2.3.5).

X9: TRANSFER OF RIGHTS

A change from NEC3 is that this option was previously only included in the PSC. This option applies to intellectual property and reverses the default position of the contract. The default is that the *Contractor* or *Consultant* effectively owns any intellectual property generated, so can use it elsewhere on other contracts, but grants the *Client* the right to use it for purposes related to the contract (e.g. maintenance), often via some sort of licence. In contrast, the use of option X9 gives the *Client* full ownership rights to material prepared for the contract, except as otherwise stated in the Scope document. This clearly makes the entry in the Scope very important. In some sectors, such as pharmaceuticals, defence and IT, intellectual property can be very important.

See the section on option X15 (the *Contractor's* design) below for further comments on ownership and use of intellectual property.

X10: INFORMATION MODELLING

In NEC3, option X10 was titled '*Employer's Agent*' and only applied to the PSC (as there was no *Service Manager*). In the NEC4 PSC, there is now a *Service Manager*, so the *Employer's Agent* option is not needed. Instead, the publishers have reused the option X number for a new feature.

Readers from the construction industry will almost certainly recognise the abbreviation BIM, which stands for Building Information Modelling. Option X10 comprises a set of clauses that appear to be fully compliant with the intent of the current UK BIM protocol, without contradicting legally or undermining the good project management aspects of the NEC. This is in stark contrast to the original BIM protocol clauses, which I described as a 'contractual car crash waiting to happen'.[16] However, a lot of information still needs to be thought through for it to work, both practicably and contractually. NEC has also published specific guidance on how to integrate NEC and its option X10 with the BIM protocol now offered under ISO 19650 (NEC, 2020).

So why no 'building' – just 'information modelling'? (i.e. why not BIM?). Earlier in this book I said that the ECC could be used for an asset that will be constructed in one place and then transported elsewhere, such as an oil rig or a ship in a yard, so the 'building' bit would not apply in these sectors. Casting the net even wider, structured data sets are everywhere these days, whatever sector you are in. As NEC3 was starting to be used outside of construction and civil engineering, the intention for NEC4 was to accelerate this trend by avoiding the use of industry-specific terminology.

X11: TERMINATION BY THE *CLIENT*

In NEC3, this was another option only specified for use within the PSC, whereas in NEC4 it is also available in the full Engineering and Construction Contracts and Term Service Contracts. The option allows the *Client* to terminate the *Contractor's* or *Consultant's* contractual obligations for a reason not otherwise identified in the contract (i.e. not identified in the termination table in Section 9 of the contract). However, the *Contractor* would get paid for its loss of margin – overhead and profit – as the *fee percentage* is applied to the value of the work that remains to be done

[16] My original article on this, titled 'Don't use the BIM Protocol with NEC3. Do this instead', can be found in the blogs section of my website: www.jonbroome.com

at the point of termination. This is, in fact, not substantially different to the position under the NEC3 ECC, ECS or TSC, where the provisions were included in the core clauses.

As a *Client*, it makes sense to effectively have the option of 'termination for any reason' in your contract, even though you are unlikely to even consider using it. Consequently, I would expect X11 to be a standard option to specify – which begs the question, why is it now an option?

X12: MULTIPARTY COLLABORATION

This option was called 'Partnering' under NEC3. I always thought it was imprecisely named as 'partnering' can be all things to all people, and can take many forms. Lawyers also disliked the term due to its association with the word 'partnership', which has a specific legal meaning. Instead, the publishers have updated the option to include the word 'collaboration', which can also be all things to all people. However, the change in title does make it clear that the option is for use in a 'multiparty' collaborative setting.

Most of the words within the option stay the same as in NEC3, a notable exception being *Client*. As *Client* is now used throughout the NEC4 family instead of *Employer*, the *Client* in the NEC3 option has been replaced by the *Promoter* in NEC4. Also, 'partnering' and 'working together' have been replaced by 'collaboration'.

Option X12 forms what I have called in Section 1.1.2.20 (when discussing the Alliance Contract) a 'simple alliance' or an 'X12 alliance'.

As discussed in Section 1.1.2.20, the essential feature of an alliance is that the commercial incentives for all members of the alliance are aligned to the success of the overall project or programme (which may involve a number of key players under contract), as opposed to the success of the individual contract between the Client and a supplier – be they a contractor, consultant or supplier of goods. So, how does X12 do this? Essentially, option X12 clauses are added to the individual works, service or supply contact of each of those key players. These X12 clauses have the following three main features, which are summarised in Figure 1.6.

- There is a Core Group of senior people drawn from the Partner organisations, which functions like a board of directors[17] for a project or programme. Consequently, these people are not involved in the day-to-day delivery, but equally are not too remote. They typically set direction, make key strategic decisions on delivery strategy and risk management, monitor performance against the measures and targets set for the alliance (sanctioning corrective action for underperformance), and are the final tier of dispute resolution before it gets contractual. They are also the champions for the project within their own organisations at, or close to, board level.
- They allow the *Promoter* – normally the *Client* – to set Key Performance Indicators, with the units of measurement and targets specified. Normally, these will be for the joint performance of Partners across the project or programme, but could be for an individual contracting organisation. Incentives can (but do not have to) be set against these Key Performance Indicators. Typically, these incentives are for performance against the *Promoter's* overall time and cost objectives – that is, they include all of the *Promoter's*

[17] Do note that lawyers shudder when I use this analogy, as directors' duties are laid down by law, whereas the Core Group's duties are defined in the Partnering Information.

Figure 1.6 Overview of unamended NEC4 option X12: multiparty collaboration

Unamended NEC4 X12: Multiparty Collaboration overview

Core Group

- Senior people from organisations: involved, but not on a day-to-day basis
- Set direction, monitor performance, last tier of non-contractual dispute resolution
- Maintain 'Schedule of Partners'

A **timetable** showing the proposed timing of each Partner's contributions. A change in timing is a compensation event.

Early warnings are given to other Partners.

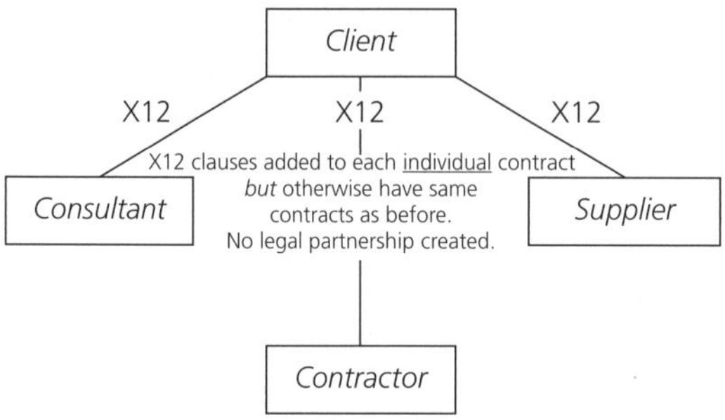

Partnering Information

- describes how Partners work together;
- similar or same in all contracts;
- can be changed by Core Group, but is a compensation event.

KEY PERFORMANCE INDICATORS

- is an apsect of performance for which a target is stated in the Schedule of Partners;
- a Partner is paid amount stated in the Schedule of Partners if the target is improved upon or achieved;
- only *Client* may add a Key Performance Indicator and associated payment and may not delete one.

costs (incuding its own internal and other costs) and are partially dependent upon the performance of the other commercial organisations within the alliance. Effectively, this creates a gain share arrangement like that in a target-cost contract, but which differs in that: (a) the reasons for changing any target are much reduced; and (b) the risks and rewards are typically shared between three or more participants (i.e. it aligns all key players' objectives to the success of the project).

- They require Partnering Information, which outlines how the players within the alliance will work (e.g. common IT systems, organisational structure, decision-making processes). If this information is too prescriptive, it will constrain how the project or programme organisation works, so it is suggested that it provides the bones of these topics, with the participants providing the flesh. (The words I use are borrowed from the field of agile project management, in that the Partnering Information needs to be 'just sufficient' for people to be on the same page about how they work together, but not overly prescriptive.) If the Core Group gives an instruction changing the Partnering Information, it is a compensation event that can reduce the Prices under the individual contracts.

Option X12 has a couple of other useful features.

- The first is a requirement to prepare a timetable that shows the proposed timings and contributions of the Partners, so that they can work in a coordinated manner. If a *Contractor* needs to change its own programme because of a change in a revised timetable, then this is a compensation event. Some of these changes may arise because opportunities due to collaboration have been spotted, where savings to the overall project exceed the cost to an individual contract. Changes that are more reactive – because one of the Partners has not met its timings – might already be a compensation event under another heading.

- The second is a requirement to give early warnings to the other Partners, not just the *Project Manager* or equivalent under the individual Partner's contract.

As can be imagined, setting one of these up from scratch means that the project or programme needs to be worthwhile in two respects.

- There needs to be sufficient leeway in what and/or how something is delivered, so it will not be worthwhile in situations where everything is prescribed.
- The size of the project and potential savings need to outweigh the investment in selecting the right partners etc. and creating a virtual organisation that works.

I have alluded to it, but will state it explicitly: the features of option X12 mean that it can be used for programmes of work – a series of similar projects over a period of time – with the incentives being set to stimulate continuous improvement.

The publishers make it clear that option X12 is an alliancing agreement, not a contract. The essential difference is that under an alliancing contract (such as the PPC 2000 and the NEC4 Alliance Contract forms), all parties sign the same contract at the same time and have joint liability for downside risks. Under option X12, the participating contractors, suppliers or consultants are not in contract with each other and have no downside liabilities to each other. Instead, each Partner has an individual contract with the *Client* only, and each contract includes option X12 as add-on clauses, thereby aligning everyone's motivations to the success of the project only.

Lastly, the option X12 heading includes the wording 'not used with option X20' (Key Performance Indicators). I do not agree with this. In general, it may well be good advice, suggesting the only appropriate incentives are ones shared between the Partners, but it might not be appropriate in some circumstances. Take for example an incentive mechanism that I helped to design for a £16.5 billion programme of works of over 10 years. The intention was that option X12 Key Performance Indicators and incentives (and its other features) would be used to create alignment and collaborative working across supply chain partners, where the management of risk and achievement of objectives required collaboration. However, option X20 would also be used when an aspect of performance was within one contractor's control.

OPTION X13: PERFORMANCE BOND

A performance bond serves the same purpose as option X4 (ultimate holding company guarantee), in that it gives the *Client* financial recourse in the event of a default by the *Contractor*, by making it easier to claw back money. Instead of it being the parent/holding company that the *Client* claims on, it is a financial institution of some sort. In theory, performance bonds have three advantages over a parent/holding company guarantee

- there does not have to be a parent/holding company in existence
- a bank or similar institution has more guaranteed security – for instance, even though a contractor subsidiary may be trading profitably, its parent company might not, so a financial institution would offer more security (but given the 2008/9 financial crisis, this is not considered as true as it once was, so the financial institution also needs to be checked out – although, as a result of the financial crisis, banks and other financial institutions are required to have much stronger balance sheets)

- willingness to pay, as a bank is less likely than a parent/holding company to take the side of the contractor subsidiary and dispute the payment.

However, the *Contractor* will have to pay a sum of money to the institution holding the performance bond, which is passed on in its tender, so ultimately the *Client* pays more for a performance bond than for a parent/holding company guarantee. In addition, depending on the economic climate, financial institutions may (a) charge significantly more than they do on a long-term average basis and (b) limit the number of bonds they issue for any particular company in order to limit their own potential liabilities. This may mean that an individual contractor is unable to take out a bond for a new contract until one for another contract can no longer be called in.

OPTION X14: ADVANCED PAYMENT TO THE *CONTRACTOR*

Under NEC3, this was for when the *Contractor* would incur significant expenditure before doing any of the physical work or, in contractually correct language, before the *starting date*, which is the enacting date for payment purposes. For instance, this might be the down payment for a specialist piece of construction Equipment needed to construct the *works*, which has to be paid out many months before the *Contractor* goes on site (e.g. a tunnelling machine or, in overseas work, transportation of Equipment to the country). Once the *starting date* had occurred, future payments to the *Contractor* have the advanced payment deducted in predetermined amounts until it is repaid. The *Client* had the option of requiring the *Contractor* to take out a bond for these advanced payments.

There is quite a significant change in NEC4, in that the advanced payment is made early on in the contract, at the first assessment date (i.e. when the contract has become 'live'), and therefore, whichever main option you are using, it can be covered by the normal payment mechanisms. However, in the same way that the amount of the advanced payment is stated in the Contract Data, so too is the rate of repayment and when repayments have to start, which could be many months after the *starting date*.

X15: THE *CONTRACTOR'S* DESIGN

This replaces the NEC3 secondary option titled 'limitation of the *Contractor's* liability for his design to reasonable skill and care'.

The NEC3 clause was publicised as being the clause to specify if the *Contractor* was going to be doing some design, which, these days, would be on virtually every contract. The problem was that it only really addressed one issue when there are a number of issues that need addressing, so Employers (under NEC3) were having to draft option Z (*additional conditions of contract*) clauses. The new NEC4 clause addresses, so far as I am aware, all bar one of the other issues as well. So what are these issues?

- The standard of liability in the ECC is generally taken to be fitness for purpose, which means that, as the *Client*, if my Scope says the asset has to do X, Y and Z and it does not, I can sue the *Contractor* for the full loss of benefit. The 'fit for purpose' liability means the *Contractor* has no defence to this failure. However, design consultants find it almost impossible to obtain insurance cover on this basis, but do have cover for exercising 'reasonable skill and care'. The exercise of reasonable skill and care would typically mean the design consultant has used a suitably experienced engineer, who followed appropriate design codes to do the work, that another experienced engineer checked the work and, generally, that an ISO 9000 certified quality-management process has been followed.

So, if the *Contractor* is responsible for design under the main contract but then subcontracts it out to a design consultant (as most do), the *Contractor* would be at risk should the asset not be fit for purpose due to a shortcoming in the design. Option X15 reduces the *Contractor's* level of design liability down to that which the design consultant is able to accept, so the *Contractor* and consultant have 'back-to-back' design liability. This reduction in liability not only applies to latent defects under law, but also to the correction of Defects, in that if there is a Defect in the *works* caused by poor *Contractor* design – but not below the standard of reasonable skill and care – then the correction of that Defect is a compensation event. Consequently, Contractors pretty much always want to have option X15 included in their NEC contracts.

Two other brief observations on liability in NEC4 ECC, compared with NEC3.

- The words surrounding the 'reasonable skill and care liability' provision have been altered to a format that insurers are apparently more comfortable with.
- Under NEC3 ECC, the burden was on the *Contractor* to demonstrate that it had used reasonable skill and care, when normally it is the claimant (i.e. the *Employer* under NEC3) who would have to demonstrate that it had not. Under NEC4, the normal situation applies, in that the *Client* would have to demonstrate, on the balance of probabilities, that the *Contractor* did not use reasonable skill and care.

- The *Contractor* effectively owns the material generated under the contract and may use it elsewhere unless either
  - the Scope says otherwise
  - 'ownership of the material has been given to the *Client*' (i.e. option X9 (transfer of rights) has been selected, which states 'The *Client* owns the *Contractor's* rights over material prepared for the design of the *works* except as stated otherwise in the Scope').
- Retention of the drawings, reports etc. developed for the contract by the *Contractor* is for 'the *period of retention*', which is normally 6 or 12 years in the UK. This is so that the timescale matches the duration of the *Contractor's* liability, which depends on whether the contract is signed 'under hand' (i.e. not as a deed) or executed as a deed.
- The *Contractor* needs to ensure it has suitable insurance in place in case the standard of reasonable skill and care is not met (i.e. the requirements for professional indemnity insurance).

What is still missing from this clause, but you could argue is a broader omission, is an explicit grant to the *Client* of the right to use the *Contractor's* – or more likely its supply chain's – intellectual property for purposes related to the contract (e.g. operation and maintenance). This applies to both any intellectual property generated in delivering the contract (foreground IP) and, more likely, intellectual property already in existence (background IP). It could be argued that clause 22.1 covers this, but it contains no mention of intellectual property or licences. Consequently, what you often see in the option Z clauses is a requirement for the *Contractor* to gain all licenses, intellectual property etc. necessary for the *Client* to use, operate, maintain, alter and add-on to the completed asset.

X16: RETENTION

This option provides another means of recourse should the *Contractor* fail to perform. In operation, at every assessment date until Completion, the *Client* retains the *retention percentage* (specified in the Contract Data) applied to the Price for Work Done to Date for the sum owed. At Completion, half of this money is paid back, with the remainder held until the Defects Certificate is issued, which is normally a year after Completion. If the *Contractor* fails to correct Defects, the cost of correcting the outstanding defects is taken out of the retention.

A difference between this procedure and those of traditional contracts is the statement of a *retention free amount*, below which no retention is retained. The idea behind this is that retention is only needed as Completion approaches, so if the *retention free amount* was specified as 80% of the Prices, then the *retention percentage* would need to be five times higher than the normal value to give the same absolute sum at Completion. In reality, I have only seen the *retention free amount* not specified as zero a couple of times.

X17: LOW PERFORMANCE DAMAGES

This secondary option states that damages are to be applied for low performance of a completed asset. This presumes that the asset's level of performance is defined in the Scope. It also presumes that the level of performance can be objectively measured, that the *Client* can supply the necessary inputs (to a stated specification) for a process and that the *Client* does nothing during the contract to undermine the *Contractor's* ability to deliver the performance, for example, specify a component that undermines performance in some respect. For these reasons alone, low performance damages are hard to enforce. Apart from that, most of what I said about option X7 (delay damages) also applies.

However, this is the obvious secondary option to use to enforce a guarantee offered on the performance of the asset, such as the power consumption of a facility. If, when tested, the facility uses more power than promised, the *Contractor* can either correct the Defect or pay low performance damages at the rate stated in the Contract Data. For more information on this, including selection on whole life costs of an asset, see Patterson and Trebes (2015b).

X18: LIMITATION OF LIABILITY

This is a secondary option that all Contractors would want included as it limits their liabilities; and they would want the levels of liability set as low as possible. As a *Contractor*, be aware that, while this option has individual caps for (a) indirect or consequential loss, (b) loss of or damage to the *Client's* property and (c) Defects due to the *Contractor's* design, the 'total liability' excludes the second of these, as well as damages for delay (option X7) or low performance (option X17) and any pain share under the target-cost payment options (see Figure 1.7). It would, however, include damages associated with not meeting Key Dates.

X19: TERMINATION BY EITHER PARTY

This option is only in the Term Service Contract (TSC) and Term Service Subcontract (TSS). In NEC3, option X19 was titled 'Task Order' and was also only in the TSC. In the NEC4 TSC and TSS, the text for Task Orders has been absorbed into the core clauses. The publishers have reused the option X number for this new option.

As it says in the text, this option is not for use with option X11 (termination by the *Client*), which means that, after a stated *minimum period of service*, either Party may terminate. For the *Contractor*, this is presumably so that it can avoid being stuck for years in an unprofitable contract. Note that, unlike option X11, the *Contractor* does not get its *fee percentage* applied to the remaining work; this is also the case if the *Client* terminates under this option.

Figure 1.7 Option X18: illustration of which liabilities are included in the 'total liability' and which are not

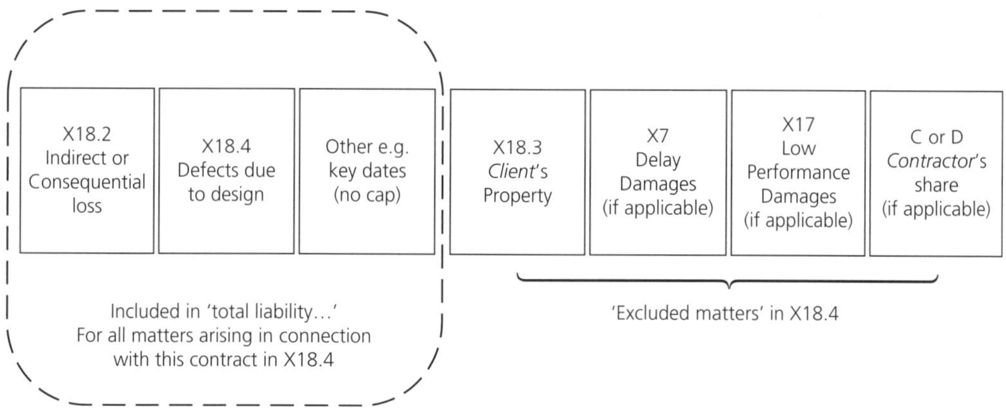

X20: KEY PERFORMANCE INDICATORS

Under this option, aspects of the *Contractor's* performance are measured and, if desired, targets can be set with incentive payments against them. What I said for option X6 (bonus for early Completion) about the benefits of using positive incentives also applies in this case. In the case of a performance specification, this could be for better performance than the contractual minimum as specified in the Scope document, assuming that better performance is of benefit to the *Client*. For instance, for a power project, the *Client* specifies as a performance specification the minimum level of output to be generated and the conversion efficiency per unit of input. Every extra megawatt it can generate and percentage increase in efficiency is worth something to the *Client*, so, in order to motivate the *Contractor*, sufficient financial benefits are allocated to the *Contractor* as incentives.

In public sector contracts, I have seen Key Performance Indicators used not just to measure the delivery of ancillary benefits for the community (e.g. local employment, number of apprenticeships), but also to reward the *Contractor* for these achievements.

Key Performance Indicators could also be used for measuring and rewarding softer metrics, such as response times under a service contract or teamworking behaviour over the duration of the project.

As I stated for option X12, I disagree that this option cannot be used with option X12 – it would just be unusual.

Note: The following secondary options, X21 to X26, are wholly new to NEC4. For simplicity, only the titles are underlined.

X21: WHOLE LIFE COST

Use of this new secondary option clause is intended to encourage the *Contractor* to come up with ideas to reduce the cost of operating and owning the asset. Having come up with an idea, the *Contractor* effectively runs it past the *Project Manager* who, if they like it, gets the *Contractor* to prepare a proposal in consultation with *Project Manager*. This proposal will include details of the capital costs involved – the proposed changes to the Prices – and a forecast of the savings to the *Client*. So far, so good. However, I do have two reservations with this clause.

- The change to the Prices is not a compensation event – and quite rightly, because the *Contractor* would only get the additional Defined Cost plus Fee and none of the whole life savings – but no guidance is given on how the change to the Prices is assessed. For example, there is no reference to the Scope, where the principles or methodology and raw data used for calculating whole life savings could be placed, or to how the net savings (operational/ownership savings minus capital costs) are to be divided up. In essence, option X21 allows the *Project Manager* (on behalf of the *Client*) to do a deal with the *Contractor*. So, given that the '*Project Manager* does not change the Scope as proposed by the *Contractor* unless the *Contractor's* quotation is accepted' (clause X21.4), the *Contractor* could hold out – or is that hold the *Client* to ransom – for a large proportion of the benefit. Alternatively, the *Project Manager* could instruct the *Contractor* to go ahead with an obviously good idea and leave the argument over the *Contractor's* share of the benefit until later. Alternatively, the *Contractor* refuses. You get the idea – silence on this matter is likely to lead to conflict!
- Modern project management, driven in part by economics, looks at benefits which, while including savings in operational costs, also include societal and environmental upsides and other intangible benefits. In a business environment, benefits could also include improved image, better customer service etc. I would like to have seen these included within the scope of this option, and for its title to have been 'whole life benefits'. A missed opportunity in NEC4.

X22: EARLY *CONTRACTOR* INVOLVEMENT

This option was first introduced to NEC3 as an addendum in November 2015, but it has been improved and is now included in the relevant NEC4 contracts. As stated earlier, the October 2020 amendments came out the week before I sent the text for this book to my internal reviewers and the impression given was that these clauses were substantially revised. So, the first part of this section is a description of early contractor involvement (ECI) in general, the second is a critique and experience of the initial June 2017 and January 2019 clauses (which are as good as identical) and the third is my opinion of the revised October 2020 clauses.

ECI IN GENERAL

So, ignoring the NEC specifics, what is ECI? In simple terms, a contractor is appointed early on to work with the client in developing the concept/outline design for the project. During the initial stage, the concept/design is progressed to a state which is sufficiently detailed that the client is confident about what they are going to get and the parties can agree the prices for the next stage. This stage also typically involves of a lot of de-risking, especially for civil engineering contracts (e.g. further site investigations, sorting out land access, early service diversions, seasonally based environmental works) and is carried out on a cost-reimbursable basis.

If the client is satisfied that the proposed design will deliver what it wants, the project can proceed to the next stage. This stage is normally done on a target-cost basis, with the costs and the fee from the first stage being included in the target prices. If the prices and other details for the second stage cannot be agreed, the contract does not proceed.

Under NEC4 option X22, this is all done under one contract. However, before I discuss the nuances of X22, a more fundamental issue is when should a Client use this option, as opposed

Table 1.5 Selection of NEC4 contract for early *Contractor* involvement

Circumstances	Suggested NEC4 contract
The *Contractor* is providing buildability and other advisory services only, with the *Client* or its own designer developing the design.	Professional Service Short Contract (PSSC)
The *Contractor* is leading the development of the design by employing other designers, under either PSS or PSSC, with no physical or investigatory work being done under this contract. The *Contractor* may supervise this work on behalf of the *Client*.	Professional Service Contract
The *Contractor* is leading the development of the design by employing other designers, under either PSS or PSSC, and is likely to be doing some early physical work, e.g. enabling or seasonal work and/or further site investigation.	Engineering and Construction Contract with option X22

to a separate Professional Service Contract for stage 1. Table 1.5 offers a starting point when using the NEC4 family.

THE JANUARY 2017 AND JUNE 2019 ECI CLAUSES

The ECC states (in the title of the option) that option X22 is only to be used with options C or E. However, ECI originally developed, in part, from the preferred contractor approach used by commercial property companies, where the detailed design and construction was let under a lump-sum approach, the equivalent of which is option A. I see no reason why option X22 could not be used with option A, which would involve simply changing the letter 'C' in clause X22.3(9) to 'A'.

An interesting variant of ECI, which I had not come across before the publication of X22, is that while there is still a Stage One and a Stage Two, the whole contract can be performed on a cost-reimbursable basis under option E, but with a higher level 'target' (compared with an option C contract) around which the gain operates. In X22, this higher level target is called the Budget, which

- includes allowances for compensation events that adjust the Prices and for other identified estimates of stated *Client* costs (e.g. land take)
- has far fewer reasons for adjustment compared with a normal NEC4 contract – these reasons are often identified by the *Client* in the contract documents.

If the Project Cost – the total Defined Cost and the Fee paid to the *Contractor* plus the other identified *Client* costs – comes in below the Budget, then the *Contractor* is paid the *budget incentive*, which is the saving multiplied by a percentage stated in the Contract Data.

Since the publication of the original ECI addendum to the NEC3 contract in November 2015, I have had a refereed article published on these clauses (Broome, 2016) and put together a number of ECI contracts. So, based on my experiences, here are the issues that typically come up.

- Public sector Clients very much like the idea of a project Budget around which the incentive mechanism works as (a) this is often what they are funded for and (b) it means that whenever there is a compensation event, the adjustment to the Prices is for forecasting purposes only. Consequently, both Parties are motivated to minimise the effects of the event, and there is much less hassle around adjusting the Prices as it does not affect the pain/gain share. However, public sector Clients do not see why the *Contractor* does not take any pain if the Project Cost exceeds the project Budget and so normally want extra clauses written in to cover this.

- Contractors very much like the idea of gain compared with the project Budget, but are much less keen on pain, especially if the *Client* costs contain items over which they have little or no influence. Consequently, such items and forecast costs should not be included in the contract Budget, because Contractors might not bid and the *Client* may be giving away unearned incentive payments. However, taking these items out reduces the attractiveness of this approach for Clients as the contractual Budget will not then match their funding budget.
- There is no formal termination if the *works* do not proceed to Stage Two. Instead, 'the work required in Stage Two is removed from the Scope' and 'the *Client* may appoint another contractor to complete the Stage Two *works*' (clause X22.5). However, I can see the potential for a legal challenge from an unhappy *Contractor* because, as there has not been a formal termination, the *Contractor* still has the contract to do the *works*. Further, because there is no termination, it is not clear how the contract is brought to a close. By this I mean, what happens about Defects and their correction? (For example, when is the *defects date?* How much does the *Contractor* gets paid, including any retention?) And the lawyers I have worked with agree. Consequently, for legal certainty, when preparing contracts that contain option X22, I write in a clause to confirm termination, the procedure to be followed and the amount due, all as per Section 9 of the core clauses.
- The preceding point opens up the question of what is in the Scope, tendered Prices and Completion Date at the initial Contract Date. Is it just the ECI work, or should it include the *works* to physical Completion? The contract does not explicitly state this and I am not sure, but I would err towards the former. This is because the contract states that 'The total of the Prices for Stage Two is assessed by the *Contractor* using the Pricing Information stated in the Contract Data' (clause X22.3(5)) and a condition for proceeding to Stage Two is that 'the *Project Manager* and the *Contractor* have agreed the total of the Prices for Stage Two'. This implies that any Prices tendered by the *Contractor* for completing all the design and construction are up for grabs, albeit using price information put forward by the *Contractor* (i.e. the Pricing Information). But absolutely nothing is stated about how this Pricing Information is used to build up the Prices for Stage Two, or how adjustments to the Completion Date are calculated based on what emerges in Stage One. Consequently, careful consideration needs to be given to how the contract documents are structured with respect to what the *Contractor* has to do in Stage One and what it *might have to do* in Stage Two, should it proceed, and how the revised Prices, Key Dates and (sectional) Completion Date(s) are arrived at.

In addition, a number of Clients I have worked with on large highways projects do not want to include ECI as per the unamended clauses. They want to give, at tender, a design that the *Contractor* can price up to give meaningful target Prices. Sometimes this design is a detailed design, but other times it will be an outline design that complies with the planning approvals etc., but which is accompanied by a more detailed 'specimen' design, that is supplied for information only. In Stage One, the *Contractor* does more research, in terms of conducting site investigations and assessing whether it can actually build this design (as well as all the de-risking). If the physical conditions are different from what was expected and/or the *Contractor* cannot build the contract design, it can propose an alternative, and the Prices for Stage Two are revised through the compensation event procedure to reflect the necessary changes. The *Contractor* also develops the design and, once Stage Two is entered into, the *Contractor* then 'adopts' the design as if it were its own, i.e. it takes responsibility for all ambiguities, inconsistencies and omissions (or, more precisely, the risk of this is within the target Prices). This has been referred to as 'optimised contactor involvement'.

THE OCTOBER 2020 ECI CLAUSES

The main changes made to the ECI clauses in 2020 were as follows.

- The ongoing submission of proposals in Stage One was expanded from including just design proposals to any proposals, so proposals can relate to, for example, construction methodology, stakeholder management proposals etc.
- The *Contractor* now prepares proposals in consultation with the *Project Manager*.
- The preparation of proposals now culminates in a separate submission to the *Project Manager* at the end of Stage One, in which everything is brought together: proposals for Stage Two; a revised programme; any changes to Access Dates (a new term), Key Dates and Completion Date (previously, these were not part of any submission); and the total of the Prices or any change to it.

All of the above changes are good and are improvements. However, none of them really address my criticisms of the January 2017 and June 2019 ECI clauses (regarding termination if the contract does not proceed to Stage Two and the content of the Scope) or the use of optimised contactor involvement. In addition, while the *Contractor* submits its proposals for Stage Two to the *Project Manager*, both on an ongoing basis in Stage One and as part of a bigger submission at the end of Stage One, the clauses are still a bit unclear about how the proposals are incorporated into the Scope. For example, if these proposals are not design proposals (but, say, commitments to a set methodology), how can they become 'Scope provided by the Contractor for its design'?

To summarise, I am enthusiastic about ECI because it gets the *Contractor* in early on, when it can add maximum value. However, whether using the original NEC4 wording or the October 2020 improvement, careful thought needs to be given to its implementation. This needs to be reflected in the contract documentation, which unfortunately means amending the words as written in the *conditions of contract*. Having said this, and in defence of the NEC4's authors, producing a one-size-fits-all set of ECI clauses is a hard ask.

X23: EXTENDING THE SERVICE PERIOD

This option only applies to the two full Term Service Contracts (TSC and TSS) and the Design Build and Operate, Contract (DBO). Basically, it does what it says on the tin. Whether by mutual agreement or because the *Contractor* has met the stated *criteria for extension*, it allows the Service Period to be extended by mutual agreement using the mechanisms of the contract, rather than by a supplementary agreement. From a client attractiveness point of view, it signals to tenderers that the *Client* is willing to extend the Service Period for good performance, and it encourages good performance once the contract is let.

X24: THE *ACCOUNTING PERIODS*

This option is exclusive to the full Term Service Contracts (TSC and TSS). It attempts to ensure that, for each stated *accounting period*, the financials for that period are wrapped up within

13 weeks of the period end, including the resolution of all disputes. If using option C, it makes sense that the end of each *accounting period* corresponds with the dates stated in the Contract Data for interim payments of the *Contractor's* pain/gain share.

X25: *SUPPLIER* WARRANTIES

This option is only included in the full Supply Contract (SC). It might also be appropriate to use X25 with the Supply Short Contract, in which case you can just cut and paste the clauses.

X26: PROGRAMME OF WORK

This final X option only applies to the Alliance Contract. It allows new *schemes* to be progressively added into the Alliance. Having said this, when working on an alliance based on this form, the lawyers and I concluded that the clauses were not adequate – mind you, this was for a contract that started with £500 million of work and was to have another estimated £6.5 billion of *schemes* added in over more than 10 years, so it involved a little bit more than adding a scheme or two.

1.2.3.5 The Y(UK) options

OPTION Y(UK)1: PROJECT BANK ACCOUNT

This was introduced to NEC3 as an addendum but is now included in the relevant NEC4 contracts. As stated earlier in this book, the October 2020 amendments to NEC4 came out the week before I sent the text to my internal reviewers. These amendments deleted the previous Y(UK)1 clauses in their entirety and replaced them with new ones, which better reflect practice. As with the X22 (early *Contractor* involvement) clauses, I will first discuss the Project Bank Accounts (PBAs) under the NEC in general, then the initial June 2017 and January 2019 clauses, and then the revised October 2020 clauses.

PBAS IN GENERAL

So how does a PBA work under the NEC? In simple terms, a trust is set up with a bank account, with the trustees being the *Client* and *Contractor*. Rather than the *Client* paying the *Contractor* and the *Contractor* then paying its Subcontractors, the *Client* pays into the PBA and the money then goes directly to the Subcontractors or even subsubcontractors (who, having signed a Joining Deed, are called Named Suppliers). In fact, the *Contractor* should also be paid via the PBA.

There are several advantages to using a PBA.

- Payments along the supply chain are faster, as there is less incentive for the *Contractor* to keep the money in the PBA than there would be if it were in its own bank account (whether to help with its cash flow and/or to reduce its interest payments). Having said this, there is some incentive for the *Contractor* to keep the money in the PBA because option Y(UK)1 explicitly states the *Contractor* gets the benefit of any interest.
- As the money is held in trust, the trustees have a duty to allocate the money appropriately. To do otherwise (e.g. to pay money that is clearly allocated to Named Suppliers into their own company) is not permitted.
- Accounting and auditing are easier, particularly under cost-based contracts. The *Project Manager* can evidence the incomings and outgoings of a PBA far more easily than it can for the *Contractor's* own bank account, which will include all the *Contractor's* other transactions.

The first two of these reasons should make the project far more attractive to the second-tier supply chain, but experience suggests that the PBA has to be explained to them in order for them to sign the Joining Deed.

A frequently asked question is: Why is this a UK option? The answer is, apparently, because trust funds in this format are unique to UK legislation. Consequently, alternative wording or a completely different vehicle are needed to create the same effect outside of the UK.

THE INITIAL JUNE 2017 AND JANUARY 2019 CLAUSES

I think it fair to say that, at the time of NEC4's publication, everyone was groping around in the dark, trying to figure out how PBAs worked in practice. This included banking people, who may have had an expert somewhere in their organisation, but those experts had to be located to help people set up PBAs.

Under the original clauses, it was the *Contractor* who had to set up the PBA. To ensure the right money goes into the PBA on a monthly basis, the *Contractor* has to identify in its payment applications the amounts due to itself and to each Named Supplier. The *Client* then pays this amount into the PBA. No problem so far.

What quickly emerged was that the Authorisation process was hopelessly bureaucratic. The *Contractor* had to prepare and sign the Authorisation (setting out the amounts due from the PBA to the *Contractor* and each individual Named Supplier), submit it to the *Project Manager* for acceptance, who would then pass it on to the *Client* for signing, who would then submit it to the bank for payment. In this age of electronic banking, this was an outdated process, even if the Authorisation were signed electronically.

In practice, what seemed to be happening (based on my limited experience) was that Contractors and banks were not only having problems with the Authorisation process (banks just do not operate this way these days), but were also having trouble setting up accounts with dual authorisation (i.e. where the *Contractor* puts in the payment details and clicks the button to say it is happy for these amounts to be paid, but, in order for the payment to be made, the *Project Manager*/*Client* must also authorise the payment, having first checked it against the application). So, what was happening instead was that the *Contractor* became the sole authoriser, with the *Project Manager* (or delegate) auditing the payments. However, even then, the *Project Manager* would often find it difficult to assess the PBA directly, instead having to use the *Contractor's* access details or, more realistically, work from a printout of payments, or even be signed in by the *Contractor's* personnel!

Fortunately, as the use of PBAs has grown, my understanding is that these issues have reduced, especially if you get through to the right person in the bank and so set up the right bank account.

THE OCTOBER 2020 CLAUSES

These amendments allow either the *Contractor* or the *Client* and *Contractor* jointly to set up the PBA, but not, for some reason, the *Client* only (as I know some Clients have done).

Instead of the Authorisation, which was passed along the participants before getting to the bank, there is now a Payment Schedule, which is prepared by the *Contractor*, copied to the *Project Manager* and supplied to the bank. In reality, 'supplied to the bank' probably means entered electronically by an employee of the *Contractor* into the bank account for payment. If the account is jointly held, then the payments have to have dual authorisation. If it is held by the *Contractor*, sole authorisation by the *Contractor* is allowed, with the *Client* checking afterwards that the right payments have been made.

To keep on top of what payments have been made, the *Contractor* also has to keep an up-to-date Project Bank Account Tracker, which is a register of all payments made to and from the PBA and the date each payment was made. This is to be submitted to the *Project* Manager within a week after each round of payments.

All this is more reflective of what, from my arm's length experience, happens in practice, so I commend the updates!

OPTION Y(UK)2: THE HOUSING GRANTS, CONSTRUCTION AND REGENERATION ACT 1996

This option incorporates the necessary terminology to comply for payment purposes with the UK Construction Acts mentioned in the explanation of the dispute resolution options (see Section 1.2.3.2). It skilfully overlays and adds to the existing words in Section 5 (Payment) of the contract, only deleting one sentence. The second Construction Act also closed out loopholes that the unscrupulous were using to avoid paying the contracting party.

Y(UK)2 was also one of the options modified by the October 2020 NEC4 amendments. The change was only very slight, and was made in response to recent case law.

In *Rochford Construction Ltd* v. *Kilhan Construction Ltd* [2020] EWHC 941 (TCC), the judges made a comment in 'obiter' (which means in passing, i.e. it was not central to the dispute) that, under the Construction Acts, the final date for payment cannot be directly linked to the provision of an invoice, as is common practice. However, the due date can be linked to the provision of an invoice, so the October 2020 amendment now links the submission of an invoice from the Party to the due date, with the due date being the later of the date of receipt of an invoice by the Party that is making payment and 14 days from the assessment date. The final date for payment is then a fixed period of time from this due date.

OPTION Y(UK)3: THE CONTRACTS (RIGHTS OF THIRD PARTIES) ACT 1999

This Act was introduced to try to reduce the need to spend large sums with lawyers on setting up collateral warranties (see option X8). Instead, under the Act, a person or organisation who is not the *Client* can enforce a term of the contract from which they would benefit, unless there is an explicit opt-out statement in the contract. Specifying this option is the opt-out, and Contractors quite definitely want it specified, provided the term and party who can enforce it are not explicitly stated in Contract Data part one. Stating the term and party in the Contract Data provides an alternative to the use of collateral warranties. However, collateral warranties are still strongly favoured by lawyers on the grounds of convention and case law and, possibly, fee generation.

To make it clear that rights of third parties are not to come out of the contract, you can select option Y(UK)3 and simply state 'none' in the Contract Data entries for term and *beneficiary*.

1.2.3.6 Use of option Z: *additional conditions of contract*

This option allows additional conditions to be added to further tailor the contract strategy and/or risk allocation. The flexibility inherent in the NEC system is intended to minimise the need to make amendments to the contract.

Unfortunately, the majority of option Z clauses that I see from live contracts, I see because the Parties are at the early stages of dispute and want a second opinion on what they mean in practice. The clauses have usually been written by lawyers, presumably in the belief that writing more increases certainty, when in reality it has the opposite effect, as, all other things being equal, ambiguity is interpreted in the light least favourable to the Party that wrote it, i.e. the *Client*. I regularly see ambiguity, not just between the option Z clauses and the standard clauses in an NEC3/4 contract, but also within and between the option Z clauses. In terms of providing contractual certainty, less is often more! The examples in the following boxes are taken from NEC3 contracts.

On one project worth about £25 million, the lawyers had got in before me and written an amendment under option C (target contract with activity schedule) to the Disallowed Cost clauses, which meant that almost any rework was disallowed (e.g. if a chippie measured twice before cutting, it was disallowed). The *Employer* had done two smaller projects using the same amendments. I asked what the *fee percentage* on these projects was, and it was 13.5%, when the going rate was 5.5–8%. I, as politely as possible, asked the *Employer* if they wanted to pay at least an extra 5% of £25 million (or £1.25 million) to the *Contractor* for a risk contingency which, due to having a small professional team, they would not be able to reclaim by enforcing the provisions and, in any case, would undermine the partnering ethos that the Parties wanted. They did not. I suggested that if they wanted to put more risk on the *Contractor*, the easiest way was to alter the pain/gain share. The amendment was taken out.

On one target-cost contract, the lawyers had done two things.

- They had deleted the physical condition and weather clauses, but then added in a page and a half of legalistic amendments that put some of the weather/ground risk back on the *Employer* if they did or did not do certain actions. It rained heavily over the summer, and the *Contractor* overran the unadjusted target Prices by some £800 000. Initially, the *Project Manager*, and then the *Employer*, refused to adjust the target.
- They had altered the contract so that Completion was only achieved at the end of a landscaping period that started on sectional Completion of the asset and went on for 3 years. What they had forgotten to do was alter the time when the *Contractor* had to pay back any share of overrun – at Completion – so the *Contractor* quite happily held onto the £800 000 plus of Defined Cost plus Fee for 3 years before the *Employer* could raise the issue of getting their money back.

Because no-one actually knew what the clauses relating to adjustments to the Prices meant and it was clear that the *Project Manager* had not followed the rather obscure procedures, the *Employer* agreed a settlement (or 'claim', as it was not a compensation event) on an 'at-cost' basis – i.e. the *Contractor* paid back the Fee on the £800 000 of Defined Cost, but not the majority of the pain that this £800 000 represented.

I could go on! I also commonly see option Z clauses cluttered up with text that should be in the Scope (or Works Information in NEC3) as it describes constraints on how the *Contractor* is to Provide the Works. As such, because the wording is part of the *conditions of contract*, it cannot be changed, whereas if it is in the Scope, it can.

Just to confirm my experiences, here is some research data from an MSc research project (Houghton and Trebes, 2011). It drew its conclusions from 28 interviews, three case studies and 58 survey responses, so it is a fairly comprehensive bit of research by Masters standards. The key points of this research were that

- 73% of respondents believed that, generally, option Z clauses create a more adversarial contract
- only 6% of option Z clauses were viewed by the researchers as 'Valid'! Of the remainder: 24% should be in the Works Information (now Scope); 20% showed a lack of knowledge of the ECC, as something similar was already in the contract (so there was risk of ambiguity or contradiction); 16% shrunk the *Employer's* (now *Client's*) obligations, thereby reducing the stimulus to good management; 16% changed language back to ICE or JCT forms and were therefore using contractually undefined language; 8% transferred risk back to the *Contractor*, typically by deleting compensation events; 8% were either onerous or unfair on the *Contractor*; and 2% were legal clarifications.

In conclusion, writing significant option Z clauses is often time-consuming and expensive (as you are likely to be paying expensive consultants, typically lawyers, to do it), but it also increases the likelihood and potential impact of 'ambiguity and inconsistency in the contract' as a project risk. Given that this risk often falls on the Party that created the ambiguity (i.e. the *Client*), I am not sure why Clients do this!

If you must write option Z clauses, I give some guidelines in Section 2.2.2.11 in Chapter 2.

1.2.4 Other aspects of flexibility

1.2.4.1 *Contractor* design

There are a number of reasons for allocating design to a contractor.

- There is a single point of responsibility for the delivery of the project. The contractor, having been given the brief, is now responsible for both design and construction, which should result in fewer compensation events.
- The design and construction periods can overlap, leading to faster delivery of the end asset.
- The contractor can utilise its ingenuity and knowledge of methods of construction and price of materials, etc., to minimise costs.
- A contractor is usually more capable of managing the design risk, which results in greater certainty of meeting the time, cost and performance objectives. A client therefore has to satisfy itself that the contractor is fully capable of managing this risk.

However, if a client, having let the contract, then changes its mind about its requirements, either because it has not thought them through properly or has expressed them unclearly and ambiguously, then this certainty is threatened, and it will cost the client dear. As one distinguished lawyer wrote, clients have to learn to 'let go' once the contract is let on any price-based design and build contract.

In the ECC, *Contractor* design is accommodated in clauses 21 and 22 and by specifying option X15. Such is the flexibility of the ECC that the *Client* can

- give a scope and outline design, a statement of the purpose of the asset to be constructed and refer to the standards that it has to be constructed to
- specify the performance requirements of the asset (e.g. if the *Client* wants a power station, the output, efficiency, control, environmental standards etc. that it must conform to, and that the *Contractor* is to design and build the facility on the specified Site)
- design the *works* itself or through a third party (i.e. a design consultant)
- use any combination of the above (e.g. if the *Client* requires an office building, it can describe the purpose of the asset and give an outline design, and state the performance requirements for the mechanical and electrical services, yet retain control for key parts of the asset, such as an architectural feature, which it can fully design itself).

Whatever course is adopted, in the ECC it is vital that the documentation (or Scope in NEC4 ECC terminology)

- states clearly what the *Contractor* is to design, otherwise the *Client* will have to pay the additional costs for that design as a compensation event
- describes what is wanted in sufficient precision to ensure that a *Contractor's* design that complies with the Scope will give the *Client* what it wants. If the *Client* insists on changing the Scope, it could give rise to compensation events, not just for the costs of redesign, but also for any resulting delay to construction.

1.2.4.2 Application across different engineering disciplines

In Section 1.1., I identified and discussed some circumstances in which the NEC4 family can be used outside of its core heavy engineering and construction origins. This section specifically discusses the flexibility inherent within the ECC. Again, many of these principles are incorporated in other contracts in the NEC4 family.

While the ECC originates from and is owned by the UK's Institution of Civil Engineers (ICE), it has been designed to be used on projects containing mechanical, electrical, civil or building elements, as well as chemical process plants. Most contracts now involve an element of work from outside the main discipline, in some cases substantially so. For example, over 50% of the cost of a building can sometimes be found in the mechanical and electrical services. A project in the heavy industrial engineering sector may well involve substantial civil engineering work for the foundations and building work to house the facility, as well as computerised control systems. Whether the management of the interfaces between different works packages is done by the client or passed down to a main contractor, who then subcontracts the various packages, it is not helpful for integration and control if each discipline has different, and often incompatible, conditions of contract. The ECC was therefore written in a way that accommodates the multi-disciplinary nature of the modern construction industry.

- The ECC omits technical information. For example, there are no detailed testing and commissioning requirements; instead, there is an outline framework for testing and quality procedures. The detail will be included in the Scope part of the contract documentation. It has been argued by some that one of the reasons why the ECC is so short compared with traditional conditions of contract is that it excludes a lot of technical information that other conditions contain, such as testing requirements. As an observation this is correct, but as a criticism it is not valid. Testing requirements will vary between sectors, because of: the maturity of the sector and the technology involved; how the project is specified

(e.g. performance specification versus fully designed); the maturity or sophistication of the individual contractor; the relationship between the client and the contractor; etc. To have in-depth testing requirements in the conditions of contract means that in many cases they are inappropriate to the circumstances.
- Some of the terminology used in the ECC is different from that used in some sectors of the industry. For instance, what are known as 'construction plant' and 'temporary works' in the civil engineering and building sectors are covered by the contractual definition of 'Equipment' (clause 11.2(9)). 'Plant and Materials' are items included in the *works* (clause 11.2(14)), i.e. 'permanent works' in traditional civil engineering terminology. This can cause some initial confusion until people become familiar with the terminology.

Having given training courses across different sectors, participants who only know a little about the ECC are often, to varying degrees, initially cynical about using it for their projects and would prefer to stick with what they know. However, by the end of the course, having understood the terminology, and realising that they have more freedom to tailor the technical aspects of the contract to their circumstances (as opposed to being restricted by what the sector conditions of contract state), as well as recognising the benefits of good project management, they are much more enthusiastic.

In practice, the ECC has been used successfully and substantially on contracts in the civil engineering, water, building, telecommunications and power industries, to name a few. In the UK, for the first two sectors it is now the dominant contract form. In the UK building sector, it is at least on par with the JCT form, tending to be used for the higher value (£20 million plus) building contracts. However, one of the conclusions drawn from my original research was that for building projects that are substantially designed at tender, the ECC may be less suitable due to the high number of small value changes that are sometimes introduced and the method by which time and cost effects of a compensation event are calculated in the ECC. This may be partly related to not having good clauses for evaluating changes in quantity (see my critique of the option B clauses in Section 1.2.2.2) or it may be that the Short Contract would be a better contract to use. Of course, if the client knows what it wants and the building is well designed, there should not be too many changes.

1.2.4.3 International use

The NEC family of contracts was written for international use. As such it does not attempt to paraphrase existing law, and generally uses simple language understandable to those whose first language is not English. In the original research, I interviewed one Hong Kong based contractor's project director, who reported how, for the first time in his experience, engineers from different countries, none of whose first language was English, would have discussions on the conditions of contract rather than leave it to the English quantity surveyors.

NEC contracts have had substantial use in Hong Kong, New Zealand and South Africa, but have also been used (to my knowledge) in Abu Dhabi, Australia, Belize, Canada, Dubai, France and Germany (for an international contract that spanned borders), Ghana, Holland, India, Nigeria, Qatar, Thailand and Vietnam.

Some of the NEC4 secondary options contain clauses that might be suitable for contracts with an international flavour, such as options X3 (multiple currencies) and X14 (advanced payment to the *Contractor*). Equally, some secondary options are only applicable where UK legislation applies, specifically options W2, Y(UK)1, Y(UK)2 and Y(UK)3. NEC also has published secondary options for use in New Zealand and, apparently, secondary options for other countries are being prepared.

1.2.4.4 Flexibility in Contract Data part one

Contract Data part one is completed by the *Client* prior to sending out the documents to the tendering Contractors. Whenever you see something in italics in the contract – as you will have seen consistently in this book – that means the specifics are stated in the Contract Data. For example, when you see '*Project Manager*', the specific individual is named in the Contract Data. One of my early clients, at the turn of the century – which makes me sound very old – was in the telecommunications industry as a management contractor. They literally had employers coming to them saying 'I want to spend £25 million on an IT network and speed to market is of the essence. How soon can you start and, more importantly, finish?' What they liked about the NEC was that the Contract Data could be agreed in a morning, whereupon the work could start. Under traditional contracts, they would be having to take out and redraft clauses over weeks and months, all while the money was being spent and they were at risk.

Examples of where flexibility can be enacted using Contract Data part one include

- being able to specify the length of time after Completion during which the *Contractor* has an obligation to correct Defects and the time allowed for correcting Defects once notified, by entering figures for the *defects date* and *defect correction period*
- altering the pain/gain shares under target-cost options C or D to fit the risk profile of the contract
- for many of the secondary options, adjusting the level of damages and caps on liability by stating different figures.

1.2.5 Conclusion on flexibility

Wherever the NEC has been used and whatever aspect of flexibility has been used, by and large it has been used again. Put simply, this implies that it works.

1.3. How clarity and simplicity are achieved

1.3.1 Overview

As has been stated, one of the principal aims of the NEC was greater clarity and simplicity compared with more traditional conditions of contract. In this section I briefly examine

- the origins of traditional conditions of contract
- the aims of the authors of the NEC
- general criticisms of the NEC and ECC since publication
- the change in drafting philosophy.

In addressing these issues, this section includes the arguments for and against the drafting style and philosophy of the NEC4 family of contracts, which answers some generic legal criticisms by giving the counter-arguments that those of a more traditional legal mindset often come up with. It should be noted that, in my experience, those lawyers who are more customer focused and/or do not have the baggage or vested interests arising from the use of more traditional forms tend to be more enthusiastic towards the NEC. Indeed, my experience is that the lawyers who are the most enthusiastic towards the NEC tend to be in-house client lawyers – think through the motivations of these lawyers in comparison with the motivations of those who make their money from disputes under the traditional forms!

Understanding the drafting philosophy behind the contract helps in the preparation and use of the NEC4 contracts.

1.3.2 Origins of traditional construction conditions of contract

Way back in 1997, Ross Hayes – a colleague at the University of Birmingham, who is still active in the NEC community – and I had a paper published that looked at the issue of clarity in construction and engineering contracts (Broome and Hayes, 1997). In it, we defined clarity as meaning 'the design and layout of the whole contract document, as well as the use and order of words within a sentence'. In addition, we extended the definition by saying that 'clarity requires that the clauses within a contract should fit together to form a logical whole, be procedurally correct and relevant to modern construction practice'.

The original authors of the NEC, in making clarity and simplicity one of their principal objectives, concluded that existing construction conditions of contract did not achieve this definition of clarity. When examining the history of these contracts (see Broome and Hayes, 1997) this is perhaps not surprising, for the following reasons.

- Their origins can be dated back to 1870, if not before. For instance
  - the Joint Contracts Tribunal (JCT) conditions are direct descendants of the Royal Institute of British Architects (RIBA) form, which was first published in 1870
  - while the original ICE conditions were first published in 1930, many of the terms found in ICE fifth edition and the International Federation of Consulting Engineers (Federation Internationale Des Ingenieurs Conseils (FIDIC)) conditions can be found in the conditions of contract used to construct the embankments of the river Thames and the main London sewers in the 1870s.
  - the Infrastructure Conditions of Contract (ICC) conditions are directly developed from the ICE conditions, and the FIDIC conditions' heritage can also be directly linked to the ICE conditions.
- The lawyers who drafted the original contracts were almost certainly Chancery pleaders, with great experience of leases, but little knowledge of the commercial and practical problems of construction. In modern procurement terminology, the contracts were developed as transactional contracts (as in one party hands over the money and the other hand over the goods or land, with no further interaction). But is this how construction projects or projects generally operate today?
- The subsequent development of these contracts has been by committee, with representatives drawn from various sides of the industry. As a number of legal commentators have noted, this has led to changes being introduced
  - reactively, in response to comments, court cases or developed practice, without any real consideration of policy or objective
  - to give one side of the industry an advantage over another.

Considering their origins and development, it is perhaps not surprising that surveys and comments of practitioners, distinguished legal commentators and the courts have consistently criticised the clarity, in its broadest sense, of traditional construction contracts. Some went so far as to call for existing contracts to be torn up and redrafted.

1.3.3 Aims of the authors of the NEC

The original authors of the NEC aimed to achieve clarity by

- using simple and commonly occurring language and avoiding legal jargon – this both avoids the need for extensive legal knowledge of case law on what the courts have interpreted the contract to say, and avoids the use of emotive phrases, which produce emotional responses (e.g. claim!)
- using identical phrases where possible
- setting out duties and responsibilities clearly and precisely – who does what and by when
- using engineering terminology common to all disciplines wherever possible – a good example of this is the clause for the acceptance of the programme (clause 31.3 of the ECC)
- not attempting to paraphrase existing law (in part because it was intended to be an international contract)
- settling for clarity above fairness in minor matters that would involve complicated text – for example, avoidance of complicated text is one of the reasons, among many, why nominated subcontractors are not mentioned in the ECC (Related to this, when I was doing my original research, one contractor, well known at the time for being contractually cute, complained that they 'Couldn't get many of the minor things we normally do, which we then use to muddy the contractual waters on a large scale'.)
- omitting matters that are more effectively covered in the technical specification (e.g. testing requirements)
- having a document with a logical structure by
  - having sections that deal with particular issues, rather than having issues scattered throughout the contract
  - treating procedures as processes
  - checking that these processes are logically complete by the use of flow charts (which are published to assist users who think that way)
  - avoiding cross-referencing between clauses
- designing the contract to reflect modern and up-to-date construction practice (e.g. regarding subcontracting as normal practice)
- having a modular system of 'bolt-on' optional clauses through which the contract can be tailored to meet the requirements of the individual project, without needing extensive redrafting
- excluding contract-specific data, so that there is less need to change, delete or add to the core conditions of contract – this data is stated in Contract Data part one (e.g. to define the length of time over which the contracting organisation is obliged to correct defects (the *defects date*), you just write the required period in the relevant place in the Contract Data, as opposed to crossing out and re-writing a contract clause).

The last two points also give the contract greater flexibility in its range of applications.

These aims and principles are extended to the rest of NEC family of contracts, which use the same (or similar, where this is not possible) terminology and interlinking processes and timescales. The benefit of this for clarity is self-evident, as compatibility should remove confusion.

1.3.4 Comments on general criticisms of the NEC and ECC since publication

Aside from detailed criticism of individual clauses, some of which are justified and some of which simply are not, the NEC style of drafting has been criticised by a number of lawyers for the following reasons.

- The drafting had not been subjected to judicial interpretation and, as a consequence, is more likely to be referred to the courts than other forms of contract. To some extent, this criticism was justified. However
  - precedent does not necessarily guarantee certainty – for instance, a series of seven cases on the conclusiveness of the architect's final certificate under the JCT form ran from 1965 to 1994, and there have almost certainly been more cases since (at what point was or is 'certainty' reached?)
  - despite its widespread use, there is still relatively little case law on the use of NEC contracts, although it is increasing.[18] The evidence would, therefore, indicate that very few disputes end up going to court. This implies that the average user of the contract thinks it is clearly written. However, David Thomas, a barrister specialising in construction law, did make a valid point at an evening talk: while the NEC is undoubtedly clearer to the average construction professional, when a problem gets to lawyers whose expertise is in case law, they have far fewer legal cases to rely on. Consequently, to the legal profession, it is more open to interpretation and hence less clear!
- The use of plain English could cause difficulties in certainty and proper legal construction. One practising lawyer has countered by stating that 'it is a matter of common sense that a contract should aim to be readily intelligible by its users and should have a logical sequence and order to it' (Herga, 1995). The research from which the first edition of this book was derived found that the majority of contract participants believed the NEC/ECC to be as intelligible, if not more so, than more traditional forms after using it on one contract. From my training experience, I would say only around 1 in 50 construction practitioners do not find this to be the case, although this is decreasing as 'old school' practitioners retire. Interestingly, a lot of the principles for plain English adopted by the NEC are now recommended by the Law Society.
- The minimal use of cross-referencing does not aid understanding. However, none of the interviewees in the original research commented adversely on the lack of cross-referencing. Indeed, several positive comments were received.
- The NEC/ECC uses new terms to describe what are well-understood terms in other contracts. This is true, but, as one prominent lawyer has commented, these traditional terms have gained a significance far beyond their contractual provisions. Indeed, this was one of the reasons for using new terminology, as old terminology produces emotive reactions (e.g. the words 'claim' and 'variation'). Another reason for the change in terminology is the multi-disciplinary application of the contract (e.g. the definitions of Plant and Materials, Equipment and acceleration). Once participants have been made aware of the changes, few problems seem to materialise in practice.
- Without familiarity, similar definitions can be confused. This has occurred in practice, although not very often. I particularly draw to the reader's attention: the differences between the date of Completion, the *completion date* and the Completion Date; the various

[18] A few months before this section was written, Matthew Garratt, a member of the NEC Board, gave a talk on case law under the NEC. He identified 40 court cases. However, 23 of these related to the enforcement of an adjudicator's decision, leaving 17 'real' cases. From Matthew's summary, by my reckoning, four of these cases related to either Z clause amendments to the *conditions of contract* or other amendments made by Clients, leaving 13 that were related to the clauses as drafted. For almost all of these (but not all!) the relevant judge sided on the common-sense interpretation of the clauses. Indeed, for some, you have to wonder how on earth they got all the way to the courts.

definitions of the Prices and Price for Work Done to Date, depending on which main option is used; and the *defect correction period* and *defects date* (see Section 1.5 for proper explanations of these terms).

- The strange use of the present tense to impose legal obligation (as one lawyer put it), although this follows practice recommended by the Plain English Campaign and has not caused problems on site. It works because of the simple statement in clause 10.1, that the actors 'shall act as stated in this contract'. I now find it much harder to write in a traditional legal style, with lots of uses of 'shall' and 'should'.
- The use of bullet points and sub-bullet points, rather than numbering subclauses, causes problems in referencing. It would be more true to say that it causes problems in the ease of referencing; for example, instead of referencing a clause as clause 31.2(f)(ii), you would have to reference it as clause 31.2, sixth bullet point, second sub-bullet point.
- The written style is extremely dense and too terse. Several of the more senior interviewees in my original research made comments to the effect that the plain English lulled participants into not giving the words proper consideration. In some instances, this terseness was addressed in the second and third editions with what, contractually, is an unnecessary expansion of clauses, but which helps those without detailed knowledge understand the contract. For instance, in NEC3, additional, but contractually unnecessary, text was added onto ECC clauses 60.1(12) and (13): if you know the contract, what this text says is stated elsewhere.
- Other lawyers have put forward the argument that, if the NEC is simple and clear, why does it need the Guidance Notes to explain it? Or, more recently, a four-day Project Manager accreditation course. The difficulty has generally been in understanding how the contract fits together and what you have to do to comply, rather than in understanding individual clauses – it is a familiarity issue. The original authors of the NEC justified the publication of the Guidance Notes on the grounds that, as an innovation, it is practical to have guidance notes to explain their intentions and the implications of each clause. I would add that having guidance notes is far more proactive than having legal columns in construction journals and magazines explaining the implications of court judgements on the standard forms, which the construction industry then has to take on board. Additionally, some other contract forms have guidance notes (e.g. the Institution of Chemical Engineers (IChemE) conditions).

Despite the counter-arguments, there is some truth in the above arguments, and many lawyers have argued that, taken together, they undermine legal certainty when using the ECC. The counter-arguments say they increase legal certainty, so the true position is probably more neutral.

However, what clients and contractors are looking for is certainty in the broadest sense – that is, certainty of outcome for their projects in terms of time, cost and performance. The prevention of disputes over interpretation at site level (because the ECC can be more easily understood at this level) helps in this respect, as does the aim of providing stimulus to good management (see Section 1.4). Indeed, during my research – and still when I do project reviews for those new to NEC – the most mentioned benefit of using the NEC and ECC was the increased certainty it gave the Parties to the contract.

1.3.5 The change in drafting philosophy

If one looks at the development of the traditional forms of contract, each revision has generally sought to reduce the risks of misinterpretation by amplification and expansion, prescribing the end result for each potential permutation of events in ever tighter detail. Thus, all the other major forms of contract have expanded with each successive edition. (In my opinion, the developers of NEC are, with each edition, also falling into this trap, albeit slowly.) One barrister asked if this would actually achieve certainty, and feared that it would not, as: 'the longer the contract, the greater the scope for ambiguities, compromises and inconsistencies. Furthermore, there will always be cases… which throw up problems the draftsmen did not envisage' as 'the possible permutation of future events is endless, whereas man's foresight is limited' (Jackson, 1986). The result is that conditions of contract are filled with many terms that are rarely read and used. The danger, also partly caused by the legalistic language employed, is that the users at site level do not read and use the relevant terms and clauses which they should, thus inadvertently placing their employers in breach of contract. Furthermore, it could be asked what the point is of employing highly qualified and experienced technical and managerial personnel if their ability to act professionally is restricted by the rules and regulations of the contract. Or, as one lawyer in the IT sector said to me, 'I know that every time I write an extra clause, I am limiting the ability of professionals to exercise their judgement in delivering the contract'. Unfortunately, my experience of reviewing many option Z (*additional conditions of contract*) clauses is that most lawyers do not get this (or, for their own commercial reasons, do not want to!).

The NEC family, on the other hand, tries to describe a generic process for the management of problems that inevitably arise during a project. In modern procurement terminology, this means it is a 'relationship-based contract', in that it describes how the contract participants will interact while delivering the project. In so doing, it attempts to tread a fine line between sufficient tightness, so that contract participants follow the procedures, and sufficient looseness, so that it can be interpreted in a way appropriate to the situation encountered. My research and consultancy experience would indicate that the authors of the NEC have been largely successful, but there is a need for fundamental change in attitudes from those who will use the contract, and from lawyers who advise and may eventually be involved in any disagreements or disputes on the contract.

A number of commentators have noted that the NEC is as much a manual of project management as a set of conditions of contract. One of the consequences of this is that NEC contracts encourage up-front discussion of how situations will be dealt with. One practising lawyer noted that: 'if the NEC is examined with a view to ascertaining whether it provides for the rights and obligations of the parties in every imaginable situation and in view of the latest developments in the law of contract and tort, then it may be [but is not necessarily] regarded as lacking. If by contrast the NEC is examined to ascertain whether its procedures stimulate good project management then it is likely to fare favourably' (Herga, 1995).

In practice, the majority of my research participants had grasped this change in emphasis, with one describing it as an 'enabling document, rather than a prescriptive document' which 'forms a framework which you interpret depending on the situation and with whom you are dealing'. However, a few interviewees within my research sample, generally with experience of less successfully implemented NEC contracts, had not realised this or realised it too late.

1.3.6 Conclusions on clarity

Despite the initial reservations of those in the legal profession, my initial research found – and my consultancy experience would concur – that, once the learning curve has been overcome, the NEC/ECC offers improved clarity, both procedurally and in terms of brevity. However, it is suggested that a potential user of an NEC contract needs to not only give each clause sufficient consideration to understand its meaning and implications, but also to reconsider how the clauses can best be implemented in practice and the purpose of conditions of contract. In this way the user can define more clearly the relationship and processes between the Parties, as opposed to their rights and obligations in every conceivable circumstance. One of the more enlightened lawyers that I work with described the NEC as 'an enabling "how to" guide to get it right document, whereas traditional forms [referring to JCT] are more like a warranty to allocate liability for when it's gone wrong'.

Or, as one client said, 'the NEC is an intelligent person's contract and the problems come when it is used by unintelligent people!'

1.4. The mechanisms for stimulating good project management and how they fit together

1.4.1 Stimulus?

The third original aim of the NEC family was to engender a 'stimulus to good management'. Because of the all-embracing scope of the term 'management', in this brief overview consideration is specifically given to project management relating to the delivery phase (by which I mean, in construction terms, the design and construction phases). While achievement of 'good management' is an objective both of the NEC and, presumably, of the organisations that contract under it, it is an enabling objective in that 'good management' should lead to a higher likelihood of the project, and hence organisational, objectives being fulfilled.

Therefore, before looking at what constitutes good management and how conditions of contract can contribute to its achievement, it is worth asking what the 'stimuli' to good management are, i.e. what are clients' and contractors' objectives? The reason for considering this is so that both Parties understand what the other wants from a contract, so that potential users of an NEC4 contract can appreciate how, by its procedures and sanctions, it aims to create the right environment for good project management, including collaboration, to flourish. Several contracts in the original research sample failed to fully achieve their objectives because the participants initially failed to understand how, by working within the principles, spirit and words of the NEC contract, rather than against them or ignoring them, they could increase the likelihood of achieving their own individual objectives. Unfortunately, I still hear of participants acting in this way today, and hence the need, more than 20 years on from the first edition of this book, for this section.

1.4.1.1 Stimuli to contractors

Commercial organisations are often seen as purely profit-maximisation organisations. This is an oversimplification.

- In the short term, commercial organisations need a positive cash flow to pay their bills. If expenditure exceeds income over a sustained period of time, a company will not continue operating. The more positive cash flow is, the more profit a company makes, either through not paying interest or by doing something useful with that cash. Particularly in the civil engineering and building sectors, where profit margins are frequently less than a few per cent, a short delay in payment can have serious effects on profit margin, and a huge effect on return on investment. Virtually all the Contractors interviewed in the original research commented on how their cash flow was more positive with NEC contracts than with contracts let under traditional forms. The more they embraced the spirit of the NEC, the more positive their cash flow seemed to be: this is because issues are wrapped up (including their financial effects) and the *Contractor* is paid as the contract progresses, as opposed to at the end (when the contractor mounts a claims offensive).

- Commercial organisations aim to produce a profit on any contract they enter into. Unsurprisingly, given the low profit margins, they prefer profit levels from a contract to be predictable, rather than fluctuating and unpredictable. For any business, not knowing when or how much you will be paid for completed work does not aid the running of a business, yet it would appear to be almost accepted practice in much of the construction industry. A few Contractors within the research sample increased their expected profit on turnover by 3–4% (i.e. to 5–6%, compared with 2–3%), and these were the ones where both Parties not only fully embraced the spirit of the ECC, but also operated it pretty closely to how it is written. A commonly expressed view by Contractors was that there is more certainty under the ECC that a reasonable profit would be achieved, provided they comply with the requirements of the contract. Representatives of both Clients and Contractors thought that there was no potential within the ECC to make 'silly money'.
- Commercial organisations may want a satisfied client, so that they win repeat-order or referred business. The 'may' depends on the nature of the client. For instance, if the next contract will be awarded on lowest cost criteria, whatever the contractor's performance on the previous contract, or if the client is unlikely to place another contract, then this motive may be subjugated in order to pursue the first two given above.

1.4.1.2 Stimuli to clients

A project has three prime objectives: time, cost and function (or performance), the last of which may have a multitude of facets by which it can be judged. The traditional view is that to improve any one of these parameters, you have to eat into another one. (This is generally true once into the delivery phase of a project, but, if you work smart, I do not believe this is absolutely true. See for instance Section 3.6.8 on managing in opportunity.) For example, to decrease the timescale of a project, it will generally be necessary to increase expenditure, and vice versa. However, within each of these objectives there are also some trade-offs, which are now briefly examined.

- *Function/performance*. A company, in commissioning a project, is doing so to satisfy a perceived need or desire, which should be expressed positively as functions or levels of performance. In turn, this ultimately should translate into business benefit. I argue, therefore, that a client's prime objective is to have an asset that performs the functions identified for it, i.e. that it is fit for purpose. It is not worth paying, for example, £999 for an asset that does not perform the function required, when you could pay £1000 for one that does. As a project progresses into the delivery phase, change inevitably happens, including clients introducing some changes to ensure or increase functionality or performance. This may mean enhancing functionality or performance in one respect, while reducing it in another. Either way, some flexibility is often required. As change normally impacts on time and cost, there is a need for a rational, consistent, fair and accepted method to evaluate these effects.
- *Cost*. There is a potential trade-off between certainty of cost and potential minimum cost. For instance, if a client is liable for the majority of risks that may eventuate, the risk premium in the tender will be low, resulting in a low tender. If none of the risks occur, the client will not pay anything additional to the tendered price, but if they do occur then

it may pay a great deal more. If, however, a lot of risk is allocated to a contractor, then the contractor will charge a premium to bear that risk and its tender will be higher, but the client will have greater certainty.

- *Time*. To a lesser extent, what applies to cost also applies to time.

What comes out of these three brief glimpses is that clients generally want an element of certainty over the time, cost and functional/performance parameters when they sanction a project and let an individual contract. To gain certainty, first, they have to have effective monitoring procedures for time, cost and function/performance – so they have knowledge of when a project or contract is deviating from the desired or expected parameters – and second, they have to be able to exercise influence in order to bring the project back within the planned parameters. Having knowledge of where a contract is heading, in terms of time, cost and function/performance, and the ability to influence it leads to greater certainty. In terms of benefits, these were the advantages of using the ECC most often cited by clients and their representatives.

However, this is not the whole story. A number of studies have found little correlation between satisfaction expressed by clients and project outcomes expressed in absolute terms of cost and time. Other factors, such as the smoothness and efficiency with which their needs are fulfilled by the industry play a large part in clients' judgements of the level of success. One research project summarised these factors as, at a personal level, 'a lack of hassle'. In my opinion, this equates to level of service, and could include ease of working with and agreeing issues with a contractor, as well as speed of settling the final account.

In my original research, on one contract subject to major changes the final account was settled on the dot of Completion, despite the changes costing £7 million over the original tender Price of £6.5 million (and this was a while ago – for today's prices, you could probably double these amounts). The *Employer* thought it had got value for money on these changes, and both Parties thought they had saved a lot in legal fees! In my research sample, it was not unusual for the final account to be more or less settled within a month of Completion, and only three out of the original sample of 29 contracts took more than 3 months. The contract teams that took less than 3 months had embraced, to varying degrees, the spirit and wording of the contract (and, as a generalisation, the more they embraced it, the quicker the final account was settled; thus the more positive the cash flow for the *Contractor*, as the sooner a compensation event was agreed, the sooner they were paid for it). One of my former contractor clients expected the final account on their contracts to be agreed within 1 month of Completion. Again, a generalisation, but the quicker the final account is settled, the more positively the *Client's* people view the *Contractor's* performance.

In fact, when I followed up, on the three contracts where it took more than 3 months to settle the final account, the account had still not been settled after 6 months. During operation, the *Contractors* had, by and large, put the contract in a drawer and carried on as per a traditional contract. Consequently, they got a traditional result, and this was reflected in what the *Employers'* people thought of them!

Minimum hassle also means that both Parties can concentrate on their core businesses. For a contractor, I argue this should be building the client a functional asset within its time and cost parameters by managing the construction process efficiently (i.e. programming with good risk management), and as a result it makes a profit. It should not be devoting considerable time and overhead to clawing back money through claims in order to make a profit.[19] For a client, I suggest that this is not spending time and money fighting these claims, whether through its own employees or through expensive consultants.

1.4.2 A brief rant *about* 'collaboration'

In June 2018, I went to the NEC Users' Group seminar in London and came away thoroughly irritated. Why? Because every second sentence uttered by the speakers seemed to contain the words 'collaboration', 'collaborative' or 'collaborative working'. This appeared to reflect the current vogue. A similar trend occurred 15 or so years ago, but that time it was 'partnering'. So, why do I want a quick rant *about*, but not against, collaboration and collaborative working? Here are my reasons.

- Collaboration seems to be taken as an end in itself, as in 'we are doing this project to be collaborative' or 'we chose NEC because we wanted to be collaborative', with little or no mention of what it is that the project is to achieve, in terms of time, cost or performance and the certainty around this. We do not do projects to achieve 'collaboration', although collaboration has benefits: collaboration generally means happier, more productive people, so the project is more likely to be delivered within its time, cost and performance parameters. In other words, collaboration is a means to an end, not the end in itself (although happier people is, in project management terms, an ancillary benefit and of worth in itself).
- There seems to be presumption that, if people are being collaborative, the result will automatically be a successful project or contract. Consequently, the focus becomes being 'collaborative', rather than delivering the project. The danger then – and I have seen this happen quite often – is that hard issues are swept under the carpet or ignored, because to raise them would potentially cause confrontation, thus undermining collaboration. As a result, the issues fester, get worse, in terms of their likelihood and potential impact, and more and more suppressed emotion becomes attached to them. Consequently, when they become so big that they cannot be ignored, the detrimental effect on collaboration, as well as on the project objectives, is far bigger. For me, collaboration is a means for enabling hard issues to be raised and difficult conversations to be conducted early on, through using all parties' expertise positively and respecting their interests.
- There is little mention of the features of the NEC that encourage collaboration: it is just taken as read that the NEC promotes collaboration. Yet, when you ask people what the features in the NEC are that promote collaboration – as we did in a workshop at the 2019 NEC People Conference – they really have to stop and think.
- If you then go on and ask people 'Well, how do you actually operate an NEC contract in accordance with the specific words while being collaborative' – which was my second question at the 2019 NEC People Conference – they really scratch their heads!
  (An edited version of the outputs from this workshop session is presented in Appendix 2.)

[19] One of my regular contractor clients did a review of their contracts and found a clear correlation between the amount of change and profitability, where contracts with little change had high profitability and those with lots of change had decreased profitability. As their operations director said, this flies against the accepted wisdom that contractors make their money on changes. He thought the reason for this was that contractors rarely get full recovery on the disruption aspects of changes, because it is so hard to demonstrate. Meanwhile, clients always think they are being overcharged for this aspect. The middle ground is that neither party is happy!

- The claim that NEC was designed for collaboration is downright false! If you go back to the original specification documents, you will not find collaboration mentioned among the principal aims or objectives. The aims were greater: flexibility, clarity and simplicity, and stimulus to good [project] management. However, Dr Martin Barnes – a (retired) project management guru and the co-instigator of the NEC, along with Professor John Perry – has publicly stated that successful projects exhibit collaboration. It is therefore something to aim for when setting up a project or contract, but, I would argue, is also a result of having a well set up project. Or put another way, collaboration is a subset of good project management.

And on that note, I will finish my rant *about*, not against, collaboration and we can get back to considering what good project management looks like and how the NEC stimulates it.

1.4.3 Good project management

This section identifies the ways in which the procedures or processes within the ECC aid effective project management of the contract in order to deliver projects that meet their time, cost and performance/functional objectives. Many of these features are common to other members of the NEC4 family, albeit perhaps simplified in the short forms of contract.

When giving seminars on the NEC, I occasionally ask participants what factors increase the likelihood of the objectives of a project being achieved. Below is a list of some factors that are likely to increase the likelihood of a project achieving its objectives

- clearly defined objectives, derived from the benefits that you wish from the project after it has been delivered
- quality personnel
- quality organisations
- good stakeholder management
- clear, unambiguous and complete information supplied to the contractor
- minimal change introduced during the delivery phase
- appropriate contract strategy to align motivations (this has been discussed in Section 1.2, including the selection of the appropriate main and secondary options and the level of *Contractor* design)
- clear and appropriate definitions of roles and responsibilities
- sufficient testing requirements to ensure fitness for purpose
- integration of actual progress and expenditure compared with planned progress and expenditure aided by
  - a risk register that identifies and contains responses to threats to achieving the project's objectives
  - early and tangible actions to minimise the impact of issues and emerging risks
  - seeking out opportunities to improve the project in time to implement them
  - proper and clear allocation of risk to reduce argument over ownership and liability
  - an effective, equitable and rapid procedure for evaluating the time and cost effect of changes
  - options for dealing with the unexpected
  - transparency and communication leading to cooperation and collaboration between the contract participants
  - an inexpensive and fair procedure for rapid resolution of disagreements.

The first six of these points are relatively independent of the type of contract used, and are discussed briefly below.

- *Clearly defined objectives*. Professor Hackman of Harvard University identified this as the number one factor, based on all the research that he is aware of (Hackman, 2002). It is a matter of common sense that, to achieve the project objectives, it helps if you know what they are. It is therefore worth spending time defining them, including any constraints, before rushing into delivery. For example, if the asset has to be fit for purpose, what is its purpose? What is the balance between cost (or time) certainty and the desire for minimum cost (or time)? From this flows a proper definition of the project, from which flows the technical requirement (in NEC4 speak, the Scope), which states what is required (the *works*) and any constraints on how it is to be provided. An awful lot of detailed work can be wasted later if the objectives are not thought through properly at the beginning.
- *Quality personnel*. Ultimately, it is not conditions of contract that build things, but organisations, which consist of individuals. If the people involved are not of sufficient quality, in terms of technical, managerial and people skills, or have incompatible values and attitudes, then the process of delivery will be adversely affected, which in turn will affect the achievement of objectives. All NEC4 long forms – by which I mean those without 'Short' in the title – require the supplier to offer the names (in Contract Data part two) of the '*key persons*' it will provide to carry out the key roles, which should be identified by the *Client*.
- *Quality organisations*. While the quality of individuals is important, if they are not backed up by a quality organisation, they could be like an Olympic swimmer floundering in a pool of mud – not going anywhere fast! By quality organisation, I mean the supporting processes (including IT), organisational structure, policies, senior people (who set direction and tone), support staff and general culture that are brought to bear on the project. At its most extreme, if quality individuals are having to 'fight' against their own organisation, as opposed to being supported and helped by it, little constructive energy will be devoted to the successful delivery of the project.
- *Good stakeholder management*. This has become more and more important, in part because
  - we live in an ever more complex and interdependent world, both socially and technically, and so more people and organisations are stakeholders in a typical project, whether through being a receiver of the project outputs or having a part to play in delivering it

  and, particularly in developed countries, because
  - the liberal culture means that seemingly everybody has a voice and a right to be heard, which is amplified by social media, and
  - many projects, particularly in civil engineering and building, are to refurbish, upgrade or build over existing live assets, thus disrupting existing users.

  Stakeholder management requirements might be included in the brief for the *Client's Project Manager* and their organisation and/or as part of the *Contractor's* Scope.
- *Clear, unambiguous and complete information* supplied to the contracting organisation. In ECC terms, good quality Site Information and Scope, so that the *Contractor* knows what it is to do to complete the contract, so that it can be priced and programmed properly. Major users found that the ECC more clearly highlights (not exaggerates) the time and cost effects of individual changes closer to the time when they occur, giving better feedback. This compares to the common situation under traditional contracts, where the indirect costs of changes are generally assessed together at the end of the contract as a claim for delay and disruption. Under the ECC, shortcomings in the documentation are

more easily identified and the lessons learnt can be, and have been, incorporated into the documentation for future contracts. It has made some experienced NEC Clients quite critical of their design team or consultant's performance!

- *Minimal change introduced during the delivery phase.* A study done in Canada (Revay, 1992) drew data from over a billion dollars' worth of work (at 1992 tender prices) from a number of heavy engineering and construction sectors. It concluded that changes introduced after a contract is let cost, on average, in the region of three times as much as if they were included in the original specification. Approximately, a third of this extra cost is on the direct costs (i.e. what it would have cost anyway), a little under a third is spent on the prolongation element (i.e. the additional time on site) and a little over a third on the loss of productivity – which in the UK we would probably call 'disruption' – caused by the change. The implication is that, if those managing a project wish to increase the likelihood of it meeting its time, cost and performance objectives, they need to have strict change control procedures, and only allow changes, whether in scope or design, that are necessary for the project to achieve its functional or performance objectives. In the almost 30 years since this study, timescales have, if anything, got tighter and projects become more complex, and so the effects of delay and disruption on the cost of a change will have increased. An implication of this is that, if change is necessary, you want to do it in a planned way to minimise the disruption element.

The following factors are relevant to the conditions of contract. In order to build familiarity with the contract, I suggest you have an ECC document alongside you, so that when I mention a clause, you can easily turn to it and have a quick read.

1.4.3.1 Clear and appropriate definitions of roles and responsibilities

The formulation of clear and appropriate definitions of roles and responsibilities is important so that participants know what function they are expected to perform within the contract team. The ECC has divided the traditional role of the engineer, architect or contract administrator into four separate roles: the designer, the *Project Manager*, the *Supervisor* and the *Adjudicator* (see Section 1.5.2), although the *Client's* designer is not mentioned in the contract, as all their contractual communications have to come through the *Project Manager*. The intention is that each individual can specialise in their particular roles, without conflicts of interest between those roles. For example, in a JCT contract, the Contract Administrator is required to make an impartial judgement over a previous decision they have made, say on the detail of a design, when that design may have been initially done by them or their company, while being paid by the Employer and with the fear of a potential claim on their professional indemnity insurance in the background. Hardly impartial!

However, in NEC4 contracts, there are only two general statements of how the players, namely the 'Parties, the *Project Manager* and *Supervisor*', are to act, which are that they

- 'shall act as stated in this contract' (clause 10.1), i.e. as specifically stated in the individual clauses of the contract
- 'act in a spirit of mutual trust and co-operation' (clause 10.2), an overarching obligation, more on which later (see Section 1.5.2).

The *Project Manager* has stated criteria by which they make a decision, either 'accepting' a communication from the *Contractor* or notifying it of the reasons – given in the relevant clause –

for not doing so 'in sufficient detail to enable the *Contractor* to correct the matter' (which is new to NEC4 and in clause 13.4). The initial research found that roles and responsibilities were well defined in the ECC, with few, if any, gaps or overlaps in responsibility, and this has been borne out by my consultancy experience.

However, it is worth highlighting some case law from May 2005 on the NEC. The case of *Costain Ltd* v. *Bechtel Ltd* [2005] TCLR 6 effectively confirmed that the *Project Manager* in administrating the contract has to interpret the reasons for non-acceptance 'impartially', with Judge Jackson stating, with regard to the specific issue over certification of payment, that

> I am unable to find anything which militates against the existence of a duty upon the project manager to act impartially in matters of assessment [of compensation events] and certification [of payments and more generally] when the project manager comes to exercise his discretion in those residual areas [of subjectivity in the NEC]... it would be a most unusual basis for any building contract to postulate that every doubt shall be resolved in favour of the employer and every discretion shall be exercised against the contractor.

What this means in practice is that the *Project Manager* interprets the contractual reasons for non-acceptance in the way that a reasonable person, experienced in that field, would do based on the information available to him or her, which includes, but (as lawyers would say) is not limited to, that supplied by the *Contractor*. Here are a couple of examples to illustrate this.

- Under a cost-based contract, the *Contractor* is missing a delivery note for, say, reinforcing bars. It would not be reasonable to refuse payment if the *Project Manager* – or more likely the quantity surveyor – could step outside and onto the Site and see a large pile of rebar clearly marked up as being that for which the *Contractor* is claiming payment. Other forms of supporting information, such as photos, are also relevant.
- A personal one this: a consultant is called in by the *Contractor*, with agreement from the *Project Manager,* to run a workshop specific to that contract. The largest item on the consultant's invoice is for his time, followed by mileage, followed by bed and breakfast at a very reasonable £40. The *Employer's* quantity surveyor – it was under NEC3 – refuses to reimburse the *Contractor* for the whole invoice because there is no receipt for the consultant's bed and breakfast, and the consultant cannot provide one because he is now in New Zealand for 3 months. This was despite the *Contractor's* personnel staying in the same accommodation, being charged the same amount and producing receipts.

To illustrate the difference between administrating the contract and project managing it, let me give you a simple example. In this case it is a school project, and the school has to be open by term time. A compensation event occurs, and the *Project Manager* asks the *Contractor* for a number of quotations based on different ways of dealing with it. Let us say one is for least time (crashing the programme with people working on top of each other), one is for least construction cost (which, when worked through, would mean the school opens late) and one is somewhere in between (if there is no other significant delay, the school will be ready just in time for term). In assessing each of the *Contractor's* quotations, the *Project Manager* has to be an impartial administrator, i.e. what would an impartial person of reasonable experience decide, based on the information presented in the quotation, other information readily available to them (such as the programme and Defined Cost information that they already know about, and by looking around the Site) and the contractual criteria in clause 63. Having done this, the *Project Manager*, now acting in the best interests of the *Client*, can choose which of these quotations to implement.

Lastly, each contract participant – and this includes the *Contractor* – has clear timescales in which to respond to a communication from another, as opposed to the 'within a reasonable time' of more traditional contracts. If this timescale is not stated in the clause, then it is 'within the *period for reply*' stated in Contract Data part one – typically 1 or 2 weeks.

1.4.3.2 Sufficient testing requirements to ensure fitness for purpose

The ECC gives an outline framework for testing, which should be developed in the Scope so that it is specific and appropriate for the chosen contract strategy (e.g. performance specification or predesigned), the technology involved in the project, the maturity of the *Contractor* and sector, and the relationship between the *Contractor* and the *Client*. This helps ensure that the final product meets its quality parameters, but does require more specific thought by those putting together the documentation.

In line with the NEC panel's policy of allowing the *Project Manager* to manage, as opposed to just administrate, the contract, clauses 44.4, 45 and 46 give the *Project Manager* various management options in the event of a Defect arising, without undermining the *Supervisor's* contractual role in deciding whether or not it is a Defect in the first place.

1.4.3.3 Integration of actual progress and expenditure compared with planned progress and expenditure

Integration of actual and planned progress and expenditure aids monitoring. Without good monitoring, so that you know whether you are on schedule and cost, you cannot take effective action to get back on schedule and cost if needed. In the ECC

- what the programme is to show is stated in far greater detail than in traditional contracts (clause 31.2) and includes, for each operation, statements of how the *Contractor* is to do the work and of principal resources (in the UK, this aligns well with the requirements of the CDM Regulations 2015)
- the programme is updated at regular intervals (clauses 32.1 and 32.2)
- sanctions exist to encourage the *Contractor* to show in the programmes the stated information (clause 50.5), to keep it up to date, and for it to be practicable and realistic (clauses 64.1 and 64.2)
- if option A (priced contract with activity schedule) is used, payment is directly linked to progress; if option C (target contract with activity schedule) is used, the *Contractor* is reimbursed its Defined Cost plus a Fee; in both cases, expenditure can be compared to that anticipated in the *activity schedule* to determine trends in progress. (Having said this, the requirement to show how each activity in the Activity Schedule relates to operations in the programme has been deleted from option C of NEC4 contracts, making this comparison much harder. I think this is regressive and detracts from good project management, so would advise that the NEC3 ECC clause 31.4 – which is the same as NEC4 ECC clause 31.4 in option A – is reintroduced as an option Z clause.)

Many interviewees in the original research and subsequent users commented on the change in emphasis of the programme, from a tool used to justify and extract claims at the end of the contract, to a project management tool for the efficient and cooperative management of the work. Some Contractors partly attributed their increased profit margin to this emphasis on programming which, as one commented, 'forces us to do what we should be doing in any case'.

Unfortunately, when doing project reviews, I still sometimes see the main emphasis being on using the programme to justify the delay to the Completion Date as the contract is progressing – which is an improvement on leaving it until the end of the contract – but with hardly any emphasis on actually managing the events, whether compensation events or not.

1.4.3.4 A register that identifies and contains responses to emerging risks and issues that threaten the achievement of the contract's objectives

Risk registers are now common to the construction industry (note I do not say commonly *used*!). Often, for a project of a particular type, they contain high-level generic risks common to all projects of that type and therefore the actions or responses are fairly generic (or, as one of my reviewers commented, 'bland and useless'). The ECC, in Contract Data part one and part two, invites both the *Client* and the *Contractor* to give their 'starter for ten' (or twenty or thirty) risks that could impact on the contract. This register should then be kept active and specific throughout the contract's duration by doing what the contract says in clause 15.

For those coming from an NEC3 background, note that the defined term 'Risk Register' has been replaced with 'Early Warning Register', 'risk reduction meeting' with 'early warning meeting' and the word 'risk' with 'matters'. In fact, all references to 'risk' have disappeared from the NEC4 early warning clauses! Hopefully, this will help break the direct link in people's minds between a traditional 'project risk register' (as beloved by the risk management profession) and what was misleadingly named the Risk Register in NEC3. The truth is that the register was and always has been more akin to an 'issues register' as defined in project management literature. Much more on this and its operation in Section 3.6.

1.4.3.5 Early and tangible actions to minimise the impact of issues and emerging risks

The early warning procedure (see clause 15) is designed to ensure that issues and emerging risks are identified at the first opportunity and that prompt action is taken to minimise their likelihood and/or impact. Indeed, under clause 15.3 it is a contractual requirement to come up with actions at an early warning meeting: if it is a pity party or blame/liability party and no actions are forthcoming, then it is not an early warning meeting in accordance with the contract. The early warning procedure is the intermediate step between how risk management is normally implemented (or not in reality) – identifying what generally can go wrong and developing high-level strategies to minimise their impact – and monitoring the programme to identify something that has already gone wrong and has affected the *works*. There is a sanction should the *Contractor* not give an early warning that it should have done, which may reduce the *Contractor's* entitlement to a compensation event (see clause 63.7, provided the *Project Manager* notifies the *Contractor* at the time of asking for a quotation under clause 61.5). The early warning procedure has proved extremely effective in operation. It has been described as 'the jewel in the NEC crown' and 'the starting point for co-operation'.

1.4.3.6 Seeking out opportunities to improve the project in time to implement them

You could argue that this is the flip side to the previous two factors. Seeking out and implementing opportunities provides

- a cushion or buffer against the inevitable impacts of downside risks and issues
- the opportunity to deliver a better than anticipated result, howsoever that result is defined.

NEC3 encouraged participants to look for the upside by having target-cost options, whereby upside gain is shared between the Parties, and by including Key Performance Indicators, against which incentives can be set for good performance, in option X20 (Key Performance Indicators) and option X12 (partnering) (multiparty collaboration in NEC4). NEC4 has gone further by

- introducing the new Alliance Contract
- under the price-based contracts, options A and B, introducing a value engineering clause (clause 63.12), whereby the *Contractor* gets a share of any savings that result from changes to the *Client's* Scope suggested by the *Contractor*
- allowing the *Contractor* to propose an acceleration, under clause 36.1, to bring forward the contract Completion Date and/or Key Dates (under NEC3, it was only the *Project Manager* who could initiate the process)
- having new option X21 (whole life cost), whereby if the *Contractor* suggests a change that reduces the whole life costs of the asset, it gets a share of the savings.

This is all for the good, and for more detailed comment on these features, see elsewhere in this book. However, I think the authors of NEC4 have missed an opportunity by not encouraging the day-to-day awareness and consideration of opportunities, which they could have done by incorporating it into the early warning clauses – see Section 3.6.8 for more discussion on this and how to do it.

My last point – more an observation – is that unless you build in time to identify opportunities early enough, you will not have enough time to implement them, or you will only be able to implement them at the last minute, which creates as much or more downside as upside.

1.4.3.7 Proper and clear allocation of risk to reduce argument

'Proper allocation' generally means that the party best able to minimise the likelihood of the risk occurring and to bear it is allocated the risk. 'Clear allocation' means that there is little or no argument over ownership of the problem, and therefore who should take the lead on actions to prevent it occurring or minimise its impact if it does. Apart from the selection of the main and secondary options, risks allocated to the *Client* are clearly listed in one place as compensation events (clause 60.1). Because they are listed in one place (with the same assessment procedure for all of them), Clients have found it easier to delete or add a compensation event depending on the project circumstances. The initial research quite strongly found that the definitions of *Employer* risks were more clear-cut than under traditional forms of contract. The main advantage of this is that contract participants spend less time arguing over whose risk something is. Consequently, the party whose risk it is can devote their efforts to managing the risk, rather than trying to avoid liability for it. The other party, knowing that it is not their risk, can contribute to the management of the risk without fear of taking on any liability.

1.4.3.8 An effective, equitable and rapid procedure for evaluating contractual changes

The core clauses of the ECC contain one procedure for evaluating both the time and cost effects of a compensation event (Section 6 of the ECC). The compensation event procedure has clearly stated and tight timescales, which are achievable in most circumstances.

Sanctions create a stimulus on the *Contractor* to both

- notify the compensation event promptly (within 8 weeks from the time of becoming aware of the event under clause 61.3, otherwise entitlement to any compensation may be lost under clause 61.4)
- submit a quotation for its effects within stated timescales (clause 62.3), otherwise the *Project Manager* has first bite at assessing the effects (under clauses 64.1 and 64.2).

Sanctions (in clauses 60.1(6), 61.4 last para., 62.6 and 64.4) create stimuli on the *Project Manager* to respond within the timescales of the contract.

See Section 1.5.7 for further expansion on the above points.

The *Contractor* is compensated for the changes in the Defined Cost and the resulting Fee, with Defined Cost being drawn from the Schedule of Cost Components. The intention is that it is a more rigorous and fairer method of evaluating the costs of changes. In practice, the compensation event procedure has caused some problems, partly due to unfamiliarity and the changes in practice it requires, and partly because, in previous editions, the Schedule of Cost Components did not match how the *Contractor* incurred additional costs in certain circumstances. The NEC3 ECC Schedule of Cost Components was a considerable improvement, and the NEC4 version is a further improvement. Note also that all the NEC4 long forms – i.e. those without the word 'Short' in the title – now contain a Schedule of Cost Components, setting out what is Defined Cost. Previously, it was only included in the ECC. A key to the successful operation of the ECC is in the administration of the compensation event procedure.

1.4.3.9 Options for dealing with unexpected events

When an unexpected event occurs, there may well be more than one way of dealing with it. Regardless of whether it is or is not a compensation event, the place to identify, discuss and evolve options is an early warning meeting. If it is not a compensation event, the cost of dealing with it is within the Prices and the *Contractor* takes the lead on how to manage it. If it is, or results in, a compensation event, provided the *Project Manager* has had a discussion about different practicable ways of dealing it, they can instruct the *Contractor* to submit a number of quotations for the different options (clause 62.1). The *Project Manager* then selects the option that best matches the *Client's* objectives, which might be the least time solution or the least cost solution. This choice has been exercised in practice, although perhaps not as often it could have been.

1.4.3.10 Transparency and communication leading to collaboration between the contract participants

Participants to a contract will collaborate if they perceive it to be in their professional and commercial interests to so do. The alternative is confrontation and dispute. In has been estimated that, in 1992, confrontation cost the UK construction industry 7% of its turnover (Doyle, 1993) – and that ignores the physical cost of not managing the events out on site – although my guess is that it has reduced since then. The traditional 'rules of the game' make it necessary for this money to be spent by contractors to recover their costs in order to make a profit, and for clients to then fight these claims. However, fighting claims is an activity of the construction industry that adds nothing to the value of the end product. The ECC aimed to change the contractual rules and to stimulate collaboration through two interlinked principles.

- *Transparency.* The ECC's programming requirements require more information to be communicated, the early warning procedure requires the *Contractor* and the

Project Manager to inform the other of any potential problem at the earliest opportunity, the Schedule of Cost Components provides an indisputable list of what the *Contractor* can include as Defined Cost in its quotation for a compensation event, and the tendered *fee percentage* prevents any argument over what the *Contractor's* margin is. Under the cost-based options (options C, D and E), open-book accounting gives transparency of cost.

- *Communication.* The ECC requires people to communicate in a form that can be 'read, copied and recorded', and within stated maximum timescales, rather than 'in a reasonable time'. The effect of this is not that people write more letters, but generally that people talk more and then confirm and summarise what they have agreed in writing. The author is aware of only one *Employer* (under NEC3) who abandoned use of the ECC after one contract because of the administrative burden. At the opposite end of the spectrum, others have said that, if an effective and efficient system – be it paperwork, email based or on a cloud system – is set up early on, issues can be resolved sooner, with less writing involved. A good system allows more time to concentrate on value-adding or cost-reducing activities, such as programming. Indeed, one *Project Manager* stated that 'we probably spend half as much time writing contractual letters to each other and twice as much time working together to resolve issues before they hit' the project.

1.4.3.11 Rapid dispute resolution

The first edition of the ECC had adjudication as its original form of rapid dispute resolution. In the UK, as a result of the Housing Grants, Construction and Regeneration Act 1996 (as amended by the Local Democracy, Economic Development and Construction Act 2009, which came into force on 1 October 2011), adjudication is now compulsory on virtually all UK construction contracts. NEC4 has added in a pre-adjudication dispute avoidance mechanism that involves senior management representatives from the *Client* and the *Contractor*, as well as the option to use a Dispute Avoidance Board (under option W3). Other pre-adjudication dispute resolution procedures (e.g. negotiation and conciliation/mediation) can be incorporated if desired, or post-adjudication dispute procedures can be specified in Contract Data part one as part of the *tribunal*, before moving onto the expense of arbitration or litigation procedures (see also Sections 1.2.3.2 and 1.2.3.3 and Section 3.8).

1.4.4 Conclusion

This section first outlined what the main stimuli to the Parties of a contract are, so that each Party gains not only a greater appreciation of other Party's objectives, but can improve its understanding of what the original authors of NEC aimed to achieve. I then had a rant *about* collaboration because is not one of the original objectives of the NEC (although I am a strong advocate for collaboration). I wanted to differentiate collaboration from good project management as they are separate, albeit ideally overlapping, things – good project management should aim to create collaboration, but is also enhanced by collaboration.

The section then briefly outlined some of the components of good project management that affect the delivery phase of a project, showed how they are promoted in the ECC through its procedures, and gave some feedback on the success of these procedures in aiding the achievement of the Parties' objectives. Some users have stated that on the best-run contracts the factors identified in this section happen anyway. A question is: Why do they not always happen? The answer could be threefold

- they are not contractual obligations in other contracts

- the ECC, indeed the whole NEC4 family, has stimuli embedded into the contract to encourage participants to meet these obligations
- other contracts, in certain instances, unwittingly have stimuli not to undertake them, or even do the opposite (e.g. to keep quiet about an issue in order to maximise entitlement).

The NEC family aims to provide these stimuli, and my initial research, followed up by success stories over the past 20 years or so, has indicated that, by and large, it has been successful. (Interestingly, some contractors, having run an NEC contract and learned its procedures, have implemented the NEC procedures in non-NEC contracts. However, they have not passed the information that these procedures give them onto those representing the client, because it is not necessarily in their interests to do so under these other forms!)

Conditions of contract are, however, only part of the 'rules of the game'. Other factors, such as appropriate contract strategy, selection of the right contractor and personnel, a realistic contract sum/Prices, good quality contract documents, use of lean construction principles, and the various principles and techniques that come under the banner of partnering and collaborative working, all help to reduce cost and time and increase certainty and quality. In this respect, use of the NEC4 family should not be seen as *the* answer, but rather as part of the answer.

1.5. A quick run through of the NEC4 Engineering and Construction Contract

The purpose of this overview is to give a reader new to the ECC – and to a lesser extent the NEC4 family – a rapid understanding of the terminology used, of the principal clauses and how the contract fits together. It is not a legal interpretation of the contract. The ECC uses some terminology that is different from traditional conditions of contract. This is partly to move away from old terminology that has emotive connotations (e.g. 'compensation events' replaces 'claims' and 'variations' – although it would be fair to argue that the term 'compensation event' has now gained some emotive association in many people's minds), partly because of the multi-disciplinary application of the contract, and partly because in achieving its objectives of flexibility and stimulus to good management, the ECC has terms and processes that other contracts simply do not have.

In giving this overview, I am, to an extent, simplifying or glossing over the detail. It is preferable that you read this section with the ECC contract open beside you. If desired, the detail can be investigated, with help of the ECC Guidance Notes and/or legal commentaries, as well as some of the other chapters in this book. For instance, see Section 3.5 on the Accepted Programme.

1.5.1 Contract Data and defined terms

The Contract Data is contract-specific data and can be found after the *conditions of contract* in the published document. Part one is filled out by the *Client* prior to putting the contract out to tender, and part two is filled out by the *Contractor* and returned with its tender. It can be likened to expanded appendices to the form of contract/tender in other standard forms, although as one *Project Manager* commented: 'it is a very useful checklist of issues which you should consider before putting any contract out to tender, and if you haven't thought about them by then, then you should do'. Information in the Contract Data is brought into the contract either by the ECC expressly stating that the information is in the Contract Data or, more commonly, by the use of italics. **Terms identified in the Contract Data** are set in italics (clause 11.1), so where italics are used in the contract clauses, you should be able to substitute in the specific name, timescale, date or number (as appropriate) with that stated in the Contract Data. This convention has been and is followed in this book when mentioning a Party or term in a context specific to the ECC.

Defined terms – i.e. words or phrases that have a specific contractual meaning – have capital initials (clause 11.1) and their definitions can be found in clause 11.2. Therefore, whenever you come across a term with capital initials, you can refer to clause 11.2 to find out its specific meaning within the ECC. This prevents repetition of long phrases throughout the contract and, in doing so, keeps sentences short enough that they can be understood – provided you know what the definitions actually mean! So, to properly understand the contract, potential users need to become familiar with these terms. Hence, I give further explanation of some of them below. Where the ECC uses capital initials, then so does this book.

Note that some of these defined terms, particularly those related to payment – i.e. Prices, Price for Work Done to Date, Defined Cost and Disallowed Cost – have different meanings depending upon which main payment option is chosen. So, if you are using the Black Book (as opposed to one of the green ECC books[20]), to find the specific contractual meaning you have to look in the definitions for the particular main payment option.

1.5.2 Parties and participants in the contract

The Parties to the contract are the ***Client*** and the ***Contractor*** (clause 11.2(13)).[21] As mentioned before, a difference between NEC3 and NEC4, and which applies throughout the NEC4 family, is that the term '*Employer*' has been replaced with '*Client*.' Other participants mentioned in the contract are the *Project Manager* and the *Supervisor*. There are only two general statements of how these individuals act.

- Clause 10.1 states: 'The Parties, the *Project Manager* and the *Supervisor* shall act as stated in this contract'. So how they act in specific circumstances and the basis for their actions are defined in the individual clauses of the contract. Some lawyers take comfort from the

[20] When the original New Engineering Contract came out there was only one contract, which subsequently became the Engineering and Construction Contract (ECC). Currently, the ECC is printed in seven versions. The main published version contains all the main payment options – this version was originally printed with a black cover (it currently has a largely black cover) so was known as the 'Black Book', and the name has stuck. In the Black Book, all option A clauses are collected together in one section, all the option B clauses together in another, etc., with these sections appearing in the middle of the document, after the core clauses (the clauses that apply regardless of which payment option is used). However, six other versions are also printed, which correspond to the six main payment options – these originally had green covers (and the current covers are mainly green). So, the option A version only contains the option A clauses, and not the option B to F clauses. Further, the option A clauses are positioned throughout the contract, in line with their numbering, rather than being grouped together. Option-specific clauses are so identified by being printed in **bold**. Most of the other long forms also have different payment options, but it is still only the ECC that is printed in different versions.
If you know that you are going to be working under a particular ECC main payment option for any length of time, it is worth getting that version, rather than having to locate the option-specific clauses in the middle of the Black Book.

[21] I have already stated this, but it is worth repeating here. I had to scratch my head a fair few times in order to come up with a consistent use of the words 'client' and 'contractor' – whether they have a capital initial or not, and whether they should be in italic or not. This is the protocol that I settled on.

- 'client' (no capital, no italics) is used for any client across the engineering, construction or any other industry, regardless of whether they are using NEC. I also use it when referring to my own company's clients.
- 'Client' (capital C, not in italics) is used for general comments about NEC Clients and is therefore typically in the plural.
- '*Client*' (capital C and in italics) is used in the specific context of an NEC contract, either before the contract is entered into or once it has been entered into. It is therefore generally in the single.

I have used a similar convention for contractors, Contractors and the *Contractor* and, when telling stories about the NEC3 edition, for Employers and the *Employer*.

traditional legal convention of using modal verb 'shall' to express a legal obligation, arguing that without it the rest of the contract – which uses the present active tense – would not make legal sense. Others think that it is utterly irrelevant! But all agree that this general statement, in whatever tense it is expressed, is necessary to make the contract work.
- Clause 10.2 states: 'The Parties, the *Project Manager* and the *Supervisor* act in a spirit of mutual trust and co-operation'. This obligation was introduced into the second edition in response to a recommendation in the Latham Report (Latham, 1994).

I will say more on what acting in a spirit of mutual trust and cooperation means in a few moments. In the second and third editions, the two obligations regarding actions – 'as stated in this contract' and with 'mutual trust and co-operation' – were connected together by the use of the word 'and', whereas in the fourth edition they are two separate sentences. None of the lawyers I know think this change makes any difference to legal interpretation. But note that it is not an 'or' – i.e. you operate the contract as per the individual clauses in a spirit of mutual trust and cooperation. Many people seem to act as if these obligations are mutually exclusive: they either act as stated in the specific clauses, with a view to getting one over you, **or** they abandon the mechanisms of the contract, but are collaborative (and I have explained my feelings on collaboration in Section 1.4.2). How about following the contract by doing what it says in a collaborative manner?

Back to what acting 'in a spirit of mutual trust and co-operation' means. Until a few years ago, lawyers loved debating what it meant, but there was little case law to help them come to a conclusion. Interpretations ranged from 'absolutely nothing' to 'a duty to act in good faith'.[22] More recently, there has been some case law on what it or similar words mean[23] and which narrows down its meaning, so here is quick summary of what it does and does not mean.

- First, it quite clearly means something! The *Northern Ireland Housing Executive* v. *Healthy Buildings Ltd (Ireland) Limited* [2017] case concerned a dispute over whether historical records should be used to price a compensation event or whether it should be based on a forecast. The contract was under the NEC3 Professional Services Contract, which has similar words in its clause 10.1. The judge stated

  > This particular contract begins with the agreement that the employer and the consultant shall act 'in the spirit of mutual trust and co-operation' (10.1). It seems to me that a refusal by the consultant to hand over his actual time sheets and records for work he did during the contract is entirely antipathetic to a spirit of mutual trust and co-operation.

- In legal interpretation, detailed or specific statements override general statements, so when push comes to shove you act as stated in the clause relevant to that specific circumstance.

[22] 'Good faith' is a high-level legal concept, in which there is a general presumption that the parties to a contract will deal with each other honestly and fairly, so as to not undermine the other party's or parties' ability to receive the benefit of being in that contract. In the UK, there is no presumption to act in 'good faith' when entering into a commercial contract. This differs from the position in many other countries, including France, Germany, the USA and Australia, which, to varying extents, recognise some form of overriding principle that, in agreeing and performing contracts, the parties should act in good faith.

[23] See in particular: *Mid Essex Hospital Services NHS Trust* v. *Compass Group UK and Ireland Ltd (t/a Medirest)* [2013] EWCA; *Chelsfield Advisers LLP* v. *Qatari Diar Real Estate Investment Company* [2015] EWHC 1322 (Ch); *Mears Ltd* v. *Shoreline Housing Partnership Ltd* [2015] EWHC 1396 (TCC); *Costain Ltd* v. *Tarmac Holdings Ltd* [2017]; *Northern Ireland Housing Executive* v. *Healthy Buildings Ltd (Ireland) Limited* [2017] NIQB 43.

In the *Mears* v. *Shoreline* [2015] case, the judge summarised by saying that he was 'not satisfied that … any such … obligation to act in a spirit of mutual trust and cooperation or even in a "partnering way" would prevent either party from relying on any express terms of the contract freely entered into'.

To give a personal story that illustrates this, I had the dubious pleasure of being an expert witness in Ghana for 4 days on an NEC3 Term Services Contract that had been terminated by the *Employer*. One of the arguments put forward by the *Contractor's* lawyer was that the *Employer* could not terminate, even though there are specific provisions in all NEC contracts for doing so, because it went against the obligation to 'act … in a spirit of mutual trust and co-operation'. The judge, in line with the UK judge, dismissed this argument.

- *Costain Ltd* v. *Tarmac Holdings Ltd* [2017] considered what these words could mean in the most detail, making substantial reference to *Keating on NEC3* (Thomas, 2012), which is considered the most authoritative legal book on NEC3. This book makes an analogy with what 'good faith' means by reference to the Australian case of *Automasters Australia PTY Ltd* v. *Bruness PTY Ltd* [2002] WASC 286. *Keating* sets out four bullet points on what it could mean, but the second and last are the most relevant.

  Taking the second bullet point first, it states 'good faith obligations do not require the parties to put aside self-interest; they do not make the other parties fiduciary', i.e. make the other party's interests central to how they behave and act, as a trustee would.

  The fourth bullet point states 'the duty is one "to have regard to the legitimate interests of both the parties in the enjoyment of the fruits of the contract as delineated by its terms"'. I attended an evening talk by the barrister who led and edited this book, who summed this up as 'having due regard to the other party's business'. So, to go back to my original question, what does and does not 'having due regard to the other party's business' mean? It means that

  - you do not deliberately mislead another party, either by your words or by keeping quiet, if you realise that they are pursuing a course of action that may have unfortunate consequences for them (and there is a good argument for saying that the early warning procedure expressly requires you to notify them)
  - you should try to find mutually beneficial solutions to issues and problems (or early warnings in NEC speak), although this might not be possible. Clause 15.3 (in the early warning clauses) actually directs you to do this, i.e. 'co-operate in… seeking solutions that will bring advantage to those affected'
  - if you can take a relatively small action that has minimal cost to you but which helps another party, you should do it. For instance, if, as a *Project Manager*, you are not accepting a communication from the *Contractor*, then tell the *Contractor* why in sufficient detail that it can correct it (i.e. as per NEC4 clause 13.4).

What I have hinted at is that there are some specific clauses that tell you to act in this way in certain circumstances. But the current legal position seems to be that, generally, you should act in a trusting and cooperative way regardless, albeit without putting the other Party's interests ahead of, or even on par with, yours.

Another way of looking at it is that acting 'in a spirit of mutual trust and co-operation' lubricates the express provisions (the 'engine of project management'), as the contract operates so much more smoothly and effectively if the participants adopt this mindset and reflect it in their actions.

The *Client's* interests are represented by the ***Project Manager***, who manages the contract on the *Client's* behalf and administers it impartially (see Section 1.4.3). I emphasise the word 'manages', because the ECC places a much greater onus on the *Project Manager* to manage than traditional conditions of contract do. This implies that

- the person selected should be more management orientated than technical orientated, although it is desirable that they have, at a minimum, a technical appreciation of the *works*
- they must be delegated sufficient powers to take the actions and decisions required within the timescales of the contract
- they know and understand the business drivers behind the project, so they can make timely decisions that best reflect the *Client's* objectives
- they are in regular contact with the *Client* and are clear if and when the *Client* wants to be consulted before a *Project Manager's* decision.

The ***Supervisor*** is responsible for monitoring whether the *Contractor* satisfies the quality standards stated in the specification, or 'Scope' in ECC terminology. Some have suggested that the role would better have been called the '*Inspector*', as the *Supervisor* does not 'supervise' any construction. The role has been described as somewhere between the traditional architect/resident engineer/employer's representative role and the clerk of works/inspector role. It has, in some cases, been seen as a demotion for the former roles, but a promotion for the latter as they have formal actions relating to testing, Defects and quality under the contract. Most of the *Supervisor's* actions are in Section 4 (quality management) of the contract. On smaller projects, which only warrant one *Client's* representative having day-to-day involvement in the contract, much of the day-to-day duties of the *Project Manager* have been delegated to the *Supervisor*, who is resident on the Site or visits the Site frequently.

Designers and **architects** are not mentioned in the ECC. Therefore, any design changes originated by the *Client's* designer have to come through the *Project Manager*. This is so that the *Project Manager* can exercise control over the extent of changes issued by the designers in order to control time and cost. This does not stop works (sub)contractors and designers talking in order to determine the best technical solution, but does mean that the *Project Manager* can evaluate the effects of the proposed technical solution on time and cost before instructing the *Contractor* to proceed with the work.

Quantity surveyors and **programmers/planners** are also not mentioned in the contract. On a project of any size, one would traditionally expect a project manager to be supported by a quantity surveyor, at the very least in a part-time role. Increasingly, because of the emphasis on programme in the NEC family, this is also the case with a programmer/planner, although they seem to be a rarer breed than quantity surveyors. This is even more the case for contractors, where the site agent is often supported by a quantity surveyor and programmer.

The ***Adjudicator*** is a third party who is called upon to decide on any dispute between the Parties that cannot be settled by themselves. Adjudication is intended to be a relatively quick, inexpensive and less procedural way of settling disputes compared with arbitration or litigation. The *Adjudicator's* costs are allocated by the *Adjudicator* once they have made their decision, which normally means being paid by the losing party. As a result of the two Construction Acts, it is now compulsory to have adjudication provisions on virtually every construction contract let in the UK. Other countries have also legislated for similar adjudication procedures.

A **Subcontractor** is defined in clause 11.2(19) and is what you would expect it to be. The definition has been refined in NEC4 to exclude certain organisations from being counted as Subcontractors: those which only design and supply Plant and Materials (i.e. do not install it); those from which the *Contractor* hires construction Equipment; and those from which it hires people. Clause 26 refers to subcontracting and clause 26.1 makes it clear that the main contract applies as if the Subcontractor's employees and resources are those of the main *Contractor*.

Lastly, **Others** are defined as 'people or organisations who are not the *Client*, the *Project Manager*, the *Supervisor*, the *Adjudicator* or a member of the Dispute Avoidance Board, the *Contractor* or any employee, Subcontractor or supplier of the *Contractor*'. This would include a party in another contract with the *Client* to do some work on the same project, such as a utility company. Others would also include third parties and stakeholders.

1.5.3 Defined terms for things and locations

The **Scope** and the **Site Information** (clauses 11.2(16) and (18)) together effectively form what are traditionally referred to as specification and drawings. Project management is sometimes described as the management of change, and the Site Information describes the starting point for that change (i.e. the Site and its surroundings). If unexpected physical conditions are encountered that satisfy the criteria in clauses 60.1(12) and 60.2 (i.e. the Site Information is sufficiently inaccurate), then a compensation event results.

The Scope describes the end point of change (i.e. it describes the *works* in detail) and any constraints on how the *Contractor* is to achieve this end point. The drawing and documents that make up the Scope and Site Information are referenced in the Contract Data. In the unamended contract, there are potentially two Scope documents in place on entering into the contract:

- the *Client's* 'Scope' referenced and incorporated into the contract via Contract Data part one – this could be a performance or functional statement or a complete design
- 'the Scope provided by the *Contractor* for its design' referenced and incorporated into the contract via Contract Data part two – if the *Client's* Scope is a complete design, there would be no need for the *Contractor's* Scope.

The former has precedence over the latter. That is, if there is an ambiguity or inconsistency between the two sets of information, the *Client's* set 'trumps' or overwrites the *Contractor*'s set, meaning that the risk is on the *Contractor* to ensure it has met the *Client's* requirements, and not the other way around (see the last bullet point of clause 60.1(1)). The Scope can be changed by an instruction from the *Project Manager*, in which case it becomes a compensation event, except in two circumstances (again, see clause 60.1(1)).

The **Site** (clause 11.2(17)) is the area within the *boundaries of the site* as identified in Contract Data part one by the *Client* for use by the *Contractor* to Provide the Works. This is often by reference to a drawing showing a plan view with the Site marked out.

The **Working Areas** (clause 11.2(20)) are the Site and any additional areas identified by the *Contractor* in Contract Data part two which are needed to Provide the Works, such as areas for offices and a storage yard. The importance of this is that, when the *Contractor* is incurring Defined Cost, either for a compensation event (main payment options A to E) or being reimbursed it under the main payment options C to E, then it is reimbursed costs directly incurred within the

Working Areas. (This is a little bit of a white lie, in that costs for manufacturing or fabrication and design outside of the Working Areas are reimbursed.) The *Contractor* can propose additions to the Working Areas during the contract, but the *Project Manager* can reject these for two reasons stated in clause 16.3.

'**Plant and Materials** are items intended to be included in the *works*' (clause 11.2(14)), i.e. they form the permanent works, while **Equipment** (clause 11.2(9)) is, in conventional building and civil engineering terms, construction plant and includes 'temporary works'. The reason for this non-traditional terminology is the multi-disciplinary application of the ECC. For instance, in the heavy engineering sectors, such as the power and process sectors, plant is machinery that is delivered to the Site, bolted in and connected up (e.g. generators and turbines).

Under the ECC, to be contractually valid, all **communications** have to be in a form that can be 'read, copied and recorded' (clause 13.1). For most purposes, this is in writing, which can include email, or it could be through the use of one of the NEC cloud-based communication systems outlined in Section 3.3.2. In fact, one of the changes between NEC3 and NEC4 is that, if such a system is specified in the Scope, contractual communications have to go through that system (clause 13.2). Unless stated elsewhere in the contract, all contractual communications that require a response must be responded to within the *period for reply* (clause 13.3), which is stated in Contract Data part one. The most suitable *period for reply* will vary between types of contract (e.g. on a design and build contract, where the *Project Manager* has to accept the *Contractor's* design, compared with a traditional construction-only contract), or it could be varied for different types of response. A detailed change in NEC4 is that any different *periods for reply* have to be stated in the Contract Data and cannot now be included in the Scope. Failure by the *Project Manager* or *Supervisor* to respond within the period required by the contract is a compensation event under clause 60.1(6).

1.5.4 Terminology for time and programming

The ***starting date*** is the date when the contract comes alive, in that three things hang off it.

- The *Project Manager* is required to prepare the first Early Warning Register within 1 week of the *starting date*, which is done by amalgamating the matters that the *Client* and the *Contractor* have entered into the Contract Data (see clause 15.2).
- It is the reference point for payment, with the first assessment date within the *assessment interval* after the *starting date*.
- Insurances need to be in place, but if the Party responsible for an insurance has not proved that the insurance is in place, then the other Party can take out the insurance and charge the responsible Party (see clauses 84, 85 and 86).

An ***access date*** is simply when the *Contractor* can have access to the Site. Note that I say 'Site', not Working Areas. So, for instance, prior to the *access date*, the *Contractor* could start to set up its cabins and yard etc. in the areas not identified by the *Client* as the Site. There could be a number of *access dates* for access to different areas or sections of the *works*. There could even be a number of different types of access given, such as to carry out surveys while an existing asset is still in use and before any physical work is done.

The **Completion Date** (clause 11.2(3)) is the date by which the *Contractor* has to complete the *works* or, in ECC speak, achieve Completion. The original Completion Date is stated in the

Contract Data as the '*completion date*'. The Completion Date may later be changed as a consequence of compensation events, but it can only be put back (delayed) by compensation events, not brought forward. The relevant clause is 63.5. If the *Project Manager*, acting on the *Client's* behalf, wishes to bring forward the Completion Date, then the **acceleration** procedure in clause 36 is followed. 'Acceleration' under the ECC means bringing forward the contractual Key Dates and/or Completion Date, not speeding up the work to meet the current ones. If, as a *Project Manager*, you wish to do the latter, then you may be able to though the compensation event process. The only other way that the Completion Date may be changed is as the result of a 'deal' to accept a Defect (see clause 45.2).

Completion (clause 11.2(2)) is the equivalent of substantial or practical completion in some other conditions of contract. However, it is more tightly defined in the ECC conditions, without the need to refer to case law. The first bullet point of clause 11.2(2) states that the *Contractor* has to do 'all the work which the Scope states is to be done by the Completion Date', and a paragraph below the bullet points clarifies what has to be achieved if there is no such explicit statement in the Scope. The second bullet point states that the *Contractor* has to correct notified Defects that prevent the *Client* from using the *works* or Others from doing their work.

It is the *Project Manager* who decides the date of Completion and certifies it under clause 30.2. If the secondary option for delay damages is selected (option X7) – which it normally is – and the *Contractor* fails to achieve Completion by the Completion Date, the *Contractor* will have to pay damages at the rate prescribed in Contract Data part one.

Note that if secondary option X5 (sectional Completion) is selected, there can be a number of different sectional Completion Dates and hence Completions.

'The **Accepted Programme** is the programme identified in the Contract Data or is the latest programme accepted by the *Project Manager*. The latest programme accepted by the *Project Manager* supersedes previous Accepted Programmes' (clause 11.2(1)). Revised programmes are submitted at intervals no longer than that stated by the *Client* in Contract Data part one.

The **programming requirements** are in clauses 31 and 32. What the first and all subsequent programmes have to show is stated in clause 31.2, and this includes statements, for each operation, 'of how the *Contractor* plans to do the work identifying principal Equipment and other resources which will be used', as well as float and time risk allowances (the difference between best productivity and average productivity, or allowances for specific identified *Contractor*-owned risks). Until a programme is submitted that shows the information required by the contract, a quarter of the Price for Work Done to Date is retained from amounts due to the *Contractor* (clause 50.5). Once such a programme has been submitted, this sanction cannot be applied but, if the *Contractor* never submitted such a programme, it could in theory be applied until the *Contractor* submits an as-built programme at the end of the contract. To be accepted, the programme has to satisfy the criteria stated in clause 31.3. Any revised programme has to show the information in clauses 31.2 and 32.1, and to be accepted has to again satisfy the criteria stated in clause 31.3. In addition to the sanction in clause 50.5 for a programme that does not contain the information required, it is mandatory for the *Project Manager* to assess compensation events if various criteria for the programme are not satisfied (see clauses 64.1 and 64.2). The *Contractor* should know more about the effect of a compensation event, particularly in terms of disruption within an operation, than the *Project Manager*; and even if they knew the same amount, the

Contractor will almost certainly be more generous in its assessment than the *Project Manager*. Consequently, all Contractors I have explained this to want the first bite at assessing compensation events, and hence have a motivation to keep the programme up to date in accordance with the contract.

The reason for incorporating these contractual incentives in the ECC is that the Accepted Programme is the base document from which both the time and cost effects of compensation events are assessed. Thus, if a compensation event occurs that results in a change in method, then the resulting changed resources are compared with those in the current Accepted Programme. Provided these resources are identified as Defined Cost, then costs are assigned to these changed resources. Any change to the Completion Date is evaluated by comparison with the Accepted Programme current at the 'dividing date', in accordance with clause 63.5.

The programme is a document vital to the administration and management of the contract and, as such, it is worthwhile

- for the *Contractor* to spend time developing the programme so that the *Project Manager* can understand what will cause the *Contractor* delay and where its costs come from, and to avoid the sanctions
- for the *Project Manager* to spend time understanding the *Contractor's* programme, ensuring that it both satisfies the requirements of the contract and does not commit the *Project Manager*, the *Client* or Others to any deadlines that they cannot fulfil (e.g. delivery to Site of something to be incorporated into the *works*). If the *Project Manager* is unhappy with the programme, they should not be afraid to exercise the powers available to them.

New to NEC4 is an express sanction on the *Project Manager* for not doing what they should do in relation to the programme. So, under the last paragraph of clause 31.3, if the *Project Manager* does not respond correctly to the *Contractor's* programme submission within the 2 weeks allowed, the *Contractor* may give a reminder notification. From the time of this reminder notification, the *Project Manager* has a further 1 week to respond. If the *Project Manager* still does not respond, the programme is treated as being accepted.

A **Key Date** is a date during the construction period by which the *Contractor* has to meet a stated Condition (see clause 11.2(11)). Typically, this would be to enable the *Client*, another contractor or a third party (an Other in ECC speak) to do something within the Site. For instance, prepare an area so that a service diversion can be carried out by a statutory service provider, such as a water board or electricity supplier, who is contracted directly with the *Client*. The initial *key dates* and associated conditions are stated in Contract Data part one, but Key Dates can be changed by the *Project Manager* during the contract (see clause 14.3), although doing so will be a compensation event (under clause 60.1(4)). Failure by the *Contractor* to meet a Key Date entitles the *Client* to deduct the direct costs for this failure (i.e. not the consequential or indirect costs; see clause 25.3). This deduction has no individual cap, either in the core clauses or even with the use of secondary option X18 (limitation of liability), although it would be limited by the *Contractor's* total liability under this option.

1.5.5 Terminology for quality

A **Defect** is described in clause 11.2(6) and is pretty much, on a superficial reading, what you would expect it to be: a part of the *works* that is not in accordance with the Scope or not in accordance with *Contractor's* design that the *Project Manager* has accepted. It can therefore apply to both the physical works and the design. However, note that what would conventionally be called a 'snag' (i.e. minor bits of work that are not yet done, but which do not affect Completion) would also fall within this definition of a Defect. Consequently, the same timescales for correcting them, once notified, apply.

The ***defects date*** is the date when the *Contractor* ceases to be obliged to remedy Defects in the completed *works*. The date itself is not specified in the contract, but is instead taken as the end of a set period of time after the Completion of the whole of the *works*. This period – loosely equivalent to what is called a 'maintenance period' in some other contracts – is entered into Contract Data part one, typically being 52 weeks. For Contractors with sectional Completion in their contracts, note that the liability to correct Defects in a particular section lasts from the date of Completion of the whole of the *works,* not just the date of Completion of that section. Until the *defects date*, the *Supervisor* notifies the *Contractor* of any Defects that they find, and vice versa (clause 43.2). Note that a Defect is related to the '*works'* (the finished article) and not to the way in which the *Contractor* is 'Providing the Works'.

If a Defect is notified on or before Completion – and provided the Defect does not prevent the *Client* from using the *works* and thus affect Completion – the *Contractor* has, from the time of Completion, the ***defect correction period*** (stated in Contract Data part one) in which to correct any notified Defects. So, if at Completion the *Supervisor* had notified all incomplete work as Defects, then the *Contractor* would have the *defect correction period* to finish off the outstanding work. For any Defects notified after Completion and before the *defects date*, the *Contractor* would have the time between the notification and the end of the *defect correction period* to correct those Defects (see clauses 44.1 and 44.2).

The **Defects Certificate** is either a statement that there are no Defects or a list of Defects that the *Contractor* has not corrected (clause 11.2(7)). It is issued by the *Supervisor*, either at the *defects date* (if there are no Defects), or at the end of the last *defect correction period* or the date when all notified Defects have been corrected (whichever is earlier) (see clause 44.3).

Figure 1.8 illustrates these concepts. To help you to fully understand the timelines, I suggest you re-read the above text while looking at the figure.

Figure 1.8 Timelines for the correction of Defects

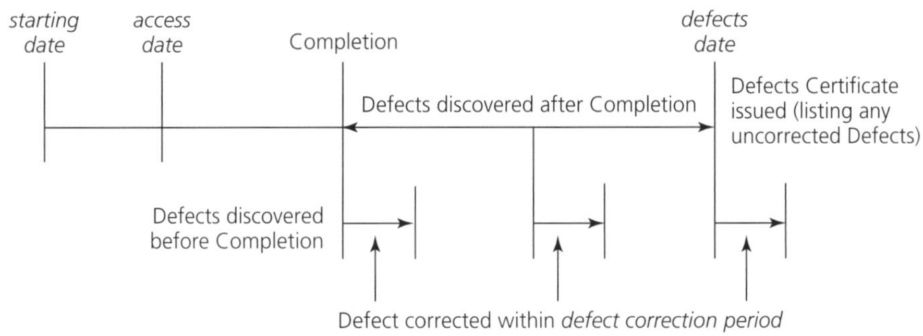

If the work done fails to meet the quality parameters (i.e. it is a Defect), then

- the procedure for correcting the Defect is specified in clause 44
- the *Project Manager* has the option of trading quality off against time and/or cost by accepting the Defect in return for reduced Prices and/or an earlier Completion Date under clause 45
- the *Project Manager* has a sanction available, in clause 46.1, to encourage the *Contractor* to correct Defects within the *defect correction period*, whereby the cost of getting someone else to correct the Defect is paid by the *Contractor*.

Note that the *Project Manager* has no power under the contract to overrule the technical decision of the *Supervisor*. Instead, the *Project Manager* manages the consequences of the *Supervisor's* technical decision, leaving the *Supervisor* free to focus on technical issues.

1.5.6 Terminology for payment

Various definitions exist for the Prices, the Price for Work Done to Date, Defined Cost and Disallowed Cost depending on which main option is used. Hence the explanations given below. If you are already using one option (or even if you are not), you might find it easier to follow through the definitions for one main payment option at a time.

1.5.6.1 The Prices

Under the **priced contracts** (options A and B), the Prices are the sum of either

- under Option A (clause 11.2(32)), the individual prices against each of the activities in the *activity schedule* referenced from Contract Data part two (which you know is in the Contract Data because it is in italics)
- under Option B (clause 11.2(33)), the lump sums and the sum of rates multiplied by quantities in the *bill of quantities* referenced from Contract Data part two.

If there are no compensation events, the original Prices are the amount the *Contractor* will receive for Providing the Works. As the contract progresses, some compensation events will inevitably occur, and so the Prices will be changed to reflect the effect of these compensation events.

Under the **target-cost contracts** (options C and D), the total of the Prices is the target, which is either

- under Option C (target contract with activity schedule), the initial Prices are the sum of the individual prices against each of the activities in the *activity schedule* (clause 11.2(32)), and is the fulcrum around which the incentive or pain/gain share mechanism operates
- under Option D (target contract with bill of quantities), the initial Prices are the sum of lump sums and rates multiplied by quantities at tender in the *bill of quantities* (clause 11.2(33)). However, once the contract is entered into this does not matter, as it is the quantity of work done, whether as a proportion of each lump sum or a quantity times rate, that matters and which forms the fulcrum around which the pain/gain share works or, in contractual speak, the Total of the Prices (clause 11.2(35).

Once again, as the contract progresses some compensation events will inevitably occur, and the target Prices will change to reflect the cost effect of these compensation events. However, the Prices are the target, not what the *Contractor* is paid.

Under the **cost-reimbursable** and **management contracts** (options E and F), the Prices are defined as 'the Defined Cost plus the Fee'. So, while the Prices are modified by compensation events, their only function is as an estimate of the final monetary sum that will be paid to the *Contractor*.

Under **all options**, the principal clause governing the evaluation of the cost effects of compensation events is clause 63.1. It means that the change in the Prices, due to the compensation event, is effectively equal to the change in the Defined Cost already incurred or forecast and the resulting Fee. Therefore, to work out the change in Defined Cost, it is necessary to first evaluate the Defined Cost of the original work item that is going to change, then the Defined Cost of the revised work item, and then subtract one from the other to give the change in Defined Cost, to which the tendered *fee percentage* is applied. Both in practice and contractually, short cuts can be taken by agreement. For instance

- you only need to evaluate the bits that are changing, not the whole operation or activity
- core clause 63.2 states: 'The *Project Manager* and the *Contractor* may agree rates or lump sums to assess the change to the Prices'. For instance, if everybody is happy, use existing Subcontractor rates to pay the Subcontractor, on top of which the *Contractor's fee percentage* is applied. But note, this is by agreement. (Some people – and, from experience, it appears to be those on the *Client* side – seem to think that agree means 'If I tell you that I will be using a rate, then you have agreed to it'. That is not agreement!)

1.5.6.2 Price for Work Done to Date

Under the **priced contracts** (options A and B), the Price for Work Done to Date is the total of the Prices at any point in the contract for

- under option A, each group of completed activities or each completed activity in the Activity Schedule (see clause 11.2(29) in option A)
- under option B, the quantity multiplied by the rate in the Bill of Quantities plus a proportion of the lump sum for each completed work item (see clause 11.2(30))

which are without notified Defects, the correction of which would delay following work.

Notice that here I have used capital initials and not italics for Activity Schedule and Bill of Quantities, whereas in describing the initial Prices I used italics and not capital initials. This is because, under clause 11.2(21), 'The [current] Activity Schedule is the [initial] *activity schedule* [referenced in Contract Data part two] unless changed in accordance with these *conditions of contract* [which will typically be by compensation events]'. There is a similar statement for the Bill of Quantities in option B (clause 11.2(22)). So, if compensation event work has been done, under options A and B, the *Contractor* is not paid until the change to the Prices has been agreed and incorporated into the Activity Schedule or Bill of Quantities. The relevant clauses are 63.14 in option A and 63.15 in option B.

- Under the **target-cost** and **cost-reimbursable contracts** and the **management contract** (options C to F), 'The Price for Work Done to Date is the total Defined Cost which the *Project Manager* forecasts will have been paid by the *Contractor* before the next assessment date plus the Fee' (clause 11.2(31)). So, the *Contractor*, if the unamended timescales of the contract are used, can run the contract slightly cash-flow positive, as it will be paid by the end of week 3 following the assessment date, yet does not have to pay out until the next assessment date, which is typically at a 4-weekly or monthly interval (i.e. they can hold the cash for a week or so). These payments could be to Subcontractors, the *Contractor's* own staff etc.

Under these options, Contractors would be wise to note that

- in accordance with the first bullet point for Disallowed Cost (clause 11.2(26)), the burden of proof – which, being civil law, is on the balance of probability – is on the *Contractor* to show it has or will be making the payment, not on the *Project Manager* to disprove it
- the minimum requirements for doing this are stated in clause 52.2.

Do note that, on the **target-cost options**, if the final Price for Work Done to Date is less than the final total of the Prices, the *Contractor* will be paid its share of any saving by the *Client* following Completion. However, if the final Price for Work Done to Date exceeds the final total of the Prices, then the *Client's* share of the overrun should be paid back by the *Contractor* following Completion. On the majority of target-cost contracts that I have seen, the *Client* has changed this so that once an overrun on the target Price is reached, it only pays the *Contractor* its share of the overrun. Others have done it so that they start deducting monies once an overrun is forecast, but my experience is that this leads to complicated clauses and even more complicated administration of the contract (i.e. monthly disagreements over what the forecast final Price for Work Done to Date will be).

1.5.6.3 Defined Cost

Defined Cost is the term used in the main option clauses to reference how cost is built up.

Under the **priced contracts** (options A and B), Defined Cost is only used for assessing compensation events, and 'is the cost of the components in the Short Schedule of Cost Components' (clause 11.2(23)). The changes between NEC3 and NEC4 have included some revisions to the two schedules of cost components, which I discuss in Section 2.3.3. However, for those familiar with NEC3, a change in NEC4 is that both schedules now incorporate Subcontractor costs under a separate heading. The practical change is that, under NEC4, when a compensation event occurs, the *Contractor* is paid what it pays its Subcontractors in accordance with their subcontracts, plus a single *fee percentage* on top. (By comparison, under NEC3, the *Contractor* was paid according to the costs built up in accordance with the Shorter Schedule of Cost Components, with the *subcontract fee percentage* then applied.)

Under the **target-cost contracts** (options C and D) and the **cost-reimbursable contract** (option E), Defined Cost is both the basis for reimbursing the *Contractor* as the work progresses and for assessing compensation events. It is defined as 'the cost of components in the Schedule of Cost Components [which, in NEC4, now includes Subcontractor costs in accordance with their subcontracts] less Disallowed Cost' (clause 11.2(24)). So, at a big picture level, there is no practical change from NEC3, apart from having one *fee percentage*, rather than two.

Under the **management contract** (option F), as per options C, D and E, Defined Cost is both the basis for reimbursing the *Contractor* as the work progresses and for assessing compensation events. However, under clause 11.2(25)

Defined Cost is

- the amount of payments due to Subcontractors for work which is subcontracted without taking account of amounts to be paid or retained from the Subcontractor by the *Contractor* ... and
- the *prices* for work done by the *Contractor*

less Disallowed Cost.

Under the management contract, the presumption is that the *Contractor* is subcontracting most, if not all, of the physical work. However, if the *Contractor* is going to do some of the physical work itself, it needs to identify the work as '*prices*' in Contract Data part two (clause 11.2(25)). Note that there is no reference to any schedule of cost components. So, any internal *Contractor* costs not identified in the '*prices*' in Contract Data part two are paid for through the *fee percentage*.

1.5.6.4 Schedule of Cost Components

There are two schedules of cost components – a full Schedule of Cost Components (or full Schedule) and a Short Schedule of Cost Components (or Short Schedule) – which are both generic lists of resources that may be used to Provide the Works. They both have the following eight main headings

- People
- Plant and Materials
- Equipment
- Subcontractors
- Charges (which would commonly be called 'preliminaries' in civil engineering or building)
- Manufacture and Fabrication (for costs incurred outside of the Working Areas)
- Design (for costs incurred outside of the Working Areas)
- Insurances.

The schedules are found at the back of the contract. The main difference between the two is that the Short Schedule calls for more rates to be put forward as part of the *Contractor's* tender, while the full Schedule wants costs to be built-up in accordance with the more detailed rules under each of the main headings.

At this juncture, it may be useful to reiterate what has been said before at various points in this book.

- Under the **priced contracts** (options A and B), the Short Schedule is used for assessing compensation events only. If the *Contractor* is to be paid for a compensation event, then the change in resources is determined using the method statements and resources forming part of the Accepted Programme. For the *Contractor* to be paid more (or less), the changed resource has to be listed in the Short Schedule and the tendered *fee percentage* is then applied. The combined sum then adjusts the Prices.
- Under the **target-cost contract** (options C and D) and the **cost-reimbursable contract** (option E), the *Contractor* is reimbursed directly any costs that are listed in the full Schedule plus its tendered *fee percentage*. Under options C and D, if a compensation event occurs, the full Schedule is used to calculate the change in Defined Cost to which the tendered *fee percentage* is applied and this is the sum by which the Prices are adjusted.

- Under the **cost-reimbursable contract** (option E), the Prices are simply a forecast of the final total Defined Cost for the whole of the *works* plus the Fee.
- Under the **management contract** (option F), neither of the schedules is part of the contract, as the *Contractor* does very little, if any, of the physical work itself. If it is to do physical work, it needs to identify the specific activities and associated *prices* in Contract Data part two. However, assuming the subcontract works packages are let under Options A to E of the NEC4 Engineering and Construction Subcontract, then what has been stated for options A to E under the main contract applies but at a subcontract level.

1.5.6.5 Disallowed Cost

The definition of **Disallowed Cost** varies between the main options

- under the **priced contracts** (options A and B), no definition exists
- under the **target-cost** and **cost-reimbursable options** (options C to E), clause 11.2(26) applies
- under the **management contract** (option F), clause 11.2(27) applies.

In the main, Disallowed Cost results from the *Contractor* not complying with various provisions of the contract or not using the resources properly on the particular contract. The administration of Disallowed Costs is discussed in more detail in Section 3.4.

1.5.6.6 Fee

Under **all options**, 'The Fee is the amount calculated by applying the *fee percentage* to the amount of Defined Cost' (clause 11.2(10)). The *fee percentage* is tendered by the *Contractor* and stated in Contract Data part two. Having a single *fee percentage* is a change from NEC3, which had two: one for work the *Contractor* was doing directly and one for subcontracted work. The NEC3 arrangement, along with other provisions, led to gamesmanship at tender by Contractors (but, as this is no longer relevant in NEC4, I will not go into it here).

Before moving on from the payment aspects, it is worth pointing out two two new NEC4 procedures that were not in NEC3.

- Under **all options**, there is now a final assessment procedure, which is given in clause 53. This was added in to close down the contract financially, rather than contractually. Previously, the payment system worked on the assumption that the contract would end on the date of the last assessment, while under contract law in the UK it remained open for either 6 or 12 years from Completion.
- Under the **cost-based contracts** (options C to F), in which the *Contractor* is reimbursed Defined Cost, a new clause 50.9 has been added in, whereby the *Contractor* can notify the *Project Manager* that Defined Cost for 'a part of the works' has been finalised and is ready for inspection, in order to be agreed by the *Project Manager*, whereupon that part only becomes 'final accounted'. This was added to address complaints from Contractors about Employers (under NEC3), in the latter stages of a contract, opening up accounts – and sometimes paying forensic accountant on a percentage savings basis – that had been inspected after the corresponding amounts had been paid. The *Contractor* had therefore already added these amounts to their accounts as earnings and cash. A subsequent complaint from Clients is 'Why can't we do this too?' as Contractors also have a habit of producing additional Defined Cost later in a contract.

1.5.7 Terminology for early warnings and the compensation event procedure

1.5.7.1 Early warning

The early warning clauses (in clause 15) give a procedure whereby the *Project Manager* and *Contractor* have to notify each other of any matter that *could* (not necessarily will) increase the Prices, delay Completion, delay meeting a Key Date or impair the performance of the *works* in use. The *Contractor* may also notify of any other matter that could increase its total cost.

As stated previously, the early warning procedure is the intermediate or missing step between how risk management is normally implemented (or not in reality) – identifying what can go wrong and developing high-level strategies to minimise the impact – and monitoring the programme to identify something that has already gone wrong and has affected the *works*.

On the presumption that the earlier you identify a matter, the sooner you can do something about it and therefore the less impact it will have, the stimulus on the *Project Manager* to early warn is that the contract, and therefore the *Client*, may suffer if they do not. The stimulus on the *Contractor* is that, if the early warning subsequently becomes or results in a compensation event (and not all necessarily will), then 'the compensation event is assessed as if the *Contractor* had given an early warning' (clause 63.7). (The proviso to this is that, when instructing the first quotation, the *Project Manager,* in accordance with clause 61.5, tells the *Contractor* that it did not give an early warning which an experienced contractor could have given. The *Project Manager* cannot obtain the quotation, decide it is too high and retrospectively introduce this sanction.) That is, if the *Contractor* had notified early, when it first became aware of the matter, the impact of the event would have been less and the *Contractor* would be entitled to its full assessment. However, if the *Contractor* did not notify early, the impact of the event will increase, but the *Contractor* would only get the same amount as if it had given an early warning – i.e. less than the full impact of the event. All Contractors in the original research sample early warned as a matter of policy. All Contractors I point this out to decide to give early warnings as a matter of course.

1.5.7.2 Early warning meeting

Having issued an early warning notification, the *Project Manager* and the *Contractor* may instruct the other to attend an early warning meeting (previously called a risk reduction meeting in NEC3), at which the participants will consider how best to deal with the matter. Early warning meetings need not be big formal affairs. Under NEC3, risk reduction meetings were often regular events, held weekly or fortnightly for all but the most urgent or significant early warnings. A change to the NEC4 is that this regular meeting is now enshrined in the contract, occurring at the frequency stated in Contract Data part one, which would typically be weekly or fortnightly. The contract includes some constructive words that set the tone for the meetings. However, what is enforceable is that those who attend decide 'on the actions which will be taken and who, in accordance with the contract, will take them'. So, an early warning meeting that is simply a pity or blame party and if it does not result in actions it is not an early warning meeting in accordance with the contract!

1.5.7.3 Early Warning Register

It is the *Project Manager* who records the decisions made at the early warning meetings and enters them into the Early Warning Register (note the capital initials, which indicate this is a defined term under clause 11.2(8)). This was confusingly called the Risk Register under NEC3. I say 'confusingly' because people thought it should look like a traditional project risk register, as beloved by the risk management profession. But, if you read the definition, it is more akin to the 'issues register' found in project management literature. I like to think of the Early Warning Register as being an 'emerging risks and issues register'.

At the same time as adding early warnings to the register, the *Project Manager* should also be removing those early warnings whose time and relevance has passed, as well as actions that have been completed or are no longer relevant. In that sense, the Early Warning Register is, like the Accepted Programme, a dynamic entity that evolves and changes as the project progresses, as opposed to something that is stuck on a wall or remains in a quantity surveyor's or programmer's computer.

Some early warnings will be or may lead to compensation events, and some will not. Contract participants on both sides of the contractual divide – Clients, their Project Managers and Contractors – have sometimes mistakenly seen an early warning as a 'set up' for a compensation event. On the one hand, I have known a *Client* to utter in frustration 'How can we help the *Contractor* if he is not telling us his problems by giving us early warnings?', and others who openly state to their *Contractor* and *Project Manager* that they want to be made aware of problems and will be disappointed if they are not. On the other hand, I know of Contractors' frustrations when Project Managers have both rebuffed their early warnings (even when it would result in minimising the effect of a compensation event) and not given early warnings themselves of, say, an imminent design change, which again costs the *Client* time and money. Indeed, if there were only one change in culture I could make, it would be for everyone involved to give, receive and cooperatively deal with early warnings in the manner intended by the authors of the contract.

1.5.7.4 Compensation events

Compensation events are events for which the *Contractor* may be entitled to time and monetary compensation. There are 21 core compensation event listed in clause 60.1, two more than in the NEC3 ECC. There are three further compensation events in the **with bill of quantities contracts** (options B and D) (clauses 60.4 to 60.6) and four in the secondary options (options X2, X12, X14 and Y(UK)2). The 'may be' in the opening sentence of this paragraph is because the compensation event is effectively a threshold that has to be crossed for the stated risk to become a *Client* risk. Below the threshold, the *Contractor* has to take the risk. Once a compensation event has occurred, the *Contractor* still has to justify any changes to the Completion Date and/or Defined Cost to obtain the additional time and/or money to which it is entitled.

The majority of compensation events are for events over which the *Client*, *Project Manager* or *Supervisor* has full or predominant control, in that they either do something they should not or do not do something they should under the contract. The exceptions, and of particular interest, are in clause 60.1(12), (13), (14) and (19).

(12) The *Contractor* 'encounters physical conditions which… an experienced contractor would have judged… to have such a small chance of occurring that it would have been unreasonable to have allowed for them'. This differs from the reasonable foreseeability test found in traditional civil engineering contracts. Further clarification to this clause is given in clause 60.2. The intended effect is to put a greater onus on the *Client* to supply more information at tender stage on the Site conditions (i.e. Site Information).
A thorough site investigation reduces risk generally, and if this information is given to tendering Contractors, then they can more accurately price this reduced risk.

(13) A weather event that would be expected to occur, on average, less frequently than once in 10 years is recorded as a *weather measurement* during the contract, with the once in 10 years weather measurements being derived from the historical *weather data*. This is a much more clear-cut test than the subjective 'exceptional(ly) adverse weather' criterion of other contracts. See Section 2.2.2.5 for further explanation of how this clause works in practice.

(14) If the NEC4 family of contracts contained cross-references (and sometimes it would be useful if it did, but not to the extent of other contracts), there would be a cross-reference here to clause 80.1, which gives a fairly fulsome list of events for which the *Client* has liability. Each is individually unlikely to happen and is caused by a third party remote from the contract.

(19) This lists events that are often referred to as force majeure/Acts of God. They are defined in the contract as, to paraphrase, events that neither Party could prevent, are extremely unlikely to happen, and stop the *Contractor* from either completing the *works* absolutely or completing by planned Completion as shown in the Accepted Programme. This clause was added into the third edition so that, if one of these showstopper-type events occurred, the *Employer* (now *Client*) could continue with the contract, rather than be forced to terminate because it was impossible to perform through neither Party's fault. Until recently, it was very rarely used. However, the COVID-19 pandemic caused, to say the least, an increased focus on this clause. See Appendix 1 for an article that looks at this and other clauses that could come into play as a result of the pandemic. Also see Section 2.2.2.6, which considers what additional compensation events you might want to write into the contract.

The procedure for notifying, quoting for, assessing and implementing compensation events is laid out in clauses 61 to 66. The same procedure applies whatever the compensation event and whichever main option has been chosen. The *Contractor* is always entitled to overhead and profit (margin), determined by applying the tendered *fee percentage* to the change in *Contractor's* Defined Cost arising from the compensation event. The maximum timescales are illustrated in Figure 1.9.

The early warning procedure feeds into the top-left box in Figure 1.9, in that it may be decided, following an early warning meeting, that an early warning is or will result in a compensation event. Equally, it may effectively (but not contractually – see the last sentence of clause 15.1) be one and the same notification.

Except for four circumstances when the *Project Manager* should notify (clause 61.1), the *Contractor's* notification of a compensation event is not valid if it is more than 8 weeks[24] since the

[24] In the original first edition 'New Engineering Contract', the time limit was 2 weeks, but the timescale was changed to 8 weeks in the second edition following legal advice. The reasons for the change included that

- the effects of the compensation event – or even that the event is a compensation event – may not become apparent within the 2 weeks
- the consequences of the time bar are so harsh – all entitlement disappears – that it needs to be balanced with a reasonable timescale.

However, I often see this timescale shortened to 4 weeks by Clients' lawyers. Given that the top legal advice was to extend it to 8 weeks, and 4 weeks is closer to 2 weeks than it is to 8 weeks, 4 weeks is of arguable legality and could possibly be successfully challenged in a dispute.

As one lawyer said to me: 'My own view is that 4 weeks is the bare minimum that client lawyers can get away with, and we all know that it is "getting away with it"!', i.e. although a challenge might not be successful, it could be.

Figure 1.9 Maximum timescales for responses for compensation events

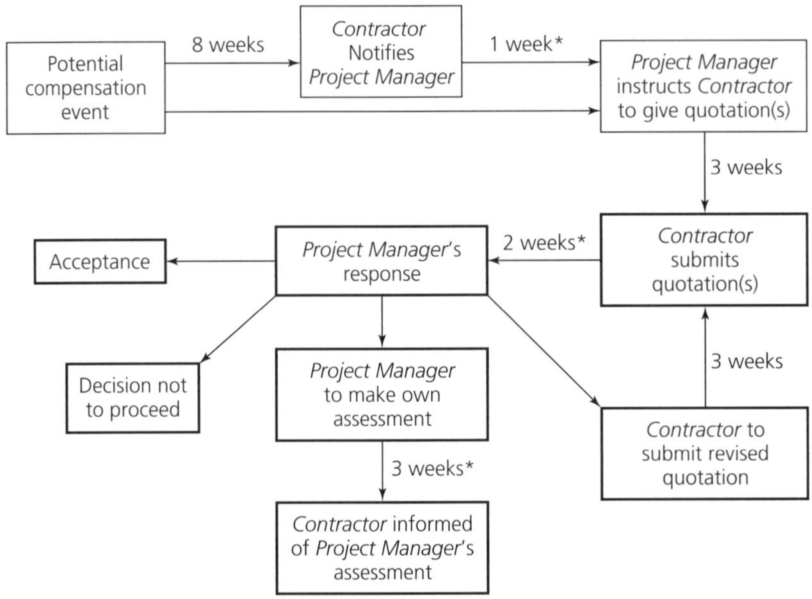

Contractor became aware of the event (clause 61.3). The intention of the authors of the ECC is to force the *Contractor* to notify compensation events promptly. If the *Contractor* does exceed the stipulated timescales, any entitlement to additional time and money is lost. The *Project Manager* has 1 week to respond to a notification, and will either instruct the *Contractor* to submit a quotation, or not to submit a quotation, for one of the five reasons stated (clause 61.4). Apart from the issue of the Defects Certificate, typically a year and a bit after Completion, there is no time bar on the *Project Manager* notifying the *Contractor* of a compensation event and instructing it to submit a quotation.

Having been instructed to give a quotation, the *Contractor* has a maximum of 3 weeks to submit one (clause 62.3) – unless a longer period is agreed (clause 62.5) – otherwise the *Project Manager* will assess the compensation event (clause 64.1), which is unlikely to be to the *Contractor's* benefit. If this is the case, then the *Project Manager* has the same period as the *Contractor* to notify the *Contractor* of their assessment, starting from when the need for the *Project Manager* to make the assessment becomes apparent (clause 64.3), i.e. the 3 weeks in clause 62.3.

On receiving the *Contractor's* quotation, under clause 62.3 the *Project Manager* has a maximum of 2 weeks – unless an extended period is agreed – either to

- accept it
- instruct the submission of a revised quotation, in which case the *Contractor* has a further 3 weeks to submit one
- notify the *Contractor* that they will be doing their own assessment, in which case the *Project Manager* has up to 3 weeks to make that assessment from the time of the notification.

A compensation event is 'implemented' once it has been accepted by the *Project Manager*, or the *Project Manager* had made their own assessment and imposed it on the *Contractor,* or a quotation has been 'treated as accepted', (see clause 66.1). After that, it cannot be re-evaluated, except as stated in the *conditions of contract* (see clause 66.3). The circumstances under which re-evaluation is possible are as follows.

- An assumption on which the forecast quotation is based and stated by the *Project Manager* proves wrong in practice (see clause 61.6 for the original assumption, clause 60.1(17) for the compensation event). This is one of the few compensation events that can reduce the Prices. Note that this does not open up the whole of the original quotation to a re-evaluation, just the part affected by the changed assumption. And contractually, it is a new compensation event, not a re-visiting of the one already implemented.
- Although not expressly stated, the *Project Manager* changes their decision, presumably in the light of an obvious error by the *Project Manager*. This is a compensation event (see clause 60.1(8)), but one that can only increase the Prices. Again, contractually, this is a new compensation event.
- The matter becomes a dispute and is resolved, preferably by the *senior representatives* or, failing that, the *Adjudicator*.

As an example of a change of decision by the *Project Manager*, I was sitting with a *Contractor's* team, trying to work out how on earth the *Project Manager* had assessed the time delay for a compensation event as 3 weeks when the logic on the programme said 17 weeks. Suddenly someone went 'I've got it': the *Contractor's* programme was two pages of A1, but the *Project Manager* had only looked at the first page, so did not appreciate the consequential effects of what was a 3-week lack of access. Once this was explained to the *Project Manager*, they literally turned the page and changed the decision to 17 weeks.

Going back to Figure 1.9, you will notice an asterisk next to three timescales that relate to communications by the *Project Manager*. Failure by the *Project Manager* to respond within these timescales, apart from being a compensation event, means

- at any time after the failure, the *Contractor* can notify the *Project Manager* that they have not responded within the timescales of the contract
- if the *Project Manager* does not respond within a further 2 weeks of the *Contractor's* notification, then the original *Contractor's* communication is treated as being accepted (That is, the compensation event, however spurious, is agreed by default, as is the quotation for a compensation event, however over the top it is. This could include where a *Project Manager* has not accepted a *Contractor's* quotation, and has notified their intention to assess the compensation event themself but has not done so. Furthermore, if the *Contractor* has submitted multiple quotations, it can choose which is accepted.)
- these deemed acceptances are contractually enforceable unless and until overturned by adjudication.

These clauses were introduced into the third edition of the NEC because of complaints by Contractors of, in the NEC Panel's words, 'indolent' project managers – in my language, substitute 'lazy', 'slack' or 'unprofessional (though Project Managers prefer 'overworked' for some reason)!

I have seen these clauses struck out in option Z clauses by Clients or Client advisors who are keen to 'protect' their Project Manager. This is misguided: prompt decision-making is part of good project management. If you are a Client, think what message this sends out to the bidders.

> I was once giving some training to a small Client and we were reviewing the changes their consultant – who would become the acting *Project Manager* on signing of the contract – was proposing. It included deleting the above clauses. The *Client* was livid, on the basis that they were paying the consultant to actively manage the contract and this undermined that obligation.

When a compensation event occurs, the change to the Completion Date is assessed following the rule set out in clause 63.5. (For the cost effects, see Prices in Section 1.5.6.1). Clause 63.5 means that the Completion Date can only be changed to a later date or kept where it is due to a compensation event, and is put back by the same amount of time that the planned Completion is delayed compared with the Accepted Programme current at the dividing date. This has several implications.

- No compensation event can result in an earlier Completion Date. In the ECC, 'acceleration' means bringing the Completion Date forward, and this has to be done through the acceleration provisions (clause 36). If a compensation event occurs that would delay Completion, then the contractual Completion Date is put back by the same amount; so, for a compensation event to affect the Completion Date, it has to be on or get on the critical path. Alternatively, the *Project Manager* can ask for the *Contractor* to produce a quotation, under the compensation event procedure, either
  - for the work to be speeded up so that the planned Completion as shown in the Accepted Programme stays the same, and hence the contractual Completion Date stays the same
  - as implied by clause 63.8, for the *Project Manager* to 'buy' some or all of the terminal float off the *Contractor* (who owns it) by including a risk allowance for time and cost in the quotation, i.e. the *Contractor* is paid additional money. (Because it is implied, but not explicit, in clause 63.8, this sum for the increased risk almost certainly needs to be accepted by the *Project Manager* in the *Contractor's* quotation, as opposed to being imposed by the *Project Manager*.)
- The *Contractor* is incentivised to show as little free float as possible in the Accepted Programme (i.e. all operations close to or on the critical path), so that any compensation event puts back planned Completion and hence the Completion Date. This is one of the reasons why it is important for the *Project Manager* to spend time evaluating the *Contractor's* programme submissions before accepting them.

For a more in-depth explanation of much of the above and other terminology related to time, see Section 3.5.4.

1.6. Conclusion

This chapter has given substantial background information on the philosophy and ethos behind the NEC4 family, with a focus on the ECC. You should now understand why the authors of the NEC chose flexibility, clarity and simplicity, and stimulus to good management (and not collaboration!) as their three key objectives and how, at a middling to high level, these objectives are put into effect. More guidance than is available in the Guidance Notes has been given on the selection of the main payment and secondary options and other aspects of contract strategy. The important clauses and main procedures of the ECC have been reviewed, to give you an overview of how the contract fits together. Feedback from the original research and gained over 20 years of consultancy experience has been given on the effectiveness of the ECC in delivering the benefits aimed for.

Given this information, you should now be able to

- make better informed decisions on whether you are willing to contract under the ECC
- make better informed decisions in the selection of the most appropriate contract strategy
- understand of how the contract fits together, in order to appreciate how the suggestions for good practice can contribute both to the efficient and effective operation of the ECC and the achievement of your organisation's objectives.

REFERENCES

APM C&P SIG (2020) *Procuring for Agile Projects*. Association for Project Management, Princes Risborough. Available at: https://www.apm.org.uk/media/42'/procuring-for-agile-projects-white-paper.pdf (accessed March 2021).

Broome JC (2002) *Procurement Routes for Partnering: A Practical Guide*. Thomas Telford, London.

Broome JC (2016) 'ECC early contractor involvement clause: a review' in Construction Law Quarterly. *Proceedings of the Institution of Civil Engineers – Management, Procurement and Law* **169**(MP4): 184–186.

Broome JC and Hayes RW (1997) A comparison of the clarity of traditional construction contracts and of the New Engineering Contract. *International Journal of Project Management* **15**(4): 255–261.

Doyle N (1993) Disputes under pressure. *New Builder* 157: 12.

Hackman JR (2002) *Leading Team: Setting the Stage for Great Team Performances*. Harvard Business School Press, Boston, MA, USA.

Herga RH (1995) A management tool rather than a clear legal statement of the parties' rights and obligations – aspects of the NEC Engineering and Construction Contract examined. MSc thesis in Construction Law and Arbitration, Centre of Construction Law and Management, King's College London.

Houghton E and Trebes B (2011) The Z factor: to what extent have Z clauses impacted on the NEC's success as a modern contract? Paper presented at NEC Users' Annual Seminar, 11 April 2011.

Jackson R (1986) Are standard forms of contract a good or bad thing? *Arbitration*, November 1986.

Latham M (1994) *Constructing the Team: Final report of the government/industry review of procurement and contractual arrangements in the UK construction industry*. HMSO, London.

NEC (2019) *NEC4 ECC – Practice Note 5: Using a Dispute Avoidance Board for contracts covered by the Housing Grants, Construction and Regeneration Act 1996*. NEC, London. Available at https://www.neccontract.com/NEC4-Products/NEC4-Contracts/NEC4-Free-resources (accessed March 2021).

NEC (2020) *NEC4 ECC – Practice Note 6: How to use the Information Protocol to support BS EN ISO 19650-2 the delivery phase of assets with NEC4*. NEC, London. Available at: https://www.neccontract.com/NEC4-Products/NEC4-Contracts/NEC4-Free-resources (accessed March 2021)

Patterson R (2017) 'NEC3 to NEC4 – evolution, not revolution; some fixes and some good ideas!', in Construction Law Quarterly. *Proceedings of the Institution of Civil Engineers – Management, Procurement and Law* **170**(MP6): 268–271. Available at: https://www.neccontract.com/About-NEC/News-Media/NEC3-to-NEC4-evolution-not-revolution-some-fix (accessed 4 March 2021).

Patterson R and Trebes B (2015a) NEC for DBFO – taking best practice procurement into PPPs. *Proceedings of the Institution of Civil Engineers – Management, Procurement and Law* **168**(MP5): 213–219. Available at: https://www.neccontract.com/About-NEC/News-and-Media/NEC-for-DBFO-paper-wins-award (accessed March 2021).

Patterson R and Trebes B (2015b) Using NEC to incentivise lowest whole-life cost. *NEC Users' Group Newsletter* **75**.

Revay SG (1992) Can construction claims be avoided? In *Construction Conflict and Resolution* (Fenn P and Gameson R (eds)). E & FN Spon, London.

Thomas D (2012) *Keating on NEC3*. Sweet & Maxwell, London.

ICE Publishing

Broome, Jon
ISBN 978-0-7277-6205-4
https://doi.org/10.1680/necaug.62054.111

Chapter 2
Good practice prior to entering the contract

2.1. Introduction

Effective and efficient implementation of any project starts with the identification of a need or desire, justifying the need or desire with a proper business case, defining the function/purpose that the project has to fulfil, and examining options to fulfil that need. Once the 'buy' option has been justified, the 'best-fit' procurement strategy needs to be developed, which includes breaking the project down into work packages, some of which may include work to be done internally, while those with work to be done externally will ultimately become contract packages. For each of these potential contracts, you need to choose the best way to select a provider, be they a contractor, consultant or supplier of goods, and choose the contract strategy that provides the best fit for each contract or category of contracts. The contract strategy includes

- selection of NEC4 as the preferred contracting vehicle (or not)
- selection of the right member of the NEC4 family to use (e.g. a long form or a short form)
- selection of the main and secondary options (in the case of a long form)
- completion of the entries in Contract Data part one – some of which adjust the level of risk and reward for the options (e.g. share of pain/gain, levels of damages and liabilities) – and preparation of option Z (*additional conditions of contract*) clauses
- selection of content for the other contract documents (e.g. the extent of the *works* that the *Contractor* is to design needs to be stated in the Scope, the results of any site investigation you do need to be included in the Site Information).

This should all result in a coherent set of contracts documents, which fit together in a logical structure.

All this is linked in with how you select the provider. For instance

- Can an individual provider carry the risk you are proposing to allocate to them?
- What is the balance between the quality and financial aspects of the assessment?
- For the quality aspects, what questions do you ask as part of the quality assessment? What weighting do you put on each question and how, if at all, do you incorporate the chosen provider's answers into the contract so they become contractual commitments (e.g. the Scope provided by the *Contractor*)?
- For the financials aspects, how do you take account of factors other than the headline Prices (e.g. the *fee percentage* and other tendered percentages and rates)?

It is only then that the clarity and stimulus to good management provisions within an NEC4 contract can aid the effective and efficient implementation of a project. Consequently, if any of the former stages are severely flawed, it is unlikely that using an NEC4 contract will save a contract from failing to meet its objectives or providing its expected benefits.

Even if all the preceding stages are completed satisfactorily, use of the NEC4 family does not guarantee success: a number of interviewees in the original research expressed the view that, in untrained hands, the ECC could be dangerous as it requires intelligence in its use. Regrettably, the original research identified some examples to justify this view, and from my consultancy experience I have also found this to be true.

Poor clients, poor client project managers and contractors who are poor at project management will not suddenly be transformed into highly competent ones by a contract that stimulates good project management. Moreover, their deficiencies will be highlighted if the other party is competent. The NEC4 family gives strong nudges and encouragement in the direction of excellence, but it does not guarantee that everyone will achieve it. Unfortunately, it is human nature for people to regard themselves as good managers, regardless of their capabilities. What I have found, typically, is that those who blame the contract for a poor outcome are the ones who do not operate it as written. Usually, those who do substantially operate the contract as written are not only the most self-critical, but also happen to be the ones who achieve better outcomes.

Consequently, while the authors of the ECC would claim that it is applicable across the whole heavy engineering and construction industry (and that the NEC4 family is applicable across a whole range of sectors), my personal view is that there are two generic situations where potential users would not gain the full benefits of using NEC4.

- If either Party cannot project manage properly, then use of the ECC may leave them exposed and the contract will suffer as a result (although probably no more than under more traditional forms). On the other hand, if your organisation has a degree of project management competence and wants to get better, then use of NEC4 should accelerate your development, provided you are also mature enough to address the issues it will highlight (not exaggerate).
- When it is thought that a client or their contractors are unable to make the cultural changes towards a more collaborative style of contracting. My original research indicated that the ECC reinforces cooperation, but cannot create the will or ability to cooperate. Using the ECC without cooperation goes against the ethos of the contract. This has implications for the selection of the *Contractor* and personnel from all sides of the industry (see Section 2.2.2.3), and implies that any training on the contract needs some emphasis on changing hearts and minds.

> Rob Gerrard, the Secretary of the NEC Users' Group, tells the story of how he went to one major client and extolled the virtues of the NEC. The client bought the virtues, but was mature enough to recognise that their project management capabilities were not sufficiently mature for them to get the best from it. Consequently, they embarked on a 3.5-year programme to improve this capability before starting to switch over to NEC.

This chapter leads you, the reader and potential user, through the changes necessary to gain the most from using the ECC in the pre-contract phases, whereas Chapter 3 leads the reader through the desired changes that need to be put in place within the first month(s) of letting the contract. As in Chapter 1, many of the comments are relevant when using any member of the NEC4 family, not just the ECC. This includes understanding the ECC's principal clauses, how it fits together and how it stimulates good management, as outlined in Chapter 1. This chapter is divided into three main sections, covering: preparing the contract, preparing the tender and evaluating the tender.

However, the guidance in this book should be taken as a whole, and I recommend that you do not use one chapter or section in isolation, as issues mentioned in later sections may have been developed from earlier sections. Then, when considering a specific aspect of implementation, the section that addresses that aspect can be referred to. For this reason, and to prevent duplication, there is cross-referencing between sections.

Examples of good, and not so good, practice from real projects, both from the original research and from my consultancy experience, are given in boxes. Much of this information was given in confidence and therefore the source matter is not attributed.

2.2. Preparing the contract

2.2.1 Overview

The main actions that the *Client* needs to undertake before to putting a contract out to tender are

- selection of an appropriate contract strategy for the contract (see Section 1.1.2 on which member of the NEC4 family to use, and Section 1.2 on the selection of the main and secondary options)
- preparation of the bid contract documentation, which includes
  - completion of Contract Data part one
  - preparation of the Scope and Site Information
  - consideration of other issues, such as how the tender will be assessed, initial risk register[1] etc.

  which will normally then be collated to form the 'Instructions to Tenderers' document.

2.2.2 Completing Contract Data part one

2.2.2.1 Introduction

Information in the Contract Data is incorporated into the contract either by the use of *italics* (clause 11.1) or by direct reference. Part one of the Contract Data ('CD part 1' for short) is the part that the *Client* fills in, and part two is filled in by the tendering Contractor. As such, CD part 1 provides a means of tailoring and fine-tuning the contract strategy, without the need to alter the *conditions of contract* clauses. In theory, this enables the *Client* to get to contract more quickly, at less expense and with less likelihood of ambiguity in the contract. One *Project Manager* described CD part 1 as 'a useful checklist of issues which should be thought about prior to putting a contract out to tender, but which often are not'.

A worked example of the Contract Data is supplied in Appendix 1 of the NEC4 ECC Guidance Notes[2] (in volume 2: *Preparing an Engineering and Construction Contract*, which is published with the contract), and earlier on in the same document there is a Contract Data checklist (at Section 2.2). Here, I provide additional advice to that given in that checklist, and in some instances in considerably more detail. It is suggested that, while reading this section, you also have open, at the minimum, the Contract Data of the contract, and ideally the Guidance Notes Contract Data checklist also, as I do not wish to replicate what is already stated.

[1] Note the small 'r's, as the NEC3 'Risk Register' is now (in NEC4) the Early Warning Register.

[2] As previously mentioned, the published guidance documents for the NEC contracts are called 'User Guides', but to avoid confusion with this book I refer to them as 'Guidance Notes' (as they were called in previous editions).

Before we start, however, it is worth pointing out that, provided you have bought a contract, then, as stated in the copyright statement at the front of the ECC contract, you 'may reproduce the Contract Data and forms for the purpose of obtaining tenders, awarding and administering contracts'. This means, for instance, that the Contract Data section can be copied into a Word document (which is also available from the publishers) and, when a specific contract arises, bits that are not needed (i.e. optional statements) can be deleted and, where more space is needed, the relevant text can be inserted before it is printed out and circulated. Alternatively, an eVersion can be purchased on an annual subscription basis.

The following subsection covers miscellaneous comments on various items where a paragraph or two covers all I think needs to be said. The remaining nine subsections then cover other aspects, which require more detail and thus are given under their own specific headings. For most of the main and secondary options, any relevant comments I have are given in Sections 1.2.2 and 1.2.3, where I go through when to use each option and the subtleties of using them. The exception is on the setting of the pain/gain share under options C and D, which is described in Section 2.2.2.10.

2.2.2.2 Miscellaneous comments

The first lines of CD part 1 are where the *Client* identifies its chosen 'big picture' contract strategy, by stating the main option and secondary options. Only in very rare circumstances would it be wise to select two main options. If this is done, there would need to be an accompanying statement identifying which parts of the *works* are covered by which payment option. Where Clients have stated two main options, typically one part of the *works* has high certainty and so is on a priced contract basis (option A or B), whereas the other part has more uncertainty and so is on a cost-based contract basis (options C to F). The problem with this arises when it comes to being paid. The *Client* naturally wants costs to fall on the part of the *works* covered by the price-based option, while the *Contractor* wants the opposite. Unless there is clear delineation between the parts, either by time or location, with the best will in the world the use of two main options will create problems. Unfortunately, I have come across situations where the *Contractor* has deliberately exploited a poorly thought through contract.

The statement of the *works* defines the extent of the contract. It is quite an important statement. Take, for instance, the example used in the Guidance Notes, in which the *works* are 'the construction of Much Binding Bypass, including the design of the bridge crossing over the River Binding'. If the *Client*, via the *Project Manager*, wanted to add in additional design work (i.e. not relating to the bridge) as a compensation event, then the *Contractor* could refuse, as this is outside of the extent of the *works*. Instead, the *Contractor* could agree to do it by a deed of variation or supplementary agreement on their terms (i.e. at whatever price is agreed to extend the extent of the *works*). For legal clarity, if the requirement to do detailed design of any part of the *works* was in the original Scope, this would not apply.

A *Client* may want to remove a portion of the *works* by reducing the *works* described in the Scope, as a negative value compensation event. If the *Client* wants to give this portion to another Contractor, the original *Contractor* can argue for a partial termination, as it has the contract to do this work. In financial terms, this means the *Contractor* would be paid its fee for the reduced value, as opposed to the difference being deducted from the Prices. In one contract, this amounted to a difference of £160 000 (two times an 8% fee percentage applied to the £1 million of work taken out).

The above example is the reason why I recommended adding termination procedures into secondary option X22 (early *Contractor* involvement). If the *works* are defined as something like 'the development, detailed design and construction of X', then, if Prices for Stage Two (the detailed design and construction) cannot be agreed and the *Client* wants to give this work to another Contractor, a similar situation could occur.

PROJECT MANAGER AND *SUPERVISOR*

For both the *Project Manager* and the *Supervisor*, I like to see named individuals as opposed to organisations. This is partly because individuals have reputations, good and bad, which affect the Prices tendered, but also because, if it is an organisation, it might not always be clear who can give instructions etc. Is it just the directors, who have legal capacity? Or is it anyone who holds themselves out as having authority and seems to have the right job title? Or is it anybody who works for that organisation? If an organisation, then it must appoint individuals on day one of the contract to be the *Project Manager* or *Supervisor*.

These people have total authority under the contract to do the actions in the contract within the timescales of the contract. As the Guidance Notes state, it is really important they be given sufficient authority from the *Client* organisation, otherwise they will constantly be having to refer back up the *Client* organisation for decisions. This will undermine the impartiality of administration and good project management of the contract, all of which ultimately can cause delay and extra cost to fall on the *Client*.

> I did some very early NEC training in Hong Kong, when the contract was first being introduced by the government. The Project Managers were often senior career civil servants. At the time, any medium to major decision had to be referred back up to a departmental committee, which sat monthly. Any 'decision' it made was actually only a recommendation, as it then had to be rubber-stamped and signed off by the head of the department. If the career civil servant broke the rules, they could be sacked and lose all their pension entitlement. Consequently, all decisions were major ones. Hardly the recipe for an empowered Project Manager!

THE *PERIOD FOR REPLY*

This is the default **maximum** timescale for responding to a communication from the other Party, which applies unless there is a specific statement otherwise. I suggest that the *period for reply* specified is no longer than two weeks. On relatively simple contracts, where for instance there is little or no *Contractor* design, there is no good reason for it to be greater than a week. However, as indicated in the Contract Data checklist, it may be useful to specify different *periods of reply* for different issues; for example, 3 working days where more information is needed in order to assess a submission for acceptance fully, but 2 weeks for acceptance of the particulars of the *Contractor's* design.

THE *BOUNDARIES OF THE SITE*

The *boundaries of the site* define what is the Site, i.e. it is the area the *Client* gives to the *Contractor* to do work on. This limits the area within which the *Contractor* can get a compensation event for encountering unexpected physical conditions under clause 60.1(12).

CLIENT IDENTIFIED RISKS TO BE INCLUDED IN THE EARLY WARNING REGISTER

I will say much more on the management of risk and issues later, in Section 3.6. This is the *Client's* chance to highlight their top 10 (or 20 or 30) risks to the contract as they see it. Usually, this is a list of risks with no further information. Sometimes, references to another document are included, often to a spreadsheet that lists the risks, together with their perceived probability, potential impact, outline mitigation measures etc. Just to be absolutely clear, risks stated here and in the equivalent statement for the *Contractor* in Contract Data part two have nothing to do with contractual ownership of them (at least in theory! – see Section 2.3.4). That is decided only by the definitions of compensation events clause 60.1 of the contract.

INTERVAL FOR EARLY WARNING MEETINGS

New to NEC4 is the statement of the maximum interval between formal early warning meetings for the duration of the contract. Again, I will say more about this in Section 3.6, but I view these as 'senior site based management' meetings, which should aim to ensure that actions generated by the much more frequent and ad hoc 'coal face' early warning meetings are being done or are effective, and to have a more proactive stance in generating actions. Consequently, the right timescale for this interval is probably weekly or fortnightly, and certainly not more than monthly.

KEY DATES

The concept of a Key Date is that the *Contractor* needs to get an area of the Site ready by a set date – the *key date* – so that the *Client* or an Other can come in and do some work, after which the area is handed back to the *Contractor* to carry on with its work. This contrasts with sectional Completion, whereby the area is taken over by the *Client* permanently. I find the two column entry in CD part 1 to be inadequate. It would be preferable to add an extra two columns, one before and one after the existing ones, stating what areas are affected and how long the *Client* or Other will be occupying the areas. Otherwise, it is not clear how the *Contractor* can programme its work. For further discussions on Key Dates, see Broome (2010), Patterson (2017) and Earnshaw (2020).

INTERVALS FOR REVISED PROGRAMMES

It is suggested that the *assessment interval* for payment is a multiple of the interval for submitting revised programmes, particularly when options A or C[3] are being used, as the programme will then tie in with the Activity Schedule, and monitoring of progress and payment can become one procedure.

QUALITY MANAGEMENT

The number of weeks stated for the period before the *defects date* is the equivalent of the defects liability period in other contracts (typically 52 weeks after Completion), while the *defect correction period* is the time that the *Contractor* has to correct a Defect that exists at or becomes apparent after Completion once the Defect has been notified (typically 2 weeks). In some cases, it may be useful to specify quite a long period between Completion and the *defects date*. For instance, on a new road or road renewal project, you could specify a timescale in which the road should not need any repairs arising from normal use. I have three comments on timescales for correcting Defects.

- As the Contract Data checklist suggests, it may appropriate to specify different *defect correction periods* for different types of work or situations. Exceptions to the standard timescales would typically apply for Defects that have health and safety consequences and

[3] As mentioned in Section 1.4.4.3, it is suggested that NEC3 option clause 31.4 (which is the same as NEC4 option A clause 31.4) is reintroduced as an option C clause, so that progress and value delivered can be tied together.

for those that hinder the operation of the asset. For instance, if a certain piece of machinery in a new factory breaks down, then the whole factory may close down until it is fixed. In both instances, 'as soon as practicably possible' is the normal timescale specified.

- It may also be appropriate, as an alternative to using warranties, to have a different *defects date* for particular types of Defect, if the connection between the *Client* and who will actually correct the Defect is direct enough, for example, if the *Contractor* is also the manufacturer of an item it is installing.
- Contractors beware if option X5 (sectional Completion) is specified! As stated in the Contract Data checklist, there is only one *defects date*, which is at the end of a period after Completion of the whole of the *works* (typically a year). So, for example, if you had a 2.5-year contract, with a sectional Completion achieved after, say, 1.5 years, for that section the *Contractor* would have an additional year in which it has an obligation to correct notified Defects.

Do note that clause 44.4 does allow the *Project Manager* to extend the *defect correction period* for an individual Defect where access is not possible. For instance, in a live school where a Defect is not dangerous, but correcting it would cause disruption and be potentially dangerous, the correction period can be extended to include the next school holiday. *Project Managers* have also used this clause to take a common-sense stance where, for instance, there is a long lead time for replacement parts.

THE *ASSESSMENT INTERVAL*

In some contracts, the *Client* has specified this as 4 or 5 weeks, in order to fit in with their accounting systems, with the *Client* being billed on, for example, the last Friday of each month. Others have used the term 'monthly'.

CLIENT WILLING (OR NOT) TO TAKE OVER THE *WORKS* BEFORE THE COMPLETION DATE

If the *Client* states it is not willing to take over[4] the *works* before the Completion Date, even though Completion may be achieved weeks or months before, but subsequently does, then it would be a compensation event for a change in the Scope via clauses 35.2 and 60.1(1). The reason for saying this is twofold.

- I have seen Clients state they do not want to take over the asset (presumably because lawyers like to keep risk on contractors for as long as possible), when a simple question ascertains that they want the asset as soon as possible!
- What happens if the contractual Completion Date has been progressively put back by compensation events so that, even though the *Client* did not want to take over the *works* before the original *completion date* in the Contract Data, it does now want to take over before the current Completion Date? Answer: it is an additional compensation event.

[4] The point of take over affects liabilities. Prior to take over, the *Contractor* is fully liable for damage to the *works* and hence security, etc. After take over, the *Client* is liable for damage, except for damage that results from actions of the *Contractor* (e.g. when coming back to fix a Defect).

> Here's a little war story from some start-up training I did.
>
> The contract was to straighten out several kilometres of country main road, and the *Contractor* had to ensure that traffic could travel between the start and finish and ensure access 'during construction', which included returning parts of the old road to countryside (i.e. not a usable road). According to the Contract Data entry, the *Employer* (as it was an NEC3 contract) was not willing to take over before the Completion Date. Consequently, on Completion, the *Contractor* could have refused to allow the *Employer*, road users and local residents to use the new road, while it was physically impossible to use much of the old road. Surely this was not what the *Employer* intended!

PRINCIPAL DESIGNER AND PRINCIPAL CONTRACTOR

If the contract is in the UK, who are the Principal Designer and the Principal Contractor under the Construction (Design and Management) (CDM) Regulations 2015? Arguably, this could be stated in the Scope document, but it is such an important obligation that many Clients have a statement allocating the roles in the Contract Data, with supporting details included in the Scope document.

2.2.2.3 Selecting the right people

PEOPLE 1: AUTHORITY AND DELEGATION

The Guidance Notes state that the *Project Manager* must be sufficiently close to the work and have time to carry out their work effectively. In a number of organisations, the named *Project Manager* has been a fairly senior person, who has delegated their powers down to someone more involved with the day-to-day running of the contract. For instance, this may be the engineer who prepared the documentation, be they an employee of the same company or an external consultant. Normally, limits on changes to the Prices and the Completion Date that the delegated individual can agree will have been specified. It is advisable that

- these limits are not set too low, otherwise the delegated individual has very little power to agree issues and so has to constantly refer back to the *Project Manager*, thus undermining the delegated individual's authority
- the *Project Manager's* delegate has quick and ready access to the *Project Manager* at all times, so that the ECC's timescales for actions can be met
- the *Contractor* is informed of any limits on authority delegated by the *Project Manager*.

Taking the first two points, I know of projects where the contract, as a project management tool, effectively became inoperable because the limits were set too low – the *Project Manager's* delegate could not make any decisions, instead being a 'postbox' to a remote (and disinterested) named *Project Manager*.

> On a recent project review, the person in the *Employer's* organisation who actually had the organisational authority to instruct change, however small, and agree to compensation events was not the *Project Manager*, nor the *Project Manager's* boss, nor even their boss, but the head of estates. Did this work? Absolutely not. In fact, in order for the project to progress, the participants in the contract were doing all sorts of informal and unrecorded deals in the belief/hope that they would eventually get signed off or rolled up into the big changes that were being signed off (after the work had been done). So, far from encouraging good governance – which I think was the intention – it created exactly the opposite effect.

In contrast, an advantage of the named *Project Manager* being one step back from the day-to-day running of the contract is that they can perhaps think more strategically and dispassionately about it. For instance, how the individual contract fits in with other contracts that make up the project, or in deciding on disagreements which might otherwise be referred straight to the *Adjudicator*.

Practice shows that whoever is involved in project managing the contract needs to be involved! This does not necessarily mean being physically close to the contract, but being sufficiently involved to be aware of what is happening on the contract and be ready and able to make decisions promptly. On smaller sites, the day-to-day management has often been delegated to the person who is also the *Supervisor*, as the amount of work carried out on site has not warranted two people being present full time. The named *Project Manager* may only be present for weekly meetings or during periods of high activity or crisis. My opinion, based on experiences on some sites in the research sample and from consultancy, is that the person given overall responsibility for delivery of the contract should be on site at least fortnightly, otherwise they will be too remote.

In the original research sample (from some 25 years ago), one company ran a number of contracts from a head office several hundred miles from the Site. The named *Project Manager* delegated his day-to-day responsibilities to the *Supervisor*, and kept in touch with both the *Contractor* and *Supervisor* by various means, such as video conferencing, telephone and email, as well as visiting the Site at least fortnightly. Both the *Project Manager* and the *Supervisor* thought that this worked well in practice, and the *Contractor* made favourable remarks on this arrangement. Indeed, he commented that the amount of communication between the *Project Manager* and the *Contractor* was unusually high.

In today's much more connected world, contracts can be run from the other side of the world, let alone a few hundred miles away!

PEOPLE 2: ATTITUDE

What sort of people should the *Project Manager* and the *Supervisor* be? Indeed, what sort of people should the *Client*, *Contractor* and Subcontractor staff the site with? Research interviewees from all sides of the industry were asked, in the light of their experience, what attitudes were desirable for people present on site on an ECC contract. The following personal attributes were mentioned, with the most commonly mentioned given first: a willingness to change, trusting, open, team minded, proactive, communicative, honest, solution orientated, commercially minded, fair, upfront, systematic, realistic, ethical and able to negotiate amicably. The following attributes were given as undesirable attitudes: adversarial, confrontational, using the contract to make money, dishonest and procrastinating. Of note is the most commonly expressed attribute: a willingness to change. A number of more senior interviewees commented that the younger members of the staff team seemed both more willing and more capable of changing.

If a Client, having decided to use the ECC, wishes for a cooperative attitude to be established, then selecting a Contractor with the above-mentioned desirable attitudes should be seriously considered. Some might wish to include attitude in their selection process for the *Contractor*. One senior member of a *Contractor's* staff noted that many of his contemporaries, who are deemed to be good at their jobs, are traditionally confrontational and, in his opinion, not suitable to oversee an ECC job. Similarly, companies participating in an ECC contract, whether as *Client*, *Contractor*, Subcontractor or consultant, would be wise to select personnel with the desirable attitudes mentioned above, and I know this happens in practice.

> The selection procedure on one contract involved pre-qualification, from which six Contractors were selected to bid in the normal manner. Each Contractor whose priced bid was within 10% of the lowest was then invited to an interview with the *Client's* site team, where each tendering Contractor's proposed site team was evaluated on attitude and how well the two teams could work together.

With the rise of collaborative working, selecting the *Contractor* partly on attitude and people, as well as on competence, is not unusual. But doing it rigorously is hard in practice, as most people can 'fake' being collaborative for an hour or so.

There is a saying in recruitment that 'you select on attitude, not skills', because you can teach skills, but attitude is within a person. While I disagree with the rather absolute nature of this, I do think there is a lot of truth in it, and I have seen people with an Accredited NEC Project Manager badge (see the following subsection) manage contracts badly, and people without one manage contracts well. Of course, having both the right attitude and the right skills is the best!

PEOPLE 3: SKILLS

As stated just above, having a suitable attitude or mindset is probably a more important starting point for a good project manager than having specific taught skills, but skill set is important. Since the publication of the NEC3 version of this book, two qualifications have emerged that are relevant to being a good NEC Project Manager. However, I will cover skills more generally first, before focusing on them.

General skills that were mentioned as being desirable for the top people on site, whether they represented the *Contractor* or the *Client*, were (with the most commonly mentioned given first) more: proactive/forward looking, commercially aware, professional, experienced and competent. A number of interviewees commented that the *Project Manager* and *Contractor's* agent need to be better all-round managers than on conventional non-ECC contracts, as they need to have an appreciation of the cost and time implications of their decisions, rather than just technical issues. This is because they can no longer make their technical decisions in a vacuum, leaving the time and cost consequences to be fought over by the claims consultants, quantity surveyors and forensic planners at the end of the contract. On the larger sites, the people in charge will be advised by others (e.g. planners, quantity surveyors and technical experts) and therefore need to be able to listen, understand and give direction to these people – i.e. speak their language.

It could be argued, as a number of individuals in the research sample did, that under the more traditional conditions of contract the programme is seen as a reactive tool to support claims while, under the ECC it is a more proactive tool for forward planning of the works. This implies a change in emphasis in how people programme and, possibly, the skills required. Thirty per cent of interviewees in the original research specifically stated that increased planning skills, as opposed to progress 'loggers', were desirable on ECC sites compared with other conditions of contract.

One of my original contractor clients, in my view, had great foresight when the ECC started to become commonplace in the civil engineering sector. In the same way that they traditionally had quantity surveyors supporting their site agents, they increased their number of programmers to an almost equivalent amount. On larger sites, the programmes might be based on site full time, in the same way a quantity surveyor might be. Likewise, on smaller sites they would visit regularly. The planners would update the programme more often and do the detailed planning in consultation with the site agent, engineers, foremen and Subcontractors, as well as the *Project Manager*. The *Contractor's* senior management view was that they should be doing this regardless of whether it is an NEC contract or not, as it means the site runs much more efficiently and profitably. Contractually, because they brought the *Project Manager* or their planning assistant with them, they could justify the time, resource and hence cost effects of compensation events much more easily. A secondary benefit was that it introduced a more technical promotion stream for their engineers outside of pure management.

In a similar vein, 15% of interviewees noted that increased estimating skills were needed on site, as opposed to reactive claims-type quantity surveying skills. A number of interviewees stated that estimating required a change from just using a bill of quantities and or calculating dayworks using time records, to a greater use of operational/resource-based estimating, where costs are assembled more from first principles. Consequently, quantity surveyors involved in estimating need a greater appreciation of programming, even though they may not be doing it themselves. A quote from a *Contractor's* agent illustrates the point:

> most surveyors do not do estimating. They are used to ... doing it retrospectively, when they know what all the costs are and can compile everything ... The danger for them, under the ECC when they are compiling it in advance, is that they have forgotten something. ... The fact that you have to build into that quote any potential extension of time and other knock on effects ... means that, at times, you have to be quite cute to pick it all up.

Another *Contractor's* chief quantity surveyor made similar comments. On his large site, not only were there more quantity surveyors and planners on site than there would have been for a non-ECC contract, but personnel were moved on and off the site depending on whether they could adapt their attitude and skills to those required of the ECC. He found that planners and estimators who were used to working from a full specification tended to require information that was not always available, and were too methodical and therefore too slow. He also found that staff who had the site experience and were willing to, in his words, 'take a view on a matter and then move on' were most appropriate. A frustration on site was that the *Client's* quantity surveyors (who happened to be paid on a time-charge basis) also did not have these skills and outlook. For example, they tended to insist that construction costs arising from compensation events had to be justified in full and in detail from first principles. While this is not wrong contractually, it does not help practical delivery. To quote Warren Buffett, the world's most famous investor, when evaluating a company's worth and what he will pay for it, he would 'rather be approximately right, than precisely wrong'. In a similar vein, when taking a view on the future effect of a compensation event, you will never be *precisely* right, so why waste time trying to be. As one *Contractor's* contract director said to me on a contract with 2700 compensation events, 'some you lose, some you gain, but if neither Party is looking to screw the other, statistically it is going to work out about right'.

Another common complaint from Contractors is that some Client-side quantity surveyors cannot even cope with true cost, but are used to – or hide behind – pricing books, which give costs per unit of quantity of a type that may (a) be only remotely related to the actual work type being done, (b) be based on different original circumstances, and (c) not take account of delay and disruption. In my experience as a trainer, this frustration at Client's quantity surveyors – or should I call them 'cost consultants' – is quite often expressed by Contractors, and leads to additional professional costs for both Parties. Consequently, the agreement of change has become a long and drawn out affair, at great cost to both Parties, while losing many of the benefits that use of the ECC can bring. On the other hand, where Clients' programmers and quantity surveyors are competent and collaborative, they contribute to the management of the contract by, for instance, making helpful suggestions on how to deal with compensation events.

In the same way that the principal people on site have to be more competent, in the early days of NEC, a number of Clients, Contractors and Subcontractors assigned their more capable staff to work on their NEC contracts. One interviewee went so far to as to state that the ECC 'demands that every contract is staffed up with competent people'. Some interviewees raised the wider point that, while planning and estimating skills were present in the construction industry, they are not common enough, and that this has training implications.

One Subcontractor's managing director stated that

> the calibre of people we needed on the job was at the top end of the range, because administratively you have got to get it right – there is no scope for you getting it wrong, for muddling through. ... That meant the job got done in the right way and I think one thing the ECC is going to do is make us more professional. ... I think a lot of the people in the industry, who maybe ought not to be there, are not going to find a hiding place. ... We learnt very quickly on this job that if we were not professional, it would hurt us.

All of the above with respect to skills was identified some 20 plus years ago, as I completed my research. While I would say it is less true today – in that there has been a general upskilling of people – I would still say it is more true than not true.

The importance of the calibre of people was recognised by one large public sector client, who I was helping to go over to both ECC and the Term Services Contract (TSC) back in 2011. They realised that the NEC would expose the shortcomings of having, in the worst-case scenario, career civil servants who knew little about construction, project management and the NEC managing projects and facilities. Consequently, they asked me to do a first draft competency framework for an ECC *Project Manager* and TSC *Service Manager*. Broadly speaking, for the ECC *Project Manager* and their team, I divided the framework into four parts.

1. General project management competence – based on the Association for Project Management's (APM's) Competency Framework.
2. General contract strategy, contract law, and contract management competence over and above that in point 1 – as I believed (as I do now, albeit to a lesser extent) the APM's competency framework is deficient in this respect for a project manager operating in the contractual sphere.

3. Knowledge, understanding and application of the NEC3 ECC or TSC.
4. Background construction knowledge and experience – which gave a framework for the ability to contribute to: the technical resolution of issues/early warnings; the evaluation of the programme, in terms of knowledge of methods, resources and durations; and cost, in terms of how costs are built up by contractors and how Schedules of Cost Components are used.

This competency framework could then be used to evaluate both their internal project managers and any consultants appointed to act on their behalf.

However, as I mentioned at the start of this section, there have been two significant developments since I drafted the *Project Manager/Service Manager* competency framework.

- The NEC Contract group, which is tasked with publishing and commercially exploiting NEC on behalf of the Institution of Civil Engineers (ICE), runs a 4-day NEC4 ECC Project Manager Accreditation course, which is claimed to 'equip delegates with the skills necessary to be able to manage a project under an NEC4 Contract, with specific reference to role and importance of the Project Manager under the NEC4 Engineering and Construction Contract'. This addresses point 3 in the competency framework (see the box above). Having qualified as an Accredited NEC4 ECC Project Manager – for there is coursework and an exam – you are entered onto the ICE Register of Accredited NEC Professionals.
  People who go on the course give good to great feedback on it, and it is undoubtedly beneficial for anybody acting in this role to gain the accreditation. Having said this
  - the accrediting body owns the only organisation allowed to run the qualification, which (while a great business model) could be construed as a conflict of interests
  - Contractors complain that some Accredited NEC Project Managers think that they know it all when (having been asked to comment on some of their decisions in accordance with the contract) they clearly do not: it is, after all, only a 4-day course
  - the accreditation is based on knowledge, not experience – to reinforce this and the preceding point, one of the trainers privately described it as 'allowing you to take off the learner-plates, but you still have a hell of a lot to learn'
  - 4 days of training does not make you an all-round project manager.
- The Association for Project Management (APM) now has a much more structured sequence of qualifications.[5] The training (certainly for the more advanced qualifications) is done by external providers, but the APM develops the syllabus and runs the exams,

[5] The hierarchy of APM qualifications is

- Project Management Fundamentals (PMF) – which can be a couple of days of e-learning and online multiple choice questions. It is about basic knowledge and terminology.
- Project Management Qualification (PMP) – traditionally a week's taught course with a written exam at the end, and is about developing understanding of application. The course and exam can now be done online. Having experience isn't necessary to pass the exam, but it helps.
- Project Management Professional (PMP) – in theory, this shows you have the knowledge and understanding required to achieve Chartered status, but you need more experience before you can demonstrate professionalism.
- Chartered Project Professional (ChPP) – this is pretty much the highest project management qualification you can get. You not only have to demonstrate experience and professionalism to get it (via a case study and an interview), but, to keep it, you also have to demonstrate ongoing professional development.

with Chartered Project Professional (ChPP) qualification being the highest standard. This addresses point 1 in the competency framework. Indeed, some Clients are starting to want not just their *Project Manager*, but also the *Contractor's* senior person, to have the education and experience of a 'Chartered Project Professional or equivalent'. It is only a matter of time before some start to insist on this. Regardless, what better way to demonstrate this professionalism than having the qualification.

So, on the one hand, my criticism of just being an Accredited NEC Project Manager is that, potentially, the teaching and qualification is not broad enough to make you a good project manager. On the other hand, my criticism of just being a ChPP – or having one of the other qualifications that can be a stepping-stone to it – is that you do not have enough specific knowledge of NEC. The ideal, if you are into qualifications, is to do and have both.

The last comment on both these qualifications is that they do not have to be for pure-bred project managers only. People who have support roles, such as cost consultants, quantity surveyors etc., can benefit from the NEC project manager course. Membership of and qualifications from the APM are for all project professionals, not just pure project managers.

PEOPLE 4: NUMBERS ON SITE

Under all NEC contracts, people should be dealing with issues as the contract progresses, as opposed to collecting records to agree retrospective quotes, albeit as the project progresses. Logically, it therefore makes sense that more professional time is needed while the contract is live. That is not to say that more professionals' time is needed over the duration of the contract, from the start to the settlement of the final account. In this respect, the original research found varying estimates, with some saying more overall and some saying less.

In my consultancy experience, I would say the amount of professionals' time needed to manage and administer the ECC depends on the following variables (which amounts to a summary of what has been said before on people).

- Choice of main payment option, with option A needing the least and option D the most. However, the choice of main payment option is partially dependent on the degree of change expected.
- The degree of change, in terms of compensation events, actually experienced.
- The team's joint knowledge and understanding of the contract and how to run it: this implies some form of joint training is advantageous (see Section 3.2 for more on this).
- The attitude and skill of the individual people and culture of the organisations to the contract.
- Whether the organisational structure and relevant delegations of authority, systems and procedures to operate the contract efficiently and effectively are in place and understood, and whether project participants have the discipline to operate them (see Chapter 3 generally).

2.2.2.4 Specifying the first programme

This subsection is about whether to specify a time within which the chosen *Contractor* has to submit a first programme for acceptance (in CD part 1, e.g. a number of weeks from the Contract Date) or whether to make tendering Contractors reference a programme in their Contract Data part two (CD part 2), which on award of the contract would become the Accepted Programme.

If a programme is not in place by the time the contract becomes operational (i.e. the *starting date*), then it is hard to evaluate the time and cost consequences for all but the smallest of changes under the ECC. The ECC allows two approaches: either a programme is referenced in CD part 2 and becomes the Accepted Programme on signing the contract, regardless of how good or bad it is; or, in CD Part 1, a time given for the *Contractor* to submit its first programme for acceptance, which is normally not more than the difference between the Contract Date and the first assessment date (i.e. within the *assessment interval*).

The advantages of the former approach are that the *Client* can assess the quality of the tendering Contractors' programmes at tender (including as part of a scored quality submission), the winning *Contractor* will know what its and Others' key commitments are, such as what things they have to supply and when, and that a programme is in place from day one of the contract. As a result, in the original research, in just under half of the contracts tendered (14), the *Employer* wanted a programme to be referenced in CD part 1 which, on signing of the contract, was to become the Accepted Programme. However, they were obviously disappointed with what they received as none of the submitted programmes, in the *Project Managers'* opinions, satisfied the criteria stated in clause 31.3, and needed further development to do so. Consequently, the reference to the programme in CD part 2 was taken out prior to signing the contract, and the *Employer* reverted to a programme being submitted within the timescale now stated in CD part 1.

From a tendering Contractor's perspective, why should it develop a full programme when it has not even won the work and, as a result, has not appointed the key Subcontractors who will actually do the work? It will often be wasted work and, even if the Contractor does win the work, things will change. In practice, Contractors will use multi-level planning, whereby planning gets more detailed as the time approaches for the work to be done, with greater involvement of the people who are going to do the work.

Since the research, many Employers/Clients have accepted this argument, so instead ask (in their Instructions to Tenderers) for a programme that 'is capable of being developed into the Accepted Programme' (often specifying more precisely what they mean) but with the first month or two's work being programmed to the level of detail required in the contract. If the *Client* is happy with this programme at the time the contract is to be signed (i.e. there may be some further discussions and clarifications), then it is referenced in CD part 2 at signing of the contract. If not, the provisions in CD part 1 are enacted. From the *Contractor's* perspective, this leaves it open to the sanction in clause 50.5 of a quarter of the Price for Work Done to Date being retained and the *Project Manager* imposing their assessment of any compensation events on the *Contractor*, neither of which the *Contractor* wants. (See Section 3.5 for a lot more on programming, and also see Patterson, 2020.)

2.2.2.5 Operating the weather clauses

This is how the weather compensation event procedure operates contractually.

- Historical *weather data* comes (in the UK) from nearby Met Office station(s)[6] for, in any calendar month, the cumulative rainfall and the numbers of days with rainfall over 5 mm,

[6] The Met Office can now triangulate from a number of different weather stations to take account of the Site's location, including altitude, to give more accurate figures compared with just taking readings from the nearest location. Furthermore, if the weather at the Site is not historically comparable with that at the nearest weather station, the Client does have the option of triangulating from a number of sites and/or making fair adjustments to reflect the differences between the locations.

a minimum air temperature less than 0 °C or snow lying on the ground. These will vary of over the year depending on the month.
Note that the standard list of *weather measurements* in CD part 1 does not include wind speed, which can affect the use of cranes (I am told that 40 km/h is the approximate figure at which crane operations have to stop) or other operations done at height. Consequently, if no additional measurements are identified in CD part 1, the *Contractor* takes the risk of high wind speed.

- For each of these historical measurements, a value is calculated which 'is shown to occur on average less frequently than once in ten years' (see clause 60.1(13)). The Met Office can both supply the base data (for a fee) and calculate a once in 10 years figure (for a larger fee).
- During the contract, *weather measurements* are collected and compared with the once in 10 years figures. If any of the measurements exceeds the once in 10 years figure within a calendar month, it is a compensation event.

So, provided the weather encountered on Site and at the existing Met Office station is historically comparable, taking *weather measurements* on Site makes this a much fairer and more clear-cut test than the 'exceptionally adverse weather' test found in non-ECC contracts.

But notice my qualification in the statement above. For the procedure to operate effectively and fairly, it is imperative that the *weather measurements* are taken on or near to the Site, preferably by a weather station operated by the Site. There are three good reasons for this.

- There are two questions to ask for any potential compensation event: (1) Is it a compensation event?; and, if 'Yes', (2) What is its quantum in terms of additional time and cost? If a *weather measurement* at a station distant from the Site exceeds the once in 10 years figure, then it may be a compensation event, but if it has not had any effect on Site, then the *Contractor* gets a zero assessment. On the other hand, if it is dry at the distant weather station, but rains cats and dogs on Site, stopping or severely hindering work (as has happened on numerous sites I know about), the lack of rain at the weather station means there was no compensation event, so the *Contractor* gets no additional time or money.

> One project involved raising the level of a road bridge over a small river. This river was prone to flash flooding due to weather conditions encountered upstream, and not in the near vicinity of the Site, which in turn flooded the road. When flash flooding occurred, work would not only have to stop, but could also involve reconstructing some of the temporary works. The *Contractor* noted that this was not a physical condition that 'an experienced contractor would have judged ... to have such a small chance of occurring' (clause 60.1(12)), nor was it a weather compensation event (clause 60.1(13)). As it was an uncontrollable risk, the *Contractor* thought that the *Employer* should, at the very least, have taken some of the risk for flooding occurring by specifying a threshold at which the river height became a compensation event. This would have resulted in a lower tender Price.

On one Site, where a new bridge was being constructed over a dual carriageway to the north of Dartmoor in the south west of England, the construction was severely hindered by higher than normal rainfall in summer. However, the *weather measurements* and *weather data* came from Torquay, some 20 miles away, where the weather had been OK. As a result, there was no compensation event. As the contract was a target-cost contract, the *Contractor* took a hit under the pain/gain share mechanism. Did the *Contractor* learn its lesson?

No! The follow-on contract was for the construction of a new bypass to the west of Dartmoor, some 40 miles from Torquay. This time the entire summer earth-moving season was a washout and, again, the *weather measurements* did not match the conditions actually encountered at the Site. Consequently, again, there was no compensation event. The standing time and subsequent delay literally cost millions of pounds in Defined Cost with no adjustment to the target Prices.

- If you get the *weather measurements* from the Met Office, you will only know the figures after the end of each month. Consequently, you will only know when the compensation event started, and hence when you should have started taking and agreeing records of its effect on time and cost, after you get these figures. This will make the assessment of the compensation event difficult.
- This is an old comparison, but having a simple weather station at the Site certainly used to pay for itself within 6 months by comparison with getting the *weather measurements* from the Met Office. You can use the weather station on subsequent contracts. You do not even have to remember to take the measurements on a daily basis, as they will be automatically recorded and fed from the site weather station (situated, for example, on top of a site cabin) to someone's computer. If the *weather measurements* are to be taken at the Site, it is sensible for the Scope to require the *Contractor* to set up the required weather station and to record the *weather measurements*.

Clients should only consider specifying additional *weather measurements* when they affect work on the Site (e.g. wind speed when tower cranes are likely to be used). Alternatively, an additional compensation event can be specified in CD part 1, as an entry for compensation event number 60.1(21). When doing so, they should ensure that what is being measured is what will affect the work.

On work being done at sea from a barge, the *Employer* initially tried to define an additional weather compensation event as wave height, but realised that there would be difficulties in measuring wave height. So they then thought about using wind speed as the criterion. Both of the tendering Contractors pointed out that swells can come in from storms out at sea several days after a storm had passed the geographical location of the Site. It was subsequently agreed that an additional compensation event would simply be when they were unable to work at sea.

2.2.2.6 Adding compensation events, including for COVID-19

Clients should only add additional compensation events when they affect the *works* and, when doing so, should ensure that what is being measured or defined is what will affect the *works*. For instance, at a port, the number of ship movements per week that the *Contractor* needs to allow for (see also the previous example on weather).

Under NEC3, if the *Employer* wanted to add in additional compensation events, they had to do it by using option Z (*additional conditions of contract*) clauses. Under NEC4, there is a specific entry in CD part 1 to do this. Additional events listed in CD part 1 would become compensation events under the new compensation event clause 60.1(21). But there is also an entry in CD part 1 for 'additional *Client's* liabilities', pointed to from clause 80.1. By reference from clause 60.1(14), the unlikely and remote events listed as additional *Client's* liabilities would also be compensation events if they occur. So, if you want to add in an additional compensation event in order to reduce the risk premium that the *Contractor* will charge, where should you enter it? As a clause 60.1(14) *Client's* liability or as a clause 60.1(21) compensation event?

In an excellent article in the *NEC Users' Group Newsletter*, Richard Patterson (2007) pointed out that if they are the latter, then, under clause 82,[7] the *Client* is also indemnifying the *Contractor* against potential claims, proceedings, compensation and costs due to a *Client's* liability. The example Richard gives is of an additional event for flooding along a river. If a bit of construction Equipment, such as a digger, is washed down the river and damages property downstream, then if the event was specified as a *Client's* liability, the *Client* would be liable for this consequential damage, but if it was an additional compensation event under clause 60.1(21), the *Client* would not be liable.

More recently, Clients have almost had to write in additional compensation events related to COVID-19, otherwise contractually-aware Contractors have just refused to contract with them. In part, this is because, at the time of finishing off this book (in early 2021), the effects of COVID-19 are part of the everyday landscape of doing business, so clause 60.1(19) has no relevance for contracts signed after mid-March 2020 (i.e. it will not qualify as a compensation event). So, the following are some of the new compensation events (with comments) that Clients are writing in – or considering writing in – as a result of the COVID-19 pandemic.

- In England and Wales, the Construction Leadership Council's Site Operating Procedures describe standard, but generic, procedures for doing construction operations safely with respect to COVID-19. I know there are equivalent and similar versions for Northern Ireland and Scotland, and it would not surprise me if most countries have a similar document. A change to using the latest version is identified as a compensation event that may reduce Prices, with the Scope document requiring the *Contractor* to comply with the version of the document current at the Contract Date.
- Option X2 (changes in the law), as written, covers all changes in the law, not just those related to COVID-19. As a result, this is changed to be a change in the law resulting from the effects of COVID-19. In the UK, new Acts of Parliament and changes to statutory instruments (which seem to be delegated to the relevant national parliaments) are changes in the law, but there seems to be a blurring when you get into any 'guidance' issued by government. As a result, I have seen 'Guidance' stated as a defined term and included within the definition of a change in law.

[7] Because Patterson's article was published way back in 2007, when NEC3 was out, the terminology here has been updated to reflect NEC4 wording (i.e. *Client* instead of *Employer*, *Client's* liability instead of *Employer's* risk).

- Early on in the pandemic, Scotland decided to have a complete lockdown, which included shutting down most construction sites. While it could be covered by the previous bullet point, a legal requirement to close construction sites has been expressly stated as a compensation event.
- Early on in the pandemic, many sites were struggling to get basic supplies: Contractors for Connect Plus (the client organisation that maintains and improves the M25 motorway) were struggling to get soap and hence meet basic health and safety needs, regardless of the additional requirements due to COVID-19. (At the time of my final review, this appears to be less of a concern for contractors.)
- Despite the *Contractor* following the protocols, an outbreak of COVID-19 causes suspension of work as people have to self-isolate. This might affect just a few people (those who have worked in the same bubble) or the whole site team (i.e. *Client*, *Contractor* and Subcontractor people). This can also result in part suspension of the *works*.
- Somewhat vaguely, I have seen 'the effects of the COVID-19 pandemic are more onerous than at the Contract Date' written in as a compensation event. Expressed this way, it means the Prices can only increase and not decrease.

2.2.2.7 Deleting compensation events

Clients should give considerable thought before deleting a compensation event, and do it by use of an option Z (*additional conditions of contract*) clause. They certainly should not delete a compensation event over which they have predominant control or responsibility, which are compensation events 60.1(1) to (11) and 60.1(14) to (18) and, arguably, 60.1(20). For physical conditions and weather compensation events (60.1(12) and (13)), the *Client* will be paying a premium to the *Contractor* if it transfers these risks to the *Contractor*. Note that it is physical conditions, not just ground conditions, so unexpected services could also count, as could structures and other items within an asset that is being refurbished.

As contractors are, hypothetically, more risk averse than clients (because they are generally less financially strong – i.e. their balance sheets generally have a fair amount of debt on them and their cash flow is not as large or as predictable as that of client organisations – especially compared with civil engineering clients), it is likely that the *Client* will pay more in the long term for allocating an uncontrollable risk to the *Contractor*, particularly in a sellers' market. Generally, the less information the *Contractor* has on a risk, the less it will be able to structure its work in order to minimise the probability of occurrence and the impact of that risk on the work, so the greater the premium required for taking it. As an absolute rule, risk should not be allocated to a party unable to sustain the consequences of that risk.

Having said this, while physical condition and weather risks might be 'uncontrollable' in terms of whether they do or do not occur, a contractor is often best placed to manage the consequences. Consequently, I have more sympathy with deleting these risks and hence allocating them within the Prices under a target-cost contract, where the risk is shared under the pain/gain mechanism. However, I would expect a 'shallower' pain/gain, in the sense that the Client takes a greater share than if these compensation events were not deleted.

Two points about deleting the physical conditions compensation event. The first is that taking out the physical condition compensation event is not an excuse not to do good site investigations for inclusion in the contract as Site Information. If you fail to get proper investigations, the premium in the Prices will be higher. The second is that, if the *Client* has provided designs based on the site investigations and these designs turn out not to be fit for purpose, they will have to be changed. This will be a change to the Scope, and therefore will still be a compensation event (under clause 60.1(1)) even though the *Client* has deleted the physical condition compensation event.

2.2.2.8 Specifying the *Adjudicator*

The primary requirement for the adjudication procedure to operate is for an *Adjudicator* to be selected when needed. The second requirement for it to work effectively is that both Parties have trust in the *Adjudicator's* independence and competence, so that the *Adjudicator's* decisions are respected and the likelihood of a dispute being taken further is decreased. The original research made the following findings.

- When no *Adjudicator* was stated in CD part 1, the likelihood of one being appointed by the Contract Date was reduced. Indeed, on two contracts, the Parties were 'squabbling' over the name of the *Adjudicator* when neither had any intention of using him. The result was that no *Adjudicator* was ever appointed.
- When only one name was suggested, tendering Contractors usually did not query or object to the choice of the *Adjudicator* (despite, in some cases, having apprehension about the named person) in case doing so was perceived as being confrontational and would count against them.
- The highest degree of satisfaction with the selection procedure was when three or more names were suggested, from which the *Contractor* could pick one in the case of a dispute. A number of Employers deliberately put forward names of ex-contractor personnel with experience in the technical aspects of the construction who they felt would be both competent and fair. This nullified potential objections from Contractors that the *Employer* was choosing someone who would just see things from their own perspective.

Employers who let a large number of contracts under the ECC (or indeed under any form in the UK since the introduction of the Housing Grants, Construction and Regeneration Act 1996) would have a large number of adjudicators on retainers who, hopefully, would have little to do. However, this could be perceived as compromising their independence.

Consequently, it has become normal practice to specify that the choice of *Adjudicator* is 'to be agreed between the Parties in the case of a dispute'. This has the advantage that the *Adjudicator* can be selected for their specialist knowledge, based on the particular matter in dispute (and most professional bodies have a list of qualified adjudicators, with location and experience, on their databases and, sometimes, website). What is more, most adjudicators are periodically peer reviewed, whereby their professional body reviews a recent case. The relatively small downsides of this approach are that the *Adjudicator* has a very limited time in which to become familiar with the contract, and the appointment of the *Adjudicator* can become contentious in itself as some adjudicators, like arbitrators, can attract reputations.

However, in NEC3 and NEC4, a third party can be named in CD part 1 as the *Adjudicator nominating body*, which will act if the Parties cannot agree on or find an *Adjudicator*. Because of the introduction of the Housing Grants, Construction and Regeneration Act 1996, many institutions – not just the ICE, as per the Contract Data checklist – that could be named as this body now have a list of trained and accredited adjudicators and, when asked for a fee, should be able to quickly select one with an appropriate contractual and sector background from its database. A starting point for the selection of a suitable adjudicator would be to choose an institution that is representative of the works that are being built. For instance, choosing the Royal Institution of British Architects (RIBA) as the *Adjudicator nominating body* for a civil engineering contract would not make sense, neither would choosing the ICE for a building project.

Another aspect to bear in mind is whether the dispute is a technical or legal one, or one over quantum. This may affect who would be the best choice for the *Adjudicator*, as you will want someone with an appropriate background. The Construction Industry Council has a list of accredited adjudicators drawn from all parts of the construction industry and from all disciplines.

In conclusion, it is now common practice for 'to be agreed between the Parties in the case of a dispute' to be inserted where CD part 1 asks for the name of the *Adjudicator*, with another party specified to choose the *Adjudicator* if the Parties cannot agree. Regardless of whether this is the case, the appointed *Adjudicator* has to be seen as competent and unbiased, have expertise relevant to the disputed matter and be familiar with the relevant NEC4 conditions. Otherwise, adjudication is just a process that has to be gone through before you can get to arbitration or litigation.

2.2.2.9 Form of *tribunal*

The tribunal is the stage of dispute resolution when all else has failed. As the Guidance Notes point out, there are two forms of *tribunal* available. The first is arbitration, where an independent but experienced and qualified person decides the dispute in a private hearing. The second is litigation, where you go court – in the UK, this would be the Technology and Construction Court (TCC), where the judges will often have technical appreciation as well as legal expertise.

The arbitration has three general advantages over litigation.

- It can be a lot cheaper and quicker. However, depending on the complexity of the issue and the behaviour of the Parties, it can also be almost as expensive. While I know arbitrators who have given examples of substantial savings compared with going to court, I also know lawyers who argue that arbitration can be more expensive.
- Arbitration is a private procedure and happens behind closed doors. Therefore, your grievances are not aired in public. Perhaps more importantly, in terms of your organisation's reputations, the other Party does not air your deficiencies in public!
- Because the arbitrator tends to have come up through the industry, the decision can be fairer and a bit less purely legally based, and the process a bit less confrontational.

However, the main disadvantage of arbitration is that the losing party can appeal on a point of law that is of general interest, such as the interpretation of unamended NEC clauses.

My general feeling is that the advantages of arbitration outweigh the advantages of litigation, so would recommend arbitration to be specified as the *tribunal*, but I know lawyers who would argue the opposite. However, in reality, you never really want to get anywhere close to either arbitration or litigation – i.e. neither is the preferable option!

2.2.2.10 *Share ranges* and *share percentages* under options C and D

These entries in CD part 1 affect the pain/gain share – or share profile – under the target-cost options. The *share range* is by reference to the final Prices (under option C) or Total of the Prices (under option D). In the example entry in the Guidance Notes' Contract Data checklist, the first entry under *share range* is 'less than 80%' and the *Contractor's share percentage* is 40%. This means that the *Contractor's share percentage* of 40% applies only to the portion of the final Price for Work Done to Date that is below 80% of the final target Prices. For the next portion of savings (or gain), which is given as 'from 80% to 90%' compared with the final Prices, the *Contractor's* share percentage is 25%, while for the initial 10% of savings, it is 15%. Likewise, for the first 10% of any overrun (or pain), the *Contractor's* share is 15%, before rising to 25% thereafter.

However, the Guidance Notes do not give any detailed advice on setting the share formulae, so here is some. I have divided the share ranges into four zones.

ZONE 1: WHAT HAPPENS IF THERE IS A SIGNIFICANT OVERRUN OF THE TARGET THAT NEITHER PARTY COULD HAVE REASONABLE PREDICTED?

In this zone, the key question is who can best bear the overrun and hence take the majority, but not necessarily all, of the overrun. For instance, if the *Client* is a large organisation that has money coming in year on year, it will have much deeper pockets than a relatively small *Contractor* that is making a relatively small profit on what for it is a large contract. In this situation, if too much risk is put on the *Contractor*, then it will fight and squeal to absolve itself of liability, to the potential detriment of the *Client's* cost and other objectives. Ultimately, the *Contractor* may go bust, in which case all the risk will revert back to the *Client*. If, on the other hand, the *Contractor* is relatively large (and for which the contract is a small one) and the *Client* is relatively small (for which the contract is a large one), the situation is reversed, and it may be advisable to cap the overrun, i.e. to set the *Contractor's share percentage* at 100%.

ZONE 2: WHAT HAPPENS IF THERE IS A SMALL TO MEDIUM OVERRUN THAT IS WITHIN THE CONTEMPLATION OF THE PARTIES?

Given that risk within the target Prices is generally within the *Contractor's* predominant (but not necessarily exclusive) influence, then generally the *Contractor* should take the lion's share of any small to medium overrun (e.g. a *Contractor's share percentage* of 50% or more). The less this is the case, the more the initial pain share should err towards a cost-reimbursable contract, with the *Client* taking a greater percentage of overrun, i.e. the *Contractor's share percentage* is smaller.

> As a 'for instance' on zones 1 and 2, on the Channel Tunnel Rail Link, the *Employer* had deleted the physical conditions compensation event on large civil engineering contracts, but instead took 75% of any overrun up to 120% of the target Prices (so the *Contractor's share percentage* was set at 25% in my zone 2) and 90% thereafter (so the *Contractor's share percentage* was set at 10% in my zone 1).

> Another example comes from one of my regular training clients, a water company – a large company with relatively secure income – that mainly contracts with small to medium-sized contractors. Rightly or wrongly, it typically deletes both the physical conditions and weather compensation events clauses to avoid arguments over whether events are or are not compensation events and what their effects are. However, it normally takes the majority of any pain share initially, and typically all of it once the Price for Work Done to Date exceeds the Prices by more than 10% or 15%.

ZONE 3: WHAT HAPPENS IF THERE IS A SMALL TO MEDIUM UNDERRUN THAT IS WITHIN THE CONTEMPLATION OF THE PARTIES?

In this opposite case to zone 2, the *Contractor* generally takes the lion's share of any small to medium savings compared with the target Prices and which are within the contemplation of the Parties. Again, the less this is the case, the more the initial gain share should err towards a cost-reimbursable contract, with the *Client* taking a greater percentage.

ZONE 4: WHAT HAPPENS IF THERE ARE SIGNIFICANT SAVINGS THAT ARE BEYOND THE REASONABLE CONTEMPLATION OF BOTH PARTIES?

This is a point that applies to all four zones: the share of any savings or overrun for each Party should, ideally, be big enough to motivate both Parties to carry on working together to minimise costs.

> In one framework let around 2011, the *Employer* was consulting with the pre-qualified contractors on the draft contract terms, including the share profile, and it was to be an early contractor involvement (ECI) process whereby the final target was negotiated. The *Employer* was proposing that Contractors took all of the share of any overrun, half of the first 5% of any saving and, beyond that all savings would go to the *Employer*. As this was being done during a recession, the Contractors were likely to tender low fee percentages/have low margins to ensure they got on the framework. Thinking through the consequences of this from the Contractors' perspective:
>
> - they would seek to add in excessive risk (from the *Employer's* perspective) prior to agreeing the target Prices to protect the downside, i.e. their 100% share of any overrun
> - they would have no motivation to seek legitimate savings beyond the 5%
> - when business picks up in a couple of years' time, the *Employer's* projects would become unattractive, causing the Contractors either not to bid or to put their 'C' team on it, having got a nice risk contingency included in the target.

However, in this Zone, I see no point in the *Client* paying any more than is sufficient to motivate the *Contractor* to carry on striving for savings. Equally, giving no gain share for significant savings does not motivate the *Contractor* to find more savings once into this zone.

Much more information and guidance on the working of target-cost contracts is given in my other book on collaborative procurement routes (Broome, 2002).

2.2.2.11 Writing option Z: *additional conditions of contract* clauses

In Chapter 1, Section 1.2.3.6, I emphasised the need and benefits of not writing lots of option Z clauses. Hopefully you will have bought into this. There are, however, situations where option Z clauses will be needed, so here are my practical tips on writing them (some of the tips are taken from Houghton and Trebes, 2011).

1. In the same way that the NEC Panel wrote a performance specification for each set of clauses when drafting the original contract, write a justification for each proposed amendment and state positively what it is intended to achieve. If you cannot state clearly and succinctly the rationale for writing a new clause, amending an existing one or deleting one, then do not do it! Be careful not to introduce ambiguity and inconsistency by

replacing standard clauses with legalistic clauses that do the same thing or, even worse, not deleting the original clause. In addition, be wary of clauses that
- undermine the stimulus to good management
- clutter up the contract with minor issues that are unlikely to ever eventuate
- undermine the drafting philosophy of the NEC by attempting to prescribe an outcome for every foreseeable eventuality, rather than a process for project managing an issue out.

> Glenn Hide, a well-known NEC name who I work with frequently, has told me a story that I think represents good practice.
>
> A new client wanted to go over to NEC for all the right reasons. Glenn gave some initial training to the senior team and their lawyer, who was invited along. The lawyer was then asked to come up with a list of clauses he wanted to change, delete or add as option Z clauses, together with the context and rationale for each change. This list was then reviewed by the *Client*, with Glenn there. Over half of the proposed amendments were dropped, many with the lawyer's consent, as following discussion he realised they were not relevant.
>
> The lawyer then wrote the clauses, which Glenn critiqued for their contractual correctness etc.
>
> This is a good example of what I keep on banging on about as (ex)chair of the Association for Project Management's Contracts and Procurement SIG: lawyers are an expensive resource and, if they are not properly briefed and/or are allowed to run amok, can have a disproportionately negative effect on a contract. So project manage them!

This list of justifications can be developed into a table format, with the version 1 document having three columns: clause reference, justification for amendment and additional comments (e.g. links with other clauses). Once it is agreed which clauses are to be amended, the rows for those that are not can be deleted. For those that are going to added, deleted or amended, then a fourth column can be added, where the wording that will go into the contract is entered. The clauses can then be evaluated and fine-tuned against the rationale for writing them in the first place.

2. Ask yourself whether it would be better to put these clauses in the Scope document.
   - If they are specifying or describing the *works* or articulating a constraint, they should usually be in the Scope document. If they are in the Scope, they can be changed if necessary, whereas *additional conditions of contract* can only be changed by agreement between the Parties.
   - If you want to be able to legally enforce them, they need to in the *conditions of contract*.

   A real issue is that the central/legal team within a Client, which has control over the Z clauses for each contract, perhaps does not trust the project team to put the information in the Scope. For that reason alone, the Z clauses tend to grow! The practical solution to this is to have a standard set of 'legal' Scope clauses.
3. Write the clauses in the plain English, present tense style of the NEC (see Section 2.2.3.3 on writing the Scope for more guidance on this).

4. Use the terminology in the contract.
5. Put the clauses in an appendix or schedule, in an order that matches the order of the contract. This will help if someone wants to see whether a particular clause has been deleted or modified, or a new one has been added.
6. Put the more legalistic ones, which do not affect the day-to-day or month-to-month operation of the contract, at the end of this schedule, especially if they do naturally fit into an existing section of the *conditions of contract*. A few additional tips are relevant here.
   - In converting what are probably 'boilerplate' clauses from the *Client's* non-NEC contracts, follow tips 2, 3 and 4.
   - Place the defined terms for these clauses next to the relevant clauses, rather than at the start of the appendix or schedule (as would be normal legal drafting practice). This means that (a) the operational defined terms are not cluttered up with lots of legalistic defined terms and (b) it is a lot easier to understand the legalistic clauses if the relevant – and only relevant defined terms – are at the start of each topic. For example, all the defined terms related to protection of data are at the start of the clause on the protection of data. What is more, having done this conversion numerous times, I have picked up where definitions are missing and found defined terms that are not used anywhere in the contract and hence can be deleted.
7. Consider making a flow chart of amendments, to ensure logical consistency – you would be surprised how many errors and ambiguities this picks up.
8. If the as-written clauses do not reflect the rationale, are not logical, are not clear or are long-winded, either rewrite the clauses or ask yourself if their complexity justifies their inclusion.
9. Provide the rationale for the amendments to tenderers, so they can understand why you have made them. I often work on complex procurements where the *Client* is trying to do something different. The best Clients I work with consult with the potential Contractors as they develop the contract, explaining what they are trying to achieve at a big picture level, preparing notes for each amendment and asking for feedback, both in generality and on specifics. They might do this when the original notice is published, so that those who do not like the contract do not waste their time pre-qualifying. They then do another round of consultation with those who do. There is no doubt in my mind that this results in a better contract and with the tendering Contractors far more bought into what the *Client* is trying to achieve. If a negotiated procedure is used for selection, it also results in far less negotiation and hence is quicker to get into contract. And there is less likelihood of challenge, as the final contract will be far closer to the conditions on which all the Contractors tendered.
10. Ask the tendering Contractors to price each amendment that transfers risk. This way, you will have an idea of the cost of risk transfer. However, it may make tender evaluation more difficult.

In reality, it will probably not be you who writes the additional conditions of contract, but lawyers. As indicated in the boxed example, do not be afraid to project manage lawyers by following the above steps, or to challenge them as they do it. If you are doing the writing, invite and do not be afraid of challenge: it is better to have a discussion now than a contractual dispute or failed project later.

See also Section 2.2.3.3 on writing good Scope and Site Information documents, as the advice on sentence length, construction and language to use in drafting more technical, management and administration clauses applies equally to writing option Z clauses.

2.2.3 Preparation of the Scope and Site Information

2.2.3.1 Overview: why does this matter and what are we trying to achieve?

One of the key factors leading to successful project implementation, under any conditions of contract, is good contract documentation. As one research report from Australia stated: 'The greatest cause of claims and disputes in the construction industry is related to problems in contract documentation, including errors, contradiction, ambiguity and the late supply of documents, which give rise to delays and inefficiencies and hence claims' (Australian Construction Services, 1988). As previously mentioned, another research project, in Canada (Revay, 1992), indicated that, on average, clients pay three times as much for work introduced after the contract is let compared with equivalent work included in the original contract documentation. The reasons for this include the additional on-site costs due to loss of productivity (i.e. disruption and resulting delays).

A number of interviewees from the initial research sample expressed the view that the ECC makes deficiencies in documentation more apparent and the effect more immediate and evident. This is because

- clause 60.1(1) makes it clear that any change to the Scope, except in two defined circumstances, is a compensation event that may entitle the *Contractor* to additional time and cost
- clause 63.10, which is an extreme statement of the legal principle called the *contra proferentem* rule – i.e. a change to the Scope due to an ambiguity or inconsistency is interpreted in the light least favourable to the Party that prepared it[8]
- the time and cost effects are calculated and agreed, ideally based on a forecast, closer to the time of the change than is the case under other more traditional conditions of contract
- the cost effects of an individual compensation event include not only the direct costs, but also the indirect costs resulting from disruption and subsequent prolongation
- a number of Clients, and at least one of the cloud-based systems identified in Section 3.3.2.3, trace the originator of the compensation event back a stage further (i.e. is it an end user originated change or delay, consultant design change or delay, due to the project manager organisation etc.).

In the words of one interviewee, 'there are no hiding places within the NEC, so poor documentation can be traced back to its originator. Designers beware!'

Unfortunately, my experience of reviewing NEC contract documents, including option Z clauses and Scope documents (whether as a trainer, in order to give contract-specific training, while carrying out contract reviews or in order try and avoid or resolve disputes), is that most are a bit of a dog's dinner. Particularly for repeat-order clients, standardising and refining your contract documents is an investment that can pay off big time. More on this latter.

[8] In fact, it is *contra proferentem* on steroids, as the legal *contra proferentem* principle is based on the balance of probability (e.g. all other things being equal). In the ECC, any ambiguity or inconsistency is assessed in favour of the Party that did not create it, so if there are five references that are to the favour of the Party that wrote it and one to the Party that did not, then it is the one reference that has precedence.

Figure 2.1 Hitting the sweet spot in contract drafting

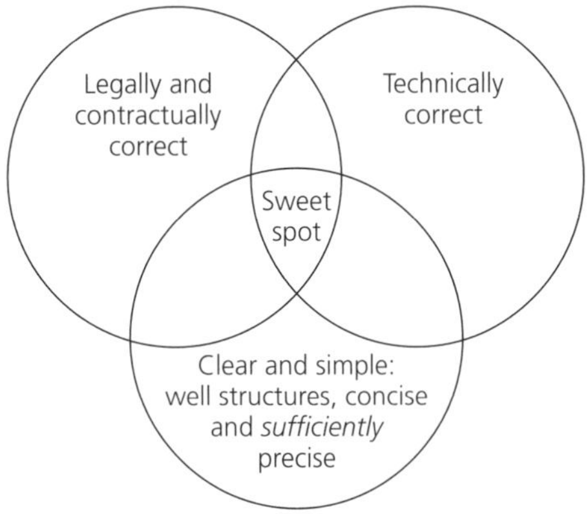

Figure 2.1 shows a diagram that I use at the start of any training session on preparing contract documentation. The following points add some explanation to the diagram.

- *Legally and contractually correct.* Ultimately these are legal contract documents, so they need to make legal sense. For example, NEC contracts, like all contracts of substance, have defined terms to ensure all participants are on the same page about what a word or set of words means. If technical authors start using different terms or referencing people who do not exist under the contract, there will, at best, be discussion or, at worst, dispute over interpretation and who should have done what by when. Similarly, the contract documents should replicate the language used in the contract. For instance, nowhere in the NEC family is the word 'approval' used when referring to an action of the *Project Manager* or their equivalent. Instead, things are accepted by the *Project Manager*, so use 'accepts', 'acceptance' etc., not 'approves' or 'approval'. (To be a more specific, you should use 'accepts' etc. for the *Project Manager*, but 'approves' etc. for any third party who has to approve something before the *Contractor* proceeds with work. This is important as clause 14.1 sets out the effects of any 'acceptance'.)
- *Technically correct.* There is no point in having a legally correct document that is technically lacking. So the technical aspects of the contract documents have to be correct as well. Unfortunately, technical and contractual people often seem to write in different languages and so need to be brought together. In practice, (a) precedence of documents is often stated within the form of agreement that the parties' directors sign to enter into the contract and (b) legal terminology is in the conditions of contract, which sit above the technical documents. Consequently, in reality, the legal usually takes precedence. So, where relevant, you want the technical people to use the legal terminology, if it fits, in preference to their own.
- *Clear and simple: well structured, concise and sufficiently precise.* Documents should be: well structured, so that people can easily find the information they need to comply with the contract; concise, so that similar information is not replicated in different places, and documents are written using sentences that are short, to the point and readily understandable; *sufficiently* precise for the *Client* to be confident it will get what it wants, but not overly prescriptive such that there is no opportunity for innovation or differentiation, both in what is provided by the *Contractor* and in how it provides it. One way of looking at this is by considering the breadth and depth of the Scope document:

- for breadth, it has to cover everything that you want the *Contractor* to do – if you leave out something and then add it in during the contract, that will be a compensation event, which may result in additional cost and time
- depth is the level of detail it goes into for any particular aspect – if you refine or add to the detail, it will also be a compensation event which, if the *Contractor* can demonstrate the change is going to cost them more than what they were originally going to provide or do, will result in additional time and cost.

It should be stressed that the way in which the Scope is written should flow from the contract strategy. For instance, it would be foolish to fully design the *works* and then, once completed, decide to allocate design responsibility to the *Contractor*.

2.2.3.2 Structuring the Scope and, if relevant, Framework Information

There are numerous ways in which the Scope and Framework Information documents can be put together. In this section, I suggest some model structures that you could use as the basis for your documentation.

MODEL STRUCTURE 1: FOR SIMPLE CONTRACTS, FOLLOW THE LAYOUT IN THE SHORT CONTRACT'S SCOPE DOCUMENT

The short contracts come with a preformatted Scope document. In the Engineering and Construction Short Contract (ECSC), the Scope is divided under the following section titles.

1. Description of the *works* – a detailed description of what the *Contractor* has to deliver, including any design that it is to do.
2. Drawings – a list of drawings that the *Contractor* has to comply with.
3. Specifications – all the specifications that the *Contractor* has to comply with. This section is laid out as a table, which has a column for these specifications to be listed. A general catch-all here (if in the UK) would be to enter 'all relevant British Standards'.

> As a little aside, when my wife and I moved into our rather dilapidated house, we had three £10–20 000 contracts with local builders for which I used the ECSC. One included replacing guttering and roof drainage, for which the *Contractor* had design responsibility. When a part of the roof drainage layout backed up and flooded due to its design, a review of the relevant British Standard revealed it was not in accordance with the standard and was therefore a Defect. The *Contractor* redid the design and replaced that part of the drainage.

4. Constraints on how the *Contractor* Provides the Works – such as access, hours of working etc.
5. Requirements for the programme – if using the ECSC, I suggest using clauses 31.2, 31.3 and 32.1 of the full ECC as a starting point for what you want to see here (i.e. simplify it down). In the ECSC, similar to the full ECC, 'one quarter of the Price for Work Done to Date is retained in assessments of the amount due until the *Contractor* has submitted a first programme ... showing the information which the Scope requires' (ECSC clause 50.8). However, unlike the full ECC, it is as required in the Scope (as opposed to the *conditions of contract*) and, as there is no 'acceptance', there is no ongoing sanction if it is not realistic, practicable or updated.
6. Services and other things provided by the *Client* – which includes when they are to be supplied by. (For example, our house – see box above – had an outside toilet, so the builders did not have to hire a portable toilet.)

While this is the given format for the ECSC, it may also do the job for simple contracts let under the full ECC, almost certainly under option A or B, where the contract has only one or two characteristics which lend itself to the full contract, such as simple *Contractor* design.

However, the downside of this format is that, with only six relatively high-level section titles, you might miss something that you should cover. An obvious one that sticks out for me is health and safety. For instance, if the work is in the UK, who will be the Principal Designer and Principal Contractor under the CDM Regulations 2015? This could fit under section 4 (constraints). You might therefore want to cross-reference both back to the ECSC clauses and the full ECC clauses (as below) to check you have covered everything you should have under these high-level section titles.

MODEL STRUCTURE 2: FOLLOW THE EXAMPLE *CLIENT'S* SCOPE STRUCTURE FROM THE GUIDANCE NOTES

Table 2.1 presents the model structure put forward by the authors of the ECC in Section 3.4 of the updated NEC4 ECC Guidance Notes (volume 2: *Preparing an Engineering and Construction Contract*).

This is a good starting point and has now almost become the industry standard. Indeed, some Clients require it to be used by their consultants when preparing Scope documents. Here are some criticisms of this structure, based on what I have found and the comments of others who have been required to use it.

There are some anomalies in terms of its order relative to the order of the clauses in the conditions of contract. For instance:

- S 400 Completion (which is defined in section 1 of the contract) comes after S 300 *Contractor's* design (which is covered in section 2 of the contract)
- S 1200 Subcontracting (which is covered in clause 26 of the *conditions of contract*) comes after S 1100 Health and safety (which is covered in the slightly later clause 27.4)
- S 1100 and S 1200, as well as S 900 Working with the *Client* and Others and S 1000 Services and other things to be provided (all of which are covered in Section 2 of the *conditions of contract*) come after S 600 Quality management and S 700 Tests and inspections (which are Section 4).

If you are to follow this format strictly – which, as I have said, some *Clients* have insisted on – there are some issues you will have to consider.

- Where do you put things that do not conform with this structure? For instance, taking payment under option A, where should you include a limit on the number of activities that a *Contractor* can have in the Activity Schedule,[9] or specify what 'form' the application is to be in (as per clause 50.2)? If you want to state in more detail (than in the *conditions of contract*) how risk and emerging issues are to be managed, where should that be included?

[9] One of my reviewers said that this should be in the Instructions to Tenderers, not the Scope, and they were partially right. If it is in the Instructions to Tenderers, then the tender *activity schedule* needs to comply with this. However, without a statement in the Scope documents, then any revised Activity Schedule – as per option A clauses 55.3 and 55.4 – need not do so.

Table 2.1 Model structure for the *Scope* as suggested in the NEC4 Guidance Notes

Section	Scope (*Client*'s)
S 100	Description of *works*
S 200	General constraints on how the *Contractor* Provides the Works
S 300	*Contractor's* design
S 400	Completion
S 500	Programme
S 600	Quality management
S 700	Tests and inspections
S 800	Management of the *works*
S 900	Working with the *Client* and Others
S 1000	Services and other things to be provided
S 1100	Health and safety
S 1200	Subcontracting
S 1300	Title
S 1400	Acceptance or procurement procedures (options C, D, E and F)
S 1500	Accounts and records (options C, D, E and F)
S 1600	Ultimate holding company guarantee (option X4)
S 1700	Undertakings to the *Client* or Others (option X8)
S 1800	Transfer of rights (option X9)
S 1900	Information modelling (option X10)
S 2000	Performance bond (option X13)
S 2100	Advanced payment to the *Contractor* (option X14)
S 2200	The *Contractor's* design (option X15)
S 2300	Retention (option X16)
S 2400	Low performance damages (option X17)
S 2500	Early *Contractor* involvement (option X22)
S 2600	Project Bank Account (option Y(UK)1)
S 2700	*Client's* work specification and drawings

- You can end up with a disjointed document. For instance
  - subcontracting requirements are generally in S 1200, but those which are related to working under options C, D, E or F are in S 1400, simply because (so far as I can tell) there is some Disallowed Cost related to Subcontracting and not following procurement procedures (second and third bullet points of clause 11.2(26) and an extra clause relating to Subcontracting procedures in clause 26.4)
  - you would be stating the bulk of requirements relating to *Contractor* design in S 300, but a few other requirements will be in S 2200, just because they are mentioned in option X15
  - if you are using a Project Bank Account, you would have these requirements in S 2600, separate from other aspects on payment (for which no 'S' section exists if under option A or B).
- You still have to decide in which sections certain things go, as they could go in two or more.

- You end up with a massive S 2700 at the end of the document, from which all the standard specifications are referenced and where drawings are included. And why is the standard specification that most large Clients have not in a separate section from the contract-specific drawings?

Consequently, those who have been told by *Clients* to use this format have found it constraining and a bit artificial, with the end result not being as user friendly as it could be.

However, I and others have found that the accompanying checklists for filling in the different sections are far more useful. They can be used to check that, whatever structure you use, you have considered all the issues detailed in the checklists (if relevant) somewhere in the Scope and that you have not missed anything.

MODEL STRUCTURE 3A: FOR MEDIUM TO MAJOR COMPLEXITY PROJECTS, FOLLOW THE *APPROXIMATE* STRUCTURE OF THE ECC *CONDITIONS OF CONTRACT*

I suggested in my book on the third edition of NEC that you could follow the *approximate*[10] structure of the ECC (the nine main sections in the *conditions of contract*). In the NEC4 ECC Guidance Notes (volume 2: *Preparing an Engineering and Construction Contract*), the authors of NEC have, in Section 3.3, helpfully provided a table that identifies all references to the Scope that appear in the contract, with a narrow column included for comments on the references. However, many of the references to the Scope are not relevant for the purposes of preparing the Scope, and most of the comments are, in my opinion, statements of the obvious and do not add much, if any, value. So I will go through the sections adding what, I hope, are value-adding comments. Consequently, many of the comments will be of value whatever structure you select.

- In Section 1 of the Scope (General), provide the following.
  - A clear general description of the *works*, including referencing general arrangement and location drawings.
  - Clear statements of general constraints on how the *Contractor* is to Provide the Works. For instance, noise levels, access, hours of working, sequencing etc. (although these could be put into section 3, on programming).

    Constraints also include generic standards that the work has to comply with. Ones that I am aware of (having helped revise technical documents in these sectors) include
    - for highways: the *Design Manual for Roads and Bridges* (DMRB) and *Manual of Contract Documents for Highways Works* (MCHW)
    - for the water sector: *Civil Engineering Specification for the Water Industry* (CESWI) and the *Water Industry Mechanical and Electrical Standards* (WIMES) and *Technical Standards 001* (TS001).

    These documents are updated by technical notes so these need to be incorporated too. I prefer to include a generic statement about this, rather than list every single update.

    A generic catch-all statement I sometimes see and incorporate on UK-based contracts is that 'all work is to be done in accordance with the relevant British Standard, unless stated otherwise'.

[10] I say '*approximate* structure' because, to me, it often makes more sense to have all the information relating to, for instance, *Contractor* design, definition of sectional Completions, payment etc. in one place, rather than scattered through the contract in subsections related to the particular main and secondary options used. As discussed earlier, this is one of the drawbacks of rigidly using Model Structure 2 as put forward by the authors of the ECC Guidance Notes.

- A clear statement of what the *Contractor* is or is not to do by the Completion Date for Completion to be achieved (see the first bullet point of clause 11.2(2)), such as provide operation and maintenance manuals, provide 'as-built' drawings for the health and safety file etc. One sentence with bullet points can highlight this (i.e. what the *Contractor* has to do), while another sentence can state what the *Contractor* does *not* have to provide to achieve Completion. A common exclusion is landscaping, because much of the work has to be done in the right season.[11]
  Where secondary option X5 (sectional Completion) is used, this also needs to be considered for each section. Note that some Clients prefer to have this statement in Section 3 (Time) for obvious reasons.
- For acceptance or procurement procedures relating to subcontracting or suppliers (under the Scope reference in the definition of a Disallowed Cost), a better place to put any acceptance or procurement procedures is in Section 2, next to any other subcontracting requirements.
- Any information relating to communications, such as pro formas to be used, or which, if any, IT system will be used or how it will be selected (see Section 3.3.2).

■ In section 2 (The *Contractor's* main responsibilities), provide the following.
  - References to all existing drawings or the current BIM model.
  - A statement of all the elements of the permanent works that the *Contractor* has to design, so that tendering Contractors can rapidly get their minds around whether, for instance, it is a design and build job or a more traditional job with a little bit of design. The statement to use from the *Client's* perspective is: 'The principal elements of the *works* that the *Contractor* is to design include, but are not limited to ...' and then, in bullet point or table format, briefly state the elements and reference the details. The way I have phrased this statement means that if an element of design has been missed in this section, but is included in a detailed section, then the *Contractor* still has to do the design. However, do make it as complete as possible as Contractors sometime complain that some items had inadvertently been left out of the summary (more through a lack of attention to detail than malice) and as a result it had affected their estimated profit margin.
  - In the table listing the elements of *Contractor* design (see above point), an indication of which of these elements are to be submitted for acceptance. In the same table, if relevant, you could also state any purposes the *Contractor's* design can or cannot be used for (as per clause 22.1). Alternatively, if option X9 (transfer of rights) is specified, in which the default is that the *Client* has copyright and hence exclusive use of any documents developed for the contract, you could state specific circumstances in which the *Contractor* can use the design or other documents.
  - Any procedures that have to be followed for acceptance, both generically and for any specific element (e.g. third party certifications and approvals), and any checks that have to be made. For checks, a common one is that a drawing, before it is submitted for acceptance, has to be internally checked and signed off by a senior Chartered Engineer or equivalent and, if in the UK, any comments from the Principal Designer under the

[11] When I was redoing the Royals Parks contract documents (for the six Royal Parks of London), I inadvertently put landscaping in as not required for Completion. When I was feeding back to the framework consultants and contractors what I had done, it was pointed out to me that, as most of their contracts were for landscaping, this was a bit daft. In contrast, on a large £100 + million contract for a new road, the requirements for landscaping maintenance – not just correction of landscaping Defects – meant that it was better to have a separate Term Service Short Contract for this work. This was signed at the same time as the construction contract, with the *starting date* being on Completion of the main construction contract.

CDM Regulations 2015 have to be taken account of. For each element of design, you may also wish to state, if relevant, any

- □ size or space limitations
- □ specific design standards or codes of practice
- □ material and workmanship specifications, referencing but not repeating relevant standards
- □ loading and capacity requirements
- □ operational or performance requirements and design life.

- References to the detailed drawings and any further descriptions of what has to be built, and to any specific constraints on an element, either on how it is built or on the end result. Having said this, the reality is that most people on site only look at drawings, so the more that can be put into the individual drawings which will be used, the better.
- Any elements of temporary works that have to be submitted to the *Project Manager* for acceptance. While not required in the Scope, I do think it is only fair that tendering Contractors know this in advance, so far as possible, rather than if and when the *Project Manager* instructs (as per clause 23.1).
- Any details of how the selected *Contractor* has to cooperate with Others in obtaining and providing information and sharing the Working Areas. The more detailed this information, the keener the pricing will be, but the more likely a compensation event, and vice versa.
- Facilities and services to be provided by the *Client* and *Contractor*. For instance, potential site yards or office locations that the *Client* can supply and the site facilities for the *Client* project team that the *Contractor* is to supply.
- Details of any criteria and procedures for accepting Subcontractors or suppliers. While these are not mentioned in the *conditions of contract*, it is often worth saying something about them. This could be stating who the *Contractor* is to use, if already selected, or giving tenderers a 'nudge' as to which subcontractors or suppliers you would like them to use. How do you do this?
  - □ If you are a private sector Client, you can just state who the *Contractor* is to use, but be aware that this might come at a cost. This is because the specialist Subcontractor or supplier knows it is effectively in a monopoly position, so can charge a high price to each tendering Contractor, which is passed on to the *Client*. Also, an individual tenderer might have had bad experiences of that specialist and prefer not to work with it. Consequently, a risk premium is added on to what the Subcontractor or supplier has quoted.
  - □ If you are a public sector Client, you will almost certainly have had to already run a compliant competition if you are to insist that particular Subcontractors, suppliers or products are to be used. And if it is a single-source supplier, you will have the same issues outlined above for a private sector Client.
  - □ If you just want to give a 'nudge' to the tendering Contractors, whether towards a single source or a small selection, then you could state which Subcontractors or suppliers are 'pre-accepted' for each category of work. Consequently, they would not have to be submitted to the *Project Manager* for acceptance under clause 26.2. While this does not stop the *Contractor* from proposing another Subcontractor, it does give a strong hint as to who the *Client* wants.
- Any health and safety requirements that are over and above either the statutory minimum or project-specific requirements in terms of how the statutory requirements are to be implemented.

- In section 3 (Time), provide the following.
  - Any additional programming requirements, as stated in clause 31.3. Common things that project managers want to see are the critical path, a monthly narrative explaining the changes since the previous submission, a complete and up-to-date list of who has to supply anything to the *Contractor* and by what date (rather than it being hidden away in the depths of numerous method statements) and histograms for the major resources. I say more on this in Section 3.5.
  - The form in which the programme is to be issued. While you could always do this in the third edition ECC, the fourth edition now specifically prompts you to state 'the form' of the programme, i.e. whether you want it in physical, pdf and/or native file format. If you want it in native file format, you will need the software to open the programme file and interrogate it, so you will also need a licence. And if you are unfamiliar with the software, you will also need to have some training on it. So, as the *Client*, you also need to decide whether you want the programming to be done using software specified by you, or if you are happy to use the *Contractor's* software (which is what the *Contractor* would almost certainly prefer). The answer probably depends on the extent to which you want to integrate the programme for this contract with programmes for other contracts, in order to have a master programme.

> For the London 2012 Olympic Games, all contractors and consultants had to apparently use the same software, so that their individual contract programmes could be assembled into an overall programme for the whole project.

  - Any restrictions on how the programme is to be prepared. In order to avoid the programme becoming unwieldy, some Clients specify a maximum allowable number of operations on the level 1 programme, which can then be broken down into activities. My experience of programmers is that they love detail – which is understandable – and have to be given a real hard metaphorical shove to produce programmes that can be readily understood (i.e. do not contain hundreds, if not thousands, of lines). Having too much detail makes it hard to get your head around the programme. Instead, from before they even get into a contract, get them to programme at different levels, where each successive level is more detailed. I would not want the level 1 programme to be more than 60 lines, but, having got my mind around this, I might well want to interrogate down into the detail, especially for the initial works in the first couple of months.
  - Any other constraints on how the *Contractor* is to Provide the Works, i.e. affects their programming. These will be more specific constraints than the headline ones identified at the start of the Scope (which you might reference, but do not duplicate).
  - Any reasons why the *Client* may wish to use the *works* before Completion, in order to avoid having to 'take over' the asset. For instance, when doing work on existing assets, so that the *Client* can continue to use it.

> According to a possible urban myth, on one of the first Highways Agency (now Highways England) contracts, the requirement to keep the motorway open was missed out … an expensive compensation event!

- In section 4 (Quality management; previously Testing and Defects in NEC3), provide the following.
  - The requirements for the quality management system. At a minimum, this will be any ISO accreditations that the *Contractor* is to have and any that the quality management system needs to comply with.
  - Procedures relating to testing and Defects, namely who is to do which test or inspection and when it is to be done (e.g. as soon as the work has been done or prior to Completion). It is very likely that the *Client* or those preparing the specification on its behalf will have to develop more detailed procedures than those found in the core conditions for testing. Testing requirements are project and technology specific, which is why detailed procedures for these are not found within the ECC *conditions of contract*. Careful consideration is needed to ensure that the procedures give sufficient comfort to the *Client* that they are getting an asset of the right quality, but without placing onerous costs on both the *Client* and *Contractor* (which ultimately the *Client* will pay for). Particularly on multi-disciplinary projects, clear statements of different testing requirements for different types of work should be provided. At its simplest, this might mean 'cutting and pasting' an old contract's testing procedures into the Scope and then changing the terminology to match that of the ECC. However, this might be a lost opportunity to review and rationalise what tests are necessary and who does them.
    For instance, most water companies require their framework Contractors do all the tests during construction, with the *Supervisor* having a watching brief (checking paperwork and that tests are done properly when on site), but the *Supervisor* does the performance tests in the commissioning phase, as this is ultimately what they are interested in.
    Related to this is the amount of notice that the *Contractor* is to give the *Supervisor* before tests and inspections, in order to avoid 'unnecessary delay' (which would be a compensation event under clause 60.1(11)). Rather than leaving it open-ended, you could define a default timescale, with different times for those tests that need a longer time to prepare for. For instance, on a pipelining contract where the *Supervisor* is always around, 2 hours' notice to inspect any work might be sufficient. However, for pressure testing of a couple of kilometres, 1 week's notice might be necessary.
    Also, do not forget about tests and inspections that need to be done on Equipment, Plant and Materials before they are brought to the Working Areas. This may be linked to payment and title, so could be covered elsewhere, in which case reference this 'elsewhere'.
  - Finally, a statement of who the *Contractor* issues any certificates to or, alternatively, where they are to be uploaded. An anomaly in the drafting of ECC is that under, clause 13.6, the *Project Manager* issues certificates to the *Client* and *Contractor*, while the *Supervisor* issues certificates to the *Project Manager*, *Client* and *Contractor*. What about the *Contractor* if it is responsible for tests and inspections that result in certificates (e.g. by a third party accreditor, who contractually would be a Subcontractor)? As many certificates will relate to quality, the most appropriate place to include this statement is this section. (Alternatively, you could add it into clause 13.6 using an option Z clause.)
- In section 5 (Payment), provide any requirements that affect payment, which will be largely dependent on the main option payment mechanism. More is given on this in Sections 2.2.4.2, 2.3.2 and 3.4.3, but a brief outline is as follows.
  - Particularly in option A contracts, any limits placed on the number of activities that the *Contractor* can have in the Activity Schedule by various means (which would need to be reflected in the Instructions to Tenderers).

- Again, particularly under option A, any extra requirements for activities to be treated as complete. For instance, all tests have to be done and the *Supervisor* notified of the results or issued with a copy of the relevant test certificate.
- Limitations placed on the value of work the *Contractor* can charge for in any budgetary period due to limits on the *Client's* cash flow. (Note that this and the previous two bullet points would need an additional bullet point added to option A clause 55.4 to be effective once the contract is awarded. It would say something like 'it does not comply with any constraints stated in the Scope'.)
- The form in which applications for payment are to be submitted. New to NEC4 is the requirement in clause 50.2 for the application for payment to be 'in the form stated in the Scope'. Volume 4 of the Guidance Notes, which goes through the contract clause by clause, gives no guidance on what this 'form' might be, but I presume it is whether it is in physical, pdf or native file format and, if native file format, which software is to be used. Alternatively, or in addition, it could be referring to how you want it presented within particular software and the level of detail or breakdown required.
- Under cost-based contracts (options C to F), amplifying the requirements for demonstrating Defined Cost and for auditing. (Much more on this is given in Section 3.4.3.)
- If using a Project Bank Account (option Y(UK)1), any requirements related to this, rather than as a separate and remote section.
- Any organisation-specific procedures for being paid. For instance, the requirement to invoice before a payment can be paid.

- In section 6 (Compensation events), provide any additional information needed for compensation events and their assessment, although this is likely to be minimal. The following three issues are worthy of comment.
  - For clause 60.1(12) (physical conditions), boundary limits could be stated. However, this is more likely to be a change to the clause enacted through the option Z (*additional conditions of contract*) clauses, with the boundary conditions stated in the Site Information, and/or as an additional compensation event under clause 60.1(21), with the wording stated in Contract Data part one.
  - Who is responsible (*Client* or *Contractor*) for sourcing the base *weather data* and taking *weather measurements* during the contract. Contract Data part one states where these are taken and who supplies them, e.g. the Met Office in the UK (see Section 2.2.2.5).
  - For value engineering suggestions under options A and B (clause 63.12) or options C and D (clause 63.13), any additional requirements for how these proposals are to be presented (e.g. process and pro formas to be used). Likewise, if option X21 is specified, requirements for proposals that reduce whole life cost.
- In section 7 (Title), provide any requirements relating to title (which effectively means ownership) – such as for moving Equipment in and out of the Working Areas, and how Equipment, Plant and Materials will be marked for this contract – and vesting (which is when a right or an interest in property, such as Plant and Materials not yet included in the *works*, is secured). These are often linked to requirements for testing before delivery, which in turn is linked to payment. Think also about the practical aspects of delivery and storage of Plant and Materials before incorporation into the *works*.
- In section 8 (Liabilities and insurance) and section 9 (Termination), any requirements relating to insurance or termination.

We now go onto the last sections, which provide any requirements related to the secondary options.

- For option W (resolving and avoiding disputes), you might want to include a section on how it is hoped issues will be resolved and disputes avoided *before* they get to the point where the dispute procedures stated in the *conditions of contract* are required. If you are using option W3, you might well want to say more about how the Dispute Avoidance Board will operate. For instance, whether the appointed people will only be called in when a dispute is developing or the Board will be expected to visit the Working Areas regularly and, if so, what it is expected to do (e.g. to coach or mentor people). (More is given on this in Section 3.8.3.)
- In the last section, present any remaining requirements relating to the secondary options. For instance, the form of an ultimate holding company guarantee (option X4), collateral warranties (X8), performance bond (X13) or advanced payment bond (X14) that may be required, although often these are added in by the lawyers, either at the end of the option Z clauses or in a separate schedule or appendices.

MODEL STRUCTURE 3B: FOLLOW THE STRUCTURE OF THE ECC *CONDITIONS OF CONTRACT*, BUT SEPARATE OUT THE DIFFERENT TECHNICAL ELEMENTS

Using this structure, you would do the same as for Model Structure 3a, but having stated the general requirements for *Contractor* design, construction and quality management and testing, you would add separate sections for the specifics. These sections could cover

- civil engineering works
- building works
- control system

in each case specifying what the *Contractor* is to design and who the drawings are to be submitted to, the standard specifications that are to be used and any unique quality management and testing requirements.

You would then revert back to the general requirements for payment etc. This has the advantage that different people can focus on the parts of the document that are relevant to them. The people responsible for putting together the documents and the people who price and do the work, including Subcontractors, both need to have an awareness of certain general requirements, but this makes it easier for them to find the specific information they need.

MODEL STRUCTURE 3C: FOLLOW THE SAME STRUCTURE AS 3A OR 3B, BUT HAVE THE MANAGEMENT AND ADMINISTRATIVE ASPECTS SEPARATE FROM WHAT THE *CONTRACTOR* HAS PHYSICALLY TO DO

In this structure, one part of the Scope document is about ‘what’ the *Contractor* has to do (in terms of design and construction) and the other part is about ‘how’ the *Contractor* has to do it. This is particularly useful when the more technical aspects are being prepared by a design consultancy and the other aspects by the *Client* or another organisation, with the two sets of information to be brought together at the last minute, as is often the way. However, even with the best pro forma templates, the two parts do need to be checked for duplication, overlaps or gaps.

This structure will be quite useful for frameworks, particularly where the *Client* has relatively fixed ways of managing and administering its contracts, but what it wants constructed varies from contract to contract and/or is prepared by different organisations.

MODEL STRUCTURE 4: DESIGN FOR FRAMEWORKS

A framework is where you have a high-level contract or agreement for a period of time – normally at least 4 years – to employ a supplier or suppliers to do projects of a certain sort, with each project being let as a separate contract. As previously mentioned, there has been an NEC Framework Contract since 2005, which gives a high-level structure and references a number of other sub-documents in which the detail is fleshed out by the *Client* organisation. This will include the conditions of contract that can be used for each project. Unless you are very familiar with the NEC Framework Contract – either the NEC3 or NEC4, as there is little difference between the versions – I suggest you go back Section 1.1.2.12, which gives an overview of the NEC4 Framework Contract.

Some characteristics of frameworks give opportunities in terms of how documents are structured.

- As well as specifying which NEC4 contracts can be used, documents need to be developed which outline the nature of the work to be done under the framework (*framework scope*), how a *Supplier* is selected for an individual Work Order (the *selection procedure*), how the Prices are built up for that Work Order (the *quotation procedure*) and what information is to be used in the quotation (*quotation information*) and, pertinent to what follows, general commitments that apply at framework level over and above any individual contract let under it. This is the Framework Information, 'which specifies how the Parties work together' (clause 11.2(2)).
- For any individual contract, much of both the technical information and the management and administration information contained in the Scope will stay pretty much the same from contract to contract, so it makes sense to develop standard documentation that contains this standard information, yet can be improved as technology advances, better ways of working are found and errors come to light.
- In much the same way that there is the Contract Data at the 'conditions of contract' level, if you have standardised the Scope documentation, it also makes sense to have standard templates at the 'Scope' level, where the contract or package specifics can be entered. Like the standardised Scope documentation, these templates can be updated and improved.

The following example – which is only intended to illustrate what is possible when given a blank canvass – is for improvement projects on and around the M25 London orbital motorway, such as lane widening at junctions, adding in cycling lanes, improving drainage to prevent flooding, technology projects to make it smarter etc. Under the Improvements Framework's terms, each project can be let as a contract up to a value of approximately £12.5 million. Figure 2.2 shows how we structured the Framework Information, Works Information (as it was an NEC3 contract) and templates.

In Figure 2.2 you will see that the structure of the Framework Information bears more than a passing similarity to the general Works Information, which exactly matches the structure of the contract-specific Works Information. So, if people are familiar with one of the documents, they will also be familiar with the structure of another. Further, if you were to read a section of the Framework Information and then the equivalent sections of the general Works Information and the contact-specific Information, you would follow a natural progression, going into more detail each time. Take the following example.

Figure 2.2 Structure of the Framework Information, general Works Information and contract-specific Works Information under Connect Plus's COFA2 Framework (using the NEC3 form)

Framework Information	General works information	Contract specific Works information
F1000 General information	W1000 General	C1000 Scope
F2000 the M25 Community	W2000 Integrating the Contract Teams into the M25 Community	C2000 Integrating the Package Teams into the M25 Community
F3000 COFA Contractor Contributions	W3000 Not Used	C3000 Not Used
F4000 Administration, Systems, and General Management	W4000 Adminstration, Systems, and General Management	C4000 Adminstration, Systems, and General Management
F5000 Design and Technical	W5000 Design and Technical	C5000 Design and Technical
F6000 Reporting and Performance	W6000 Reporting and Performance	C6000 Reporting and Performance
F7000 Contract and Commercial	W7000 Contract and Commercial	C7000 Contract and Commercial
F8000 Health, Safety and Wellbeing	W8000 Health, Safety and Wellbeing	C8000 Health, Safety and Wellbeing
F9000 Sustainability and Environmental Requirements	W9000 Sustainability and Environmental Requirements	C9000 Sustainability and Environmental Requirements
F10000 Quality Management and Assurance	W10000 Quality Management and Assurance	C10000 Quality Management and Assurance
F11000 Security	W11000 Security	C11000 Security
F12000 Network Occupancy Management NOMs	W12000 NOMS, Road Space and Traffic Management	C12000 NOMS, Road Space and Traffic Management
F13000 Not Used	W13000 Requirements for Competition	C13000 Requirements for Competition

- F12000 is about booking road space to do work on the M25, which has to be done many months in advance, often before an individual contract is signed.
- W12000 opens with a statement that you have to comply with any relevant requirements in F12000, but it is mainly about the design of traffic management arrangements and what a *Contractor* has to do, as standard, to get on and off and occupy a part of the motorway.
- C12000 might have no entries, but if there is anything that is specific or unique to that package, in terms of getting onto the motorway and traffic management arrangements, it is entered here.

But this framework should not be viewed in isolation, because Connect Plus (which runs the M25 under a DBFO (design, build, finance and operate) contract) also has the following.

- A Renewals Framework for maintaining the M25 (as opposed to improving it), where the section headings of the Framework Information, general Works Information and package-specific templates are almost exactly the same. Further, the clauses within each of these sections are pretty much the same and have the same numbering: where they are not identical, they will cover the same topic but in a slightly different way, to reflect the nuance of the framework and the work being done. If a clause is not used in a particular framework, the number is still included, but the text states 'Not used'.

- A consultancy term services contract, based on the NEC4 Professional Service Contract, which has three main levels or layers of Scope: one for what consultants are expected to do over and above any individual assignment (equivalent to Framework Information); one for what consultants are expected to do for every Task instructed (equivalent to the general Works Information); and templates for the specifics of an individual Task. As with the Renewals Framework, the structure of these layers is pretty much identical to that of the Improvements Framework documents in Figure 2.2 and the clauses within each section. While there are many more 'not used' clauses in this consultancy contract, many clauses are mirror images of the Improvements Framework or Renewals Framework clauses. What do I mean by 'mirror images'? If the clause in the framework and the renewals contract says that the *Contractor* has to submit for acceptance a document to Connect Plus's consultant and allow 2 weeks for them turn it around, then, in the consultant's contract, the same numbered clause will say that the *Consultant* has to review and accept or not accept the document within 2 weeks.

So, the documents for each arrangement are not just coherent within themselves and capable of being improved and evolved as time goes by, they are also coherent with each other. This leads to common ways of working, processes, systems, culture etc. across Connect Plus's whole portfolio of contracts.

In addition, prior to the use of frameworks, each construction contract typically comprised about 250 pages of documentation, with the specifics hidden away[12] in a mass of text. They were prepared by the designer, who would adapt the most recent similar contract by changing the relevant words, drawing references etc. In contrast, the completed package-specific documentation, excluding drawings, is now not more than 30 pages, with many entries the same from package to package. This has the following advantages.

- While the designer still has to prepare the design, the rest of the package specifics are done by completing the template, which is a much faster and more transparent process.
- The Contractors, when pricing a package, now only have to look at the drawings and the package-specific template to find out what is unique about that package. This contrasts with having to search through and draw out the specifics from a 250 page document.
- Likewise for both the on-site consultants and *Contactors'* personnel. Under the old arrangement, a change from a previous contract was sometimes not picked up, so the wrong thing was built.

Further, all the standard documentation – the Framework Information, general Works Information and templates – can be updated and improved on an annual or biannual basis, to correct errors or to reflect advances in technology and ways of working. Changes made in one framework can then easily be carried across to another.

Or put another way, our professionals – which includes *Contractors*' staff – can focus much more on the value-adding work (to reduce risk and lower whole life costs) and less on the purely

[12] This is about the same length as *all* the contract documents for the Improvement Framework. While this excludes any package specifics, it includes a preamble to explain how the framework works and the relevant documents for employing the contractor as a consultant under the Professional Service Short Contract.

administrative work. This is what Connect Plus is seeing, which is why I have used this example. Hopefully it will have opened the reader's mind to the possibilities.

CONCLUSION TO STRUCTURING THE SCOPE AND FRAMEWORK INFORMATION

When writing anything, for me, it starts with structure. So, paying someone (whether they are in-house or external) to be the 'contract mastermind' or 'contract architect' for any medium-sized or major contract or series of contracts is, to me, an absolute no-brainer in terms of investment and return. This is because a well-structured, coherent document will be shorter overall and better written, with far fewer ambiguities and repetition etc., all of which translates into less change and hence less cost and time in design, construction and contract administration. And for repeat-order clients, these benefits are magnified, because you get a return every time that format is used.

In contrast, it never ceases to amaze me how many clients are willing to pay for expensive consultants to look over failed contracts retrospectively, when getting it right in the first place would be far cheaper. Not only would you probably pay out far less in consultant fees, but you would also avoid a lot of inefficiency in construction.

2.2.3.3 Further tips on drafting good contract documents

I start this section with three tips on structuring contract documents, to reinforce points made previously.

1. Do not 'hide' contractual and commercial information in the Scope. Instead, clearly highlight it by placing it either in the option Z (*additional conditions of contract*) clauses or in the relevant part of the Scope document. For instance, on one contract a statement to the effect that the *Contractor* would not receive payment for a completed activity until all the quality assurance documentation was complete was buried away in the testing section. This information should either be in the Payment section of the Scope or stated in an option Z clause, as appropriate. Another example is not having a headline statement of what the *Contractor* is to design, meaning the design elements are scattered throughout the Scope.
2. Group similar information together, so that any ambiguities or inconsistencies can be identified and eliminated or reconciled.

> One major client, which runs the transportation services in a global city, likes earned value. However, its requirements for this are scattered throughout not just the Works Information (as the last time I saw one of its contracts, they were still on NEC3), but also the Z clauses. It did not take much to spot both contradictory requirements and requirements for similar things (but using different words). As a result, it is very hard to know exactly what the *Contractor* and, for that matter, the *Employer* have to do. For example, if the *Contractor* has to comply with each separate requirement, it would have to submit multiple similar, but slightly different, earned value documents every month.

> When developing standard Works Information for a framework that used the NEC3 Engineering and Construction Short Contract for landscaping works, I looked at past contracts, came up with categories to group text into, and then 'cut and pasted' text into these sections, so that I could spot and remove ambiguous, conflicting and overlapping

> requirements. Despite it being late 2016, one of the contracts contained references to the 1994, 2007 and 2015 CDM Regulations. The first two were clearly out of date, and so the contract required the *Contractor* to do things that were no longer needed in law.
>
> Having prepared a good first draft, I gave a 1-day course to the *Employer's* Contractors and Consultants, with the morning being more training oriented and the afternoon explaining the new format and templates and asking for feedback. There were some minor improvements as a result, but the main reaction was 'How soon can we start using this?' Compared with their normal documents, the general Works Information was much shorter and better structured, making it easier for them to find and understand the relevant information. Further, they could immediately see that the templates would make it both much quicker to put a document together and easier to understand what was unique about any contract. I am told that this materialised in practice.

3. On a similar theme, technical experts who are writing individual sections might have their own unique requirements for testing, applicable to their technical area. However, you only want to include the genuinely unique testing requirements within a particular section, to avoid having multiple repetitions of generic requirements scattered throughout the Scope. So, if information is already stated elsewhere in the documentation, cross-reference to that statement, rather than repeat it. This might appear to go against the ECC's principle of avoiding cross-referencing, but it is advisable because
   - it avoids the risk of ambiguity being introduced over time, as engineers' and other peoples' personal foibles can tend to creep into the documentation, which add cost, but little value
   - if an error is discovered in generic company-wide information that results in a compensation event, it can be rectified in the master copy, so as not to be a compensation event in a subsequent contract
   - it leads to a better structured, shorter and more coherent document.

My remaining tips relate to writing the text that fills up the sections of the Scope document. But first, there are many good tips on pages 46 and 47 of the NEC4 ECC Guidance Notes (volume 2: *Preparing an Engineering and Construction Contract*) by the authors of the contract. In other words, if you have a copy of it, read it! As I agree with all of the points made, I will not repeat them here, but will I expand upon a couple.

4. Hidden within point 13 of the abovementioned Guidance Notes is my number one point. Whether writing option Z clauses, Framework Information or Scope documents, you should construct sentences in the following way.

   Who^{noun} does **what**^{verb} to **whom**^{noun} by **when.**

   Let me unpack this a bit.
   - Both the **nouns** will generally, but by no means always, be a character who is a 'player' in the contract, i.e. the *Client*, the *Project Manager*, the *Supervisor* or the *Contractor*. If might be that a Subcontractor is actually going to be doing the work, but, as far as the *Client* is concerned, it is the *Contractor's* obligation that the work is done. Often, an undetermined 'Other' has to do something, so you would reference them as the noun. However, more

often than not, that Other can be identified either as individual entity (i.e. a named organisation) or a category of organisations (e.g. statutory undertakers). I have used '*Client's* consultant' for actions that would otherwise fall upon the *Client* and, in the UK, you might use 'Principal Designer' for actions to comply with the CDM Regulations 2015.

- For the **verb**, clause 13.1 of the third edition ECC listed out all the types of contract communication under NEC3 *conditions of contract* (i.e. instruction, certificate, submission, proposal, record, acceptance, notification and reply). Although the fourth edition wraps all these up as just 'communications', I encourage precision in terms of how the players communicate with each other. So the verbs to use between the players in the contract are 'instructs', 'certifies', 'submits', 'proposes', 'accepts', 'notifies' and 'replies'.
  Note that 'approves' is not one of the listed types of communication, so should not be used for communications between the players. However, it might be that the *Contractor* is required to get 'approvals' from third party Others, such as statutory undertakers, local authorities etc., in which case it is correct to use the words 'approves' and 'approvals'. And there may be other actions that the players have to do, but whatever action they do, use the present active tense, not the present future (i.e. do not use 'shall').
- For the **when**, be precise and state specific timescales or events.

5. Remove subjective words and phrases, such as 'reasonable', 'in the opinion of', 'to the satisfaction of', 'automatic' etc. (i.e. adjectives and adverbs). This is because, if the *Project Manager* has to issue an instruction clarifying what, for example, a 'reasonable time' is or what 'to the *Supervisor's* reasonable satisfaction' means, it is a change to the Scope and hence a compensation event under clause 60.1(1). It might not ultimately lead to an increase in the Prices or a delay to the Completion Date, but why have the argument – you might as well as be precise at the start!
   When I see words and phrases such as these written by technical people, I ask them 'By what criteria do you judge something to be "reasonable"?' and keep on pushing until I have got all their criteria to a level of precision that I am happy with.
6. Related to point 4, get your terminology right. By this, I mean make sure any references to the 'Engineer', 'Architect', 'Contract Administrator' etc. taken from other contracts are changed to '*Project Manager*' (for the actions they have to do under the contract) or '*Supervisor*' (for quality assurance). Also, be careful about referring to some commonly used documents, which may contain contractual entities that do not exist within the ECC. When you reference these documents, it is a good idea to state that, for example, the term 'Engineer' should be read as '*Project Manager*', otherwise the legal situation could be unclear. You also need to check the use of other terminology, such as Plant and Materials (which in ECC means items to be included in the *works*) and Equipment (items used to Provide the Works). All of this is quite simple to do – having identified an erroneous term, you can do a word search for it and change it to the desired term.
7. Lastly, the ECC does not include terms equivalent to 'prime cost items' or 'provisional sums'. If when writing the Scope you are unsure about what is required, you should avoid using imprecise descriptions, and instead either
   - state the requirements as a performance specification
   - find out more information and produce a precise description
   - state clear assumptions in the Scope.

   If none of these is possible, you should omit that component of the Scope and instead include a contingency in the *Client's* overall estimate for the project to cover the item (i.e. it becomes a compensation event when introduced).

> However, it is possible to have 'provisional periods', which is my expression for the *Client* stating that the *Contractor* should allow a period of time in its programme for doing an item of work that is only vaguely specified, which, once the description is firmed up, will be introduced as a compensation event changing the Scope. Provided this time is sufficient, you would then calculate the change in direct Defined Cost plus Fee from scratch, but the time and indirect costs are already built into the *Contractor's* Prices.

To summarise this section, I offer a quote from a *Project Manager* with experience of two contracts: 'You have got to be positive … and not fuzzy on the technical requirements. … You must be precise. The information to the Contractor must be complete.'

2.2.3.4 Preparing the Site Information (and Pre-construction Information)

Clause 11.2(18) states that

> The Site Information is information which
>
> - describes the Site and its surroundings and
> - is in the documentation which the Contract Data states it is in.

So, it should describe what the *Contractor* will find when it arrives at the Site, whether in the ground or in an existing asset, such as a building to be refurbished. If the *Contractor* encounters physical conditions which would 'have such a small chance of occurring that it would have been unreasonable [for the *Contractor*] to have allowed for them' in its tender by comparison with this Site Information, then it would be a compensation event under clause 60.1(12). Generally, the more information you give to potential Contractors at tender, the less likely it is there will be a subsequent compensation event.

Clause 60.2 is also relevant, as it states

> In judging the physical conditions for the purpose of assessing a compensation event, the *Contractor* is assumed to have taken into account
>
> - the Site Information,
> - publicly available information referred to in the Site Information,
> - information obtainable from a visual inspection of the Site and
> - other information which an experienced contractor could reasonably be expected to have or to obtain.

In the UK, under the CDM Regulations 2015, the definition of 'Pre-construction Information' (PCI) includes that it is information already in the client's possession or which is reasonably obtainable by or on behalf of the client. So, if you have assembled Site Information for a contract, you would be duty-bound under UK law also to forward it as PCI (i.e. the Site Information and the PCI would pretty much replicate each other). But if the PCI contained something that the Site Information did not, then, under the last bullet point of clause 60.2, the *Contractor* is deemed to have used it to judge the physical conditions regardless.

Can you see where I am heading with this? Why have essentially the same information replicated in two different documents? Indeed, when I was putting together the frameworks for Connect Plus (see Model Structure 4, above), it seemed that its health and safety people expected to see the same in the PCI as I would expect to see in the Site Information. Consequently, I merged the documents.

In a moment, I will give the headings from my standard pro forma for Site Information, but before I do, bear in mind that Site Information is static, whereas PCI is potentially fluid. What do I mean by that? Site Information is fixed in the contract and cannot change, whereas PCI should be added to as more information comes to light. So, if the *Client* or the *Contractor* is to conduct additional investigations after signing the contract – as is very likely in an early contractor involvement (ECI) relationship, but less likely in a traditional priced contract-based arrangement – then the PCI needs to be updated under the CDM Regulations 2015. However, the Site Information cannot be updated under the contract. Consequently, my standard pro forma for Site Information

- has the title 'Site Information and initial Pre-construction Information'
- has as its first clause

  The information referenced from clauses below is both

  - Site Information as defined in the NEC4 Engineering and Construction Contract June 2017 (with amendments January 2019 and October 2020) and
  - Pre-construction Information as defined under the Construction (Design and Management) Regulations 2015.

    Documents referenced from this section may be updated or replaced as the contract progresses and, as a result, may replace or update the documents referenced, in terms of the legal status, as Pre-Construction Information. However, the documents referenced below are not replaced or updated, in terms of their contractual status, as Site Information.

I would be remiss if I do not point out that the approach above is slightly different from that put forward the NEC. The NEC view is that 'Good practice is to put the information that should be works information or site information in the relevant ECC documents and refer them from the pre-construction information document' (Patterson *et al.*, 2015). I agree with any requirements for how the *Contractor* has to comply with its obligations as Principal Contractor being in the Works Information (under NEC3) or Scope, along with any other health and safety requirements. However, having a separate PCI document which (a) might have some unique information in it (i.e. is not in the Site Information) and (b) might be updated as new information comes to light (even though the Site Information cannot be updated contractually) seems daft, as it requires the preparation of two documents which might subsequently go out of sync.

The remainder of my Site Information pro forma is divided under the following headings, which I use as reminders to those putting together the contract documents. They will apply regardless of which approach you use.

- Designer's Hazard Identification and Elimination Worksheets (required under the CDM Regulations 2015)
- Environmental Impact Assessment (EIA) (required in the UK under the Town and Country Planning (Environmental Impact Assessment) Regulations 2017)
- 'As-built' drawings of existing assets
- Surveys
- Trial pit logs and/or borehole logs
- Publicly available information that tendering Contractors are directed to look at (as per the second bullet point of clause 60.2)
- Any other relevant information (as I always like to have a miscellaneous section!).

You might also want to include reports on, for instance, hydrological and hydrographic data. Regardless, if you wish to combine the Site Information with the PCC, you can use the above as a starting point.

But with all of this, you could still be relying upon interpretation as to whether the *Contractor* finds physical conditions that have 'such a small chance of occurring that it would have been unreasonable [for the *Contractor*] to have allowed for them'. Depending on the nature of the contract – for instance, a tunnelling contract through changeable or even unknown ground conditions – this could be a major source of uncertainty. And in some contracts, it might well be the worse gambler (i.e. the tenderer with the lowest Prices) who wins the contract. This would probably then result in some sort of dispute over (a) whether it is or is not a compensation event and, if it is, (b) what the *Contractor* should or should not have allowed for at tender in order to assess the change to the Prices and Completion Date.

Consequently, for such a tunnelling contract, it might well be sensible to state in a report a set of reference conditions (perhaps with ± tolerances) relating to the ground conditions that you, as the *Client*, expect the *Contractor* to find and have priced for. Have it clearly and unambiguously stated that if the actual ground conditions are found to be outside the reference conditions (± the tolerance), it is a compensation event under clause 60.1(21) (if this approach is of interest, see Patterson and Essex, 2010).

Another example could be a stated assumption, with tolerances, on the volume of asbestos that will be found on a building refurbishment contract. If the actual volume is outside those tolerances, then it is a compensation event that can both increase and decrease the Prices. (Note that for decreased Prices, you would have to explicitly state that a compensation event for this reason can reduce the Prices.)

The advantages of doing this at tender are that the assessments of risk by different tendering Contractors will be on a more common basis and there will be fewer disputes during the contract.

2.2.3.5 Conclusion

It takes a lot of work to prepare Scope and Site Information documents that are not only 'well-structured, concise and sufficiently precise', but also technically and contractually correct. This goes against what was described years ago by one interviewee in my research as 'the unacknowledged trend of getting less for less' from consultant designers.[13]

If the *Client* does not invest in this stage, it will be left exposed – under any conditions of contract – to claims for additional time and cost, which almost always outweigh the costs of any upfront investment. However, the overall time taken to prepare Scope and Site Information can be reduced, and the quality of the documents dramatically improved, by applying some forethought. That requires an investment in those documents, which, in my view, is well worth it, as it should lead to the following benefits

[13] One reason (of many) why clients want less design for less from their designers is that they are not necessarily the best placed people to do the detailed design! With the technical complexity of many building assets, it is the technical specialists of the specialist subcontractors who should do it, using the description and constraints stated in the Scope.

- savings of time in preparing the documentation (and hence fees), tender preparation by Contractors and tender evaluation
- better tenders, as Contractors are clearer about what they have to do
- less delay and increased costs, whether compensation events or not, as everybody is clearer about what they have to do and by when.

So, especially if you are a repeat-order *Client*, are you prepared to make that investment?

2.2.4 Other issues and documents at tender

2.2.4.1 Specifying tender programme requirements

In Section 2.2.2.4, I discussed how Clients can ask for a tendering Contractor's programme to be referenced in Contract Data part two with their bid, with the intention that, at the signing of the contract, this becomes the Accepted Programme. I also mentioned how Clients were refining in their Invitation to Tender what they were asking for from Contractors in terms of what they have to show in their tender programme and how it is to be shown.

Clients asking for, at tender, a fully detailed and acceptable programme presents a couple of problems for bidders, namely

- if they are one of, say, five tendering Contractors who only have a month or so to put together a bid on a speculative basis, it is a lot of work for potentially no return
- key operational staff will not be allocated and Subcontractors, who are likely to do the majority of work, will not be appointed.

Consequently, expecting a full programme that has been prepared to the standards required for acceptance has proved unrealistic.

As a result, many Clients now ask in their Instructions to Tenderers for a detailed programme, to the standards required in the contract, for the first month or so of work, and for a less detailed programme, which is 'capable of being developed into the Accepted Programme', for the remainder of work. They recognise that it is unrealistic to expect a Contractor to submit a detailed programme for the whole contract at tender.

For the more detailed programme for the first couple of months, the focus is on

- evaluating the quality of the detail in the Contractor's programme, to establish if the Contractor can programme to the high standards required by the ECC
- understanding what the *Client, Project Manager* and Others have to do or supply within the first couple of months and whether they – the *Client* or Others – can do this, as otherwise it will result in a compensation event.

For the programme for the overall contract, the *Client* often refines this by asking for a tender programme to show

- the *starting date*, *access dates*, any Key Dates, planned Completion Dates for any *sections* of the *works* and panned Completion Date for the whole of the *works*

- critical dates when the *Client* is to supply things, such as access, Plant and Materials or Equipment, and when the *Project Manager* needs to give acceptances – this could be done by a schedule with dates
- a high-level/level 1 bar or Gantt chart showing the 30–50 main operations, how they are linked to each other, including showing the critical path, the free float between operations and terminal float at the end – depending on the size of the contract and payment option, this might include a drill down into the priced activities that relate to each operation
- a general commentary or narrative on the above that adds sufficient detail on how the Contractor proposes to do the work and, if necessary for understanding, why it has proposed to adopt a certain course of action
- a resource profile showing the general level of people and Equipment on site, as well as the presence of any specialist or expensive Equipment.

The level of detail required will vary with the size, value and complexity of the contract, but its purpose should be to give the *Client* sufficient knowledge of how the tenderer intends to do the work so that, if chosen, the *Client* is confident that the *works* will be completed to time and that the *Contractor* is not looking to play games.

Clients may evaluate the programming submission, including the detail for the first couple of months, for realism, practicality and completeness as part of Contractors' scored quality submissions. The ideal is that the *Client* has no qualms about having this programme referenced in Contract Data part two upon signing the contract, whereupon it becomes the Accepted Programme. In reality, the *Client* will almost certainly want some clarifications or more detail, as well as perhaps challenges. If these cannot be resolved by the time of signing the contract, then the fallback is to state a timescale for submission of the first programme for acceptance in Contract Data part two. This would need a little explanation in the Instructions for Tenderers. From the winning *Contractor's* perspective, this leaves them open to the sanction in clause 50.5 of a quarter of the Price for Work Done to Date being retained and the *Project Manager* imposing their assessment of any compensation events on the *Contractor* (see clauses 64.1 and 64.2), neither of which it wants. (See Section 3.5 for more on programming.)

2.2.4.2 Limiting the number of activities the Contractor can tender under option A

Under option A (priced contract with activity schedule), the *Contractor* is paid for completed activities. Consequently, in order to ensure good cash flow, Contractors have broken their activity schedule down into hundreds, and sometimes thousands, of activities. Under option A clause 31.4, the *Contractor* has to provide 'information which shows how each activity in the Activity Schedule relates to the operations on each programme submitted for acceptance'. This can then result in an unwieldy and incomprehensible programme. Two ways to stop the selected *Contractor* doing this and to have a sensible number of activities in the programme are to specify in the Instruction to Tenderers

- the maximum number of activities for payment purposes that can be included in the activity schedule or
- the minimum value of any activity in the activity schedule.

However, looking at it from the tendering Contractors' perspective, I have seen Clients allow Contractors too few activities, so that their cash flow would be restricted.

> On one contract, the *Employer* (under NEC3) only allowed 30 activities in the tendered *activity schedule* for a 1.5-year contract, which naturally had more than 200 operations in the programme. Consequently, for pricing purposes, the *Contractor* had amalgamated six or more construction operations in the programme into one activity in the *activity schedule*. Because the *Contractor* naturally wanted to ensure cash flow, each amalgamated activity only stated the completed state for one construction operation, so that it would not have to complete all six operations to be paid for the activity. The net result was that the *Employer* had much reduced visibility and insight into the *Contractor's* costs and programme. Not what was intended!

In addition, Clients need to be on the lookout for the following tricks of the trade

- the usual suspect of frontloading prices to the earlier activities
- Contractors specifying the completion of an activity (in accordance with clause 11.2(29)) as the start of the next one.

Whatever is put in the Instructions to Tenderers needs to be mirrored in the Scope document to stop the winning *Contractor* changing 'a planned method of working at its discretion', as per option A clause 55.3, and then utilising these methods of ensuring cash flow. Indeed, I would suggest that an option Z clause is written which adds an extra bullet point to option A clause 55.4 (which gives the reasons why the *Project Manager* need not accept a revised Activity Schedule) to say something like 'it does not comply with any requirements stated in the Scope'.

Almost as an aside, I will mention one thing under NEC4 that has changed from NEC3 and which I think is a bit regressive. Under both options A and C in NEC3, but only option A in NEC4, the *Contractor* has to provide 'information which shows how each activity in the Activity Schedule relates to the operations on each programme submitted for acceptance' (NEC4 ECC option A clause 31.4). The easiest way of doing this is to have a level 1 programme that shows 'operations' which break down into 'activities'. If you then add the price for each activity into the programme, you can easily generate an expenditure profile. Under option A, from a Client's perspective, on top of this can be added potential compensation events to give a high and fairly superficial base for monitoring the progress of the *works* (e.g. through earned value). You could also do this under NEC3 option C, but because that is an 'open-book' contract, you could also get an idea of whether Defined Cost plus Fee or Price for Work Done to Date is above or below the curve and hence, by extrapolation, an indication of the likely pain/gain share.

I therefore think it is regressive in terms of good project management not to have these words from option A clause 31.4 in option C and, consequently, I normally add them back in as an option Z clause.

2.2.4.3 Supply of a bill of quantities when using activity schedule options

An issue that Clients may wish to consider is whether to supply a bill of quantities to tenderers, without prejudice and separate from the formal contract documentation, for contracts let under the activity schedule options and where the *Client* is supplying the majority of the design. This is suggested in the NEC4 ECC Guidance Notes, but it does have its dangers!

On all the option A contracts that I originally researched, where the *works* had been substantially designed prior to award, the tendering Contractors used a bill of quantities for estimating. This is because quantity is a significant component of cost and partly because many Subcontractors tender on the basis of a standard rate adjusted for market conditions and project circumstances (i.e. not from first principles). However, this bill has not been supplied by the *Client*, but prepared by the individual Contractors or by a quantity surveying firm jointly commissioned by them. **Important:** if you are a Contractor thinking of sharing the costs of preparing a bill of quantities, you need to gain written permission from the *Client* before approaching other Contractors. Otherwise, there is a danger that, under anticollusion legislation, you could be fined 10% of your group's worldwide annual turnover!

The following are the arguments for the supply of a bill of quantities by the *Client* under option A or C.

- The preparation of bills of quantities by the *Client* picks up many minor errors in the specification. The general move to option A and C contracts means that Clients have lost this separate internal check of their Scope document. On one NEC3 contract, the *Employer's* quantity surveyor, despite favouring the use of activity schedules, noted that a fair few minor compensation events arose from changes to the Works Information (clause 60.1(1)), which he thought would normally have been picked up in the preparation of a bill of quantities.
- It allows Contractors' estimators to quickly determine the resource needed to prepare the tender.
- Even if using the operational method of estimating (i.e. working out costs from first principles), the estimator still has to determine the quantities that make up each activity, as quantity is a significant component of cost.
- Having the total of the quantities allows a check of the calculations.
- Without having the total, the estimator has to start from a blank sheet, so the process of taking off quantities is slower than it would be when purely checking the principal quantities.
- It transfers cost to the industry, which may be replicated five or six times by all tenderers, which Clients ultimately bear. The following figures are very 'rule of thumb', but
  - if the Contractor is checking employer-prepared quantities then, compared with a contract let under more traditional conditions of contract, it costs approximately 10% more to prepare an ECC tender due to the increased emphasis on programming
  - if, as typically happens on civil engineering contracts, the Contractor is taking off a bill of quantities that correlates with how it will construct the asset – which is often different from how the Civil Engineering Standard Method of Measurement (CESMM) does it – then it is 20–30% more
  - if, as is typical on a building contract, the Contractor is taking off a full bill in accordance with the RICS New Rules of Measurement for building works (NRM 2), then it takes 50% or more additional resource.

Interestingly, none of the interviewees in the original research put forward arguments on the grounds of tradition or because of the use of estimating databases. The above factors suggest that, if no bill of quantities is supplied, the time available to tenderers for actual estimating will be less and, because there is no built-in check on quantities, their estimates may be less accurate.

The main argument against the *Client* supplying a bill of quantities is that, regardless of however many disclaimers it puts in, the *Client* is opening itself up to potential claims in law for misrepresentation if the numbers are wrong or items have been missed out. But, because they recognise the benefits of supplying some form of bill, Clients have got around this argument in a number of ways.

- Some have given the contract documents to an independent firm of quantity surveyors a few weeks before the tender comes out and told the Contractors they can purchase the bill of quantities for a set price from this firm. As the bill does not come from the *Client*, it cannot be held liable for misrepresentation. The quantity surveying firm only has to exercise reasonable skill and care in preparation of the bill. The quid pro quo for putting the work out to the quantity surveying firm is that, if any errors in the Scope are picked up, then the *Client* is told and has time to correct them.
- Some have given an estimate of quantities (with disclaimers and not referenced as part of the contract documents), but specified reasonable variances on the quantities (e.g. 10 000 $m^3 \pm 10\%$ of earth to be moved). Provided that it is within these variances, there is not any misrepresentation. If there is misrepresentation, then the *Client* would only be liable for the amount outside of these variances. However, it is important that the central estimate is sufficiently accurate and variances reasonable as, if not, it makes the whole exercise pointless.
- Some have only given quantities for the 'big-ticket' items and not in accordance with the standard methods of measurement. Consequently, as there is no attempt at completeness, they cannot be held liable for errors of omission.

2.2.4.4 A 'starter for 10' Early Warning Register

In Contract Data part one, the *Client* is invited to state its 'starter for 10' (or 20 or 30) 'matters' that apply to the project from its perspective. In NEC3 (where the Early Warning Register was called the Risk Register), this was understandably taken to mean the *Employer's* risks and hence this was often the *Employer's* initial stab at a risk register as beloved by the risk management profession. This has often been done by reference to an appendix or schedule, which is often in the form of a spreadsheet, and I see no reason why this should change under NEC4. This spreadsheet can be to varying levels of complexity (on which I will say more in Section 3.6). A number of actions often arise from this.

- The tendering Contractors are invited in the Instructions to Tenderers to comment on the *Client's* evaluation of the probability and impact of these risks and the draft responses, including adding to the responses.
- The tendering Contractors are invited to add to the list of these risks from a construction/contractors' perspective, evaluate these new risks in terms of probability and impact, and generate draft responses to avoid or minimise the overall severity.

A way to do this is to supply a spreadsheet or spreadsheet-type format for the tendering Contractors to add to. The first tab is the *Client's* list for each tendering Contractor, to which the Contractor can add its responses, and the second tab is for the Contractor to fill in from scratch, giving its additional construction risks and responses.

These responses and additional risks are assessed as part of the Contractor's quality submission. In some instances, the Contractor-owned responses – or to be precise, the avoidance or probability-reduction measures over which the selected *Contractor* has more influence – have become contractual obligations by incorporating them into the *Contractor's* Scope. However, these responses often need to be fleshed out in detail as they are too high level to be contractually enforceable, which needs to be done with contractual precision, otherwise the situation described below can happen.

One last thing on the use of the Early Warning Register: do not, whatever you do, make it or any other risk register define the contractual ownership of risks. I have seen this attempted a number of times and the result was that

- the *Client* found that it was liable for, as compensation events, many more events than the unamended contract defines as compensation events
- risks were occurring in practice that satisfied both the descriptions of *Client's* risks and *Contractor's* risks, so neither Party was sure who was contractually liable (in the end it was more often the *Client*)
- in one instance, as it was the NEC3 contract Risk Register that was referenced, it meant that new emerging risks, notified as early warnings and added to the Risk Register by the *Project Manager*, were altering the risk allocation as the contract proceeded. Unsurprisingly, the *Project Manager* stopped adding early warnings to the Risk Register.

Once again, this illustrates the point about not adding in poorly thought through option Z (*additional conditions of contract*) clauses.

2.2.4.5 Tender seminar

Prior to or at the same time as issuing the tender documentation, some *Clients* have held half-day training sessions to the preselected tenderers. This was typically when the NEC was new to a sector or country. At these sessions, as well as outlining the technical nature of the job, they have given an overview of the ECC and explained how it differs from traditional contracts and why they are using it, emphasising the principal changes needed in practice and culture. During the presentation, they have stressed that it is their interpretation of the ECC, and that it is in the tenderers' interests to receive further and independent training. The fact that the *Client* has taken the trouble to brief the tenderers has been well received by them, and taken as a sign of good faith and commitment.

Looking more broadly, a briefing to tenderers on why you are setting up a framework or contract in a particular way highlights what is important to the *Client*, which can only lead to better tenders.

2.2.4.6 Assessing tenders and making Contractors' promises contractual obligations

The criteria for assessing Contractors' tenders should be stated in the Instructions to Tenderers.

On design and build projects, the *Client* states its requirements, which are referenced and incorporated into the contract via Contract Data part one. The Contractor's proposals (in old language) for satisfying these requirements will presumably be evaluated as part of the quality assessment. If the *Client* is happy with them, they then need to be referenced in Contract Data part two as 'Scope provided by the *Contractor* for its design' to become contractual obligations.

Note for all, but especially Contractors reading this: the effect of the second bullet point of clause 60.1(1) is that the *Clients* Scope has precedence over the *Contractor's*. In practice, this means that if there are ambiguities or inconsistencies between the two, then the *Client's* Scope 'overwrites' the *Contractor's* and the *Contractor* has to work to the, normally, higher standard. Therefore, it is the winning *Contractor* which bears the cost of an ambiguity or inconsistency, and so it needs to thoroughly check its own proposals at tender.

However, for Clients it also worth noting that, if your Scope is silent on something and the *Contractor* writes something in its Scope – such as an obligation on the *Client* – which is incorporated into the contract via Contract Data part two, then it becomes a contractual obligation on the *Client*. Likewise, if the *Client* has given a performance or functional specification and the *Contractor* has said in its Scope that it will use a certain product to meet it, then to change the product would be a compensation event (provided the product specified does meet the performance or functional specification). Consequently, the *Client* still needs to properly review the *Contractor's* Scope before it becomes incorporated into the contract!

I have already mentioned the assessment of the programme and risk register as part of Contractors' quality submissions. Clients have also evaluated other aspects, such as the training and competence of the key people, the systems to be used by the Contractor to administer and manage the contract etc., much of which is covered later on in this book. If the *Client* wishes these answers to become contractual commitments, then they need to be referenced in Contract Data part two as part of the *Contractor's* Scope. (Note that I do not say 'Scope provided by the *Contractor* for its design' as per the contract, as this would only be for the end state of the *works*. Instead, I modify the contract so that where 'Scope provided by the *Contractor* for its design' is mentioned it is changed to just 'Scope provided by the *Contractor*'. I then modify Contract Data part two to have two entries, one for 'Scope provided by the *Contractor* for its design' and one for 'Scope provided by the *Contractor* for how it is to Provide the Works'. These statements then become contractual obligations, whereas if they are exclusively in the programme then they are not.)

Lastly, whichever main payment option you intend to let your contract under, the Contractors are invited to tender various percentages and rates. My early consultancy experience clearly showed that these percentages and rates come down if Contractors know they are going to be evaluated as part of their tender submission. And my subsequent experience of helping Contractors with bids clearly shows that they, shall we say, 'optimise' their bid based on how these percentages and rates are evaluated. How to do this is described later (see Section 2.4.1), as we need to understand the two Schedules of Cost Components first, but for the time being, just make sure you do it!

As stated before, statements describing how all the information discussed is going to be used to select the winning *Contractor* should, as a matter of good practice, be made in the Invitation to Tender. If you are subject to procurement legislation, it's an absolute must.

2.3. Preparing the tender

The principal differences to conventional contracts in preparing a tender under the ECC are

- accuracy
- programme and Activity Schedule
- understanding the Schedule of Cost Components
- completing the other parts of Contract Data part two.

2.3.1 Accuracy of tender

Numerous Project Managers and Contractors, and quantity surveyors acting on their behalf, have stated in the light of their experiences that it is much harder 'to make up' for a low tender under the ECC. This is for a number of reasons.

- The base from which change is evaluated – the regularly updated programme, which includes principal resources and method statements – is more transparent and hopefully agreed and up to date.
- The rigour of the whole compensation event procedure, including the use of the Schedule of Cost Components, which gives headings for which the *Contractor* can charge additional sums, means that the justification for additional costs and time is more rigorous and transparent.
- The more 'real-time' assessment of compensation events prevents the *Contractor* from waiting until the end of the contract, doing a cost/expenditure reconciliation and then claiming an inflated amount from the *Client* in expectation of receiving considerably less – the 'rule of thirds' as one *Client* referred to it.

As one Contractor's contracts manager stated: 'you can play the same games under the NEC, but you have to be a lot cleverer to play them, the rewards for playing them are less and, in playing them, you are working against the ethos of the contract'. Asked what the consequences of being found out were, he replied that 'you would probably get less payment and at a later date. Consequently, I do not think it is worth playing them'. The implication is that tendering Contractors need to have greater confidence in the accuracy of their tender and, in order to have this, need to apply greater rigour to the preparation of it.

For Clients, an advantage is that the tender Price should be more realistic and closer to the out-turn cost, which will aid financial projections and budgeting. And this is what a number of clients, both in my original research search sample and subsequently, have reported: if anything, tender prices go up – but the final amount paid goes up less by comparison with other standard forms of contract.

2.3.2 Programme and the *activity schedule*

The contractual sanctions for not having an Accepted Programme were outlined in Section 1.5.4 and suggestions for the *Client* specifying more precisely what it wants to see in the Accepted Programme and how it wants it to be presented were given in Section 2.2.4.1.

Under option A (priced contract with activity schedule), payment is linked to completion of activities or groups of activities specified in the Activity Schedule. It is essential, therefore, for the items listed in the tendered *activity schedule* to correspond with or relate to the operations in the Accepted Programme, as per option A clause 31.4. A benefit of this is that monitoring and payment become one and the same procedure because the *Contractor* is paid for completing an activity/operation. Under option C (target contract with activity schedule), the *Contractor* is paid Defined Cost plus a Fee, and during the contract planned expenditure can be compared to actual expenditure as a means of evaluating progress. Under both options, the allocation of a sum to an activity which accurately represents the true costs plus margin of doing that work aids the assessment of the change in costs due to any compensation event. As one *Contractor's* quantity surveyor stated: 'Evaluation at tender stage ... is different, because you have to think in terms of the activity schedule; ... you have to really think about preparing the programme at tender time, rather than necessarily when the contract is awarded'.

When tendering under an activity schedule option, the *Contractor* needs to establish the main activities and their quantities from the programme. Therefore, it needs to prepare a programme in sufficient detail to do this. Having derived the main activities and their quantities, some longer duration activities will need to be further split into smaller activities to ensure cash flow. Especially under option A, the priced contract, tendering Contractors should ensure the following.

- All the work that is in the Scope is included in the *activity schedule*, otherwise they will not be paid for it (under option C, the target Prices will be artificially low).

> On several NEC3 contracts I know about, the *Contractors* have lost their profit margin through omitting to price work because they did not take it off from the Works Information. The worst case I have heard of was where one *Contractor* forgot to price up about £600 000 worth of work: the contract should have been tendered at £1.3 million, but was instead awarded at £700 000!

- The **finished state** of each activity is clearly identifiable, so there is a minimum of discussion over whether the activity is complete and hence the *Contractor's* entitlement to payment. Under the legal *contra proferentem* rule, if there is ambiguity or inconsistency, then it is interpreted on the balance of probability in favour of the Party that did not create it, which in this case is the *Client*. Taking the example of an activity vaguely specified as 'completion of first fit mechanical and electrical services' in an area, the *Project Manager* might have a much expanded scope of what 'first fit' is compared with the *Contractor* and, provided their interpretation is not too extreme, it is the *Project Manager's* interpretation which would legally stand. So, in this example, the *Contractor* would be wise to break the activity down into separate activities for different disciplines (e.g. electrical, ventilation, IT). An example from a road project might be 'Chainage 0 to 100, Left Hand Side, excavate drainage down to formation level', with the following activity being 'Chainage 0 to 100, Left Hand Side, install drainage, backfill and compact to final level minus 300 mm'. For Contractors, I have heard of a trick that has been used a number of times: specify the end of an activity as the start of the subsequent activity. For Clients, you may wish to look out for this one and outlaw it!
- Operations of a long duration are subdivided into shorter activities which, at a maximum, can be completed within any given *assessment interval* – typically monthly or four weekly. An obvious example is maintaining the site establishment: if this is treated as a single contract-long activity, the *Contractor* would only be paid at the end of the contract. If broken down into monthly activities, the *Contractor* would get paid each month. However, if a month-long activity of physical work overruns the assessment date by 1 day then, by a strict interpretation of the contract, the *Project Manager* should not certify payment for it that month, leading to a cash flow risk for the *Contractor*. It makes more sense to break the activity down to a weekly or fortnightly basis, in line with any internal reporting structures. This would allow the *Contractor* to get any delayed operation/activity back on track by the end of the *assessment interval* for payment or, failing that, get at least some payment for the operation.

The other advantage of doing this is that when delays occur – which could be caused by compensation events and/or *Contractor*-owned delays – the *Contractor* is still contractually entitled to be paid for, say, three out of four of its existing activities.
Of course, some long-duration operations do not lend themselves easily to subdivision, as there is no clear cut-off point, and payment by percentage completion may be appropriate. The danger of this is that you then may have disagreements about whether a 3-month operation is, say, 35% complete at the end of the first month as per the Activity Schedule description, or 32%, in which case the *Project Manager* does not want to certify payment. What happens is that the Parties then go out and measure the work, reverting to the bill of quantities approach, which is administratively burdensome for little additional value.
In reality, the majority of Project Managers grant Contractors some leeway if this is being reciprocated. In this instance, it might be that, provided the *Contractor* is predominantly on time and has finished the activity by the time the *Project Manager* has to certify (which in the unamended ECC is 1 week after the assessment date), the *Project Manager* will certify payment for that activity. And I endorse this common-sense approach.
However, when the programme becomes seriously out of sync with the current Activity Schedule, it is time to revise it under clauses 55.3 and 55.4.

- Finishing operations are separated from the associated main operation. For instance, on a tunnelling contract, the *Contractor* divided the tunnel into segments, each of which had additional activities such as pointing, fitting the wedge block keys, initial clean, repair of Defects and final clean.

> I heard of one building site where the *Contractor* had not done this, and planned to reverse out of the building doing finishing activities. As a result, the *Project Manager* was refusing to certify rooms/activities as complete for payment. This had a serious effect on the *Contractor's* cash flow!

- Subcontractors' activities mesh with the *Contractor's* own, so that what the *Contractor* is paying out, it is also receiving.
- Delivery of major items of Plant and Materials to the Working Areas are specified as an activity when they will not promptly become part of the *works*. This may include for fabrication outside of the Working Areas where the *Contractor* will be making staged payments.

However, there are some dangers of over extensive subdivision of *activity schedules*.

- As mentioned before (see Section 2.2.4.2), under option A clause 31.4, the *Contractor* has to provide 'information which shows how each activity in the Activity Schedule relates to the operations on each programme submitted for acceptance'. Therefore, if each activity is a bar on a programme, then the Accepted Programme becomes over-complicated and, as a consequence, hard to follow and unwieldy. This has been a common problem.

> As an example, I was doing some work at a Contractor's premises and was passing a meeting room where the Contractor's quantity surveyor for a £40 million contract, which was in the final stages of negotiation with the *Client*, was having an argument with the proposed mechanical and electrical (M&E) Subcontractor. As the quantity surveyor knew

me, I was invited in. As the project was for a new-build physics laboratory, the M&E component was worth about £25 million, which the Subcontractor wanted divided into 4000 activities for an 18-month subcontract duration. That would have given about 220 activities to be measured per month, which neither of them wanted. The Subcontractor's reason was that, on a previous contract with another *Contractor*, the *Client's Project Manager* had been absolutely rigid in only certifying for payment activities that were 100% completed, and hence so had the *Contractor*. This had severely impacted their cash flow and they wanted to avoid a repeat. I suggested that rather than arguing among themselves for the same thing, perhaps it would be better to have a chat with the *Project Manager* for this contract to see how rigidly he was going to enforce the contract or whether he was going to apply common sense and a bit of give and take.

They did this and the *Client's Project Manager* said he would and so the *activity schedule* was simplified down to fewer activities. I pointed out that if the *Project Manager* did start being overly rigid, then provided the Subcontractor – and hence main *Contractor* – changed 'a planned method of working at its discretion' (as per NEC4 option A clause 55.4), the Subcontractor – and hence main *Contractor* – would be allowed to revise the Activity Schedule. With no constraints in the actual contract on the number of activities allowed, they could revert back to their 4000-odd activities.

- Under all the main options, the *Contractor* has to show the order and timings of operations, with 'for each operation, a statement of how the *Contractor* plans to do the work identifying principal Equipment and other resources which will be used' (see last but one main bullet point in clause 31.2). If each activity is an 'operation', it therefore requires an individual statement of method and principal resource schedule, which can create a great deal of paperwork.
- The simplicity is lost and there tends to be a reversion towards the bill of quantities approach, where everything is measured.

For these reasons, I suggested that, conceptually, an operation is made up of activities, as this will lead to fewer statements of method and resources. It also lends itself to the concept of multi-level programming (see Section 3.5.2), where the operations are shown in the highest level programme and activities in the lower ones. If the *Contractor* is submitting the programme in electronic format, it may simply be a matter of double-clicking on an 'operation' bar, which then expands out to show the 'activities' that make up that operation. An alternative is that a much smaller Gantt or bar chart of activities accompanies each operation's statement for how the *Contractor* plans to do the work.

My last point is a word of warning to Contractors and, for that matter, Clients who cause this game to be played and is specific to option A (priced contract with activity schedule). If you look at the wording of the contract in option A clause 11.2(29), it says that

> The Price for Work Done to Date is total of the Prices for
>
> - each group of completed activities and
> - each completed activity which is not in a group.
>
> A completed activity is one without notified Defects the correction of which will delay following work.

I know of no Contractor who wants to be paid for a **group** of completed activities: they want to get the cash in! But in terms of having a logical cost breakdown structure that is reflected in the Activity Schedule – for instance, grouping together all activities to do with digging out foundations for a building – it does make sense to group, i.e. to have activities as a subset of operations.

However, on a few rare occasions, the *Project Manager* has started, after contract award, to group activities together for payment purposes using the referencing system of the *Contractor*. So the natural response is to construct the activity schedule so that there is no natural grouping, either in description or numbering reference, although it does 'relate' to the Accepted Programme.[14] Both at tender and during the contract, this reduces the ease of relating the Activity Schedule to the programme, which undermines a number of advantages of using activity schedules (see the start of this subsection). Consequently, I do not recommend Clients or their Project Managers do this. In fact, it may well be worth Clients stating they will not do this in their Invitation to Tenderers – or even deleting the references to 'group' in clause 11.2(29) – so that tendering Contractors do not feel the need to play these games in order to protect their interests.

When evaluating tenders, Clients should also bear these points in mind. For instance, asking for simplification of the *activity schedule* prior to award. However, Clients should resist the temptation of specifying activities for all but the simplest of projects (in which case they should be using the Short Contract!). If they do, then the consequences might be that

- the *activity schedule* will not mesh with the Accepted Programme, because the *Contractor's* operations in its programme are not divided up and/or assembled in a way that directly correlates with the *activity schedule* proposed by the *Client*
- the *Contractor* will alter its programme to optimise cash flow rather than the construction process. I know this happens on other forms of contract, such as GC Works, where employers specify milestones.

In addition, the *Client* will tend to specify far fewer activities than Contractors would, again affecting Contractors' potential cash flow and encouraging games (caused by the *Client*) to secure better cash flow.

All of this undermines transparency of both cost and programme, which have consequential effects on project management generally and ease of assessing and agreeing compensation events.

> On one contract in the original research sample, the *Project Manager* did specify the activities for which the *Contractor* would receive payment. However, these did not match how the *Contractor* intended to carry out the work, so the *Contractor* distributed the costs of miscellaneous and unspecified operations across various activities. Consequently, the Accepted Programme contained resources and method statements for operations that did not relate to the activities in the *activity schedule*. Although it is the change in resources from those in the Accepted Programme that should be used to assess the financial effects of compensation events, this led to severe problems in establishing the base costs when evaluating compensation events.

[14] If you are the *Contractor* and find that the *Project Manager* is doing this, then the countermeasure is in option A clause 55.3 ('If the *Contractor* changes a planned method of working at its discretion'), which gives you the excuse to revise your Activity Schedule as I have outlined (i.e. without grouping).

Unsurprisingly, the *Project Manager* stated that in future he would want the *Contractor* to specify the activities.

On another NEC3 contract, the *Employer* went part way towards this, asking for activities to be grouped under broad headings for tender comparison purposes. The tenderers replied that they had some activities that did not fit into any of the headings specified by the *Employer*. After some discussion, a 'miscellaneous' heading was introduced, which everybody seemed happy with.

Finally, just so you do not think these potential problems are exclusive to activity schedules: the danger of the *Client* dividing the work up into segments that do not match how the tendering Contractors would do the work also applies to bills of quantities. This was identified on one contract within the research sample, and has consistently been highlighted in research reports as a problem with bills of quantities under any form of contract (see, for example, NPWC/NBCC Joint Working Party, 1990).

2.3.3 Understanding the Schedule of Cost Components

2.3.3.1 Use and application

First of all, a bit of background: my ex-Professor, John Perry, who co-instigated the development of the NEC, did a lot of research in the late 1970s and early 1980s into target and cost-reimbursable contracts, culminating in the publication of a highly regarded CIRIA Report (Perry and Thompson, 1982). In this research, it was found that, when there was a schedule of cost headings against which a contractor was directly reimbursed, with anything else deemed to be in a fee, there was a lot less disagreement about what contractors should be reimbursed. Hence one of the uses of the full Schedule of Cost Components (full SoCC), under options C, D and E, is for reimbursing the *Contractor* its own predominantly site-based costs, with Subcontractors' costs being paid as per their subcontracts. On top of this Defined Cost, a single tendered *fee percentage* is applied, which equates to the *Contractor's* margin (more on this later, in Section 2.3.4).

John Perry and Martin Barnes, the other co-instigator of NEC, then extended this logic and said something like 'Aren't there similar problems under priced contracts when agreeing the cost of a claim or variation?', as in a contractor is paid a 'fair amount', sometimes and sometimes not including an unagreed amount for profit. 'So', they continued 'why not have a defined schedule of allowable costs with a tendered percentage for margin added on for our equivalent of claims and variations', and this is what they did for compensation events.

As a result, when a compensation event occurs

- under options C, D and E, another use of the full SoCC is to assess the change in target Prices by using it to build up the change in predominantly site-based costs to which the same *fee percentage* is applied
- under options A and B (the priced-based options), the Short Schedule of Cost Components (SSoCC) is similarly used. Essentially, this contains more tendered rates and round-ups, so is more suitable for less complicated change which, in theory, should happen when a price-based contract strategy is selected.

You may notice that I have not mentioned option F (management contract). This is because a Defined Cost is what the *Contractor* pays its Subcontractors in accordance with their subcontracts, with all internal costs (unless specifically identified as *prices* in the Contract Data) being covered by the *fee percentage*. In other words, there is no equivalent schedule in option F.

With each edition of NEC, the two Schedules of Cost Components have been progressively revised and, in my opinion, improved. This trend was continued in NEC4. So, for those who are coming from an NEC3 background and are already familiar with the two Schedules of Cost Components, here is a very quick summary of key changes.

- There is now only one *fee percentage*, not two.
- Subcontractor costs are now included as part of each Schedule of Cost Components, rather than directly in the definition of Defined Cost. Consequently, there are now eight main headings, compared with seven in NEC3. The eight sections are: People, Equipment, Plant and Materials, Subcontractors, Charges (equivalent to Preliminaries), Manufacture and fabrication (outside the Working Areas), Design (again outside the Working Areas) and Insurance.
- Under options A and B, the *Contractor* is now reimbursed what it pays its Subcontractors in accordance with their subcontracts and then, as with other components of Defined Cost, has its *fee percentage* applied on top. This was the case under NEC3 for options C, D and E, and still is under NEC4. It means that under options A and B, the *Contractor* does not have to produce the 'real' cost of Subcontractor staff and is no longer at risk of losing money through compensation events if a Subcontractor's *fee percentage* is higher than its own.
- Along with there being only one *fee percentage* applied to all components of Defined Cost, there are far fewer round-up percentages within each Schedule[15]: the overhead percentages for people, design, Working Areas and manufacture or fabrication have all disappeared, leaving only the percentage for adjustment for Equipment in the SSoCC. This simplifies tender assessment and massively reduces the opportunities for Contractors to play games at tender – which is bad news for me as a trainer and consultant, but good news for the industry. However, as a result, the costs of small items of Equipment previous covered in these tendered percentages need to be agreed by the *Contractor* and *Project Manager* – even the site agent's 4-year-old laptop!
- Lastly, the SSoCC is no longer the Shorter Schedule of Cost Components, but the Short Schedule of Cost Components, i.e. they have shortened 'Shorter' to 'Short'.

For compensation events, regardless of which main option is being used, changes in the Defined Cost are, in simple terms, calculated in the following way.

1. You evaluate the change in method, if any, from that in the Accepted Programme.
2. From this, the changes in resources compared with those stated in the Accepted Programme are derived. These can be the same resources but used for longer, or new resources brought in specially for the compensation event.

[15] The original reason for having these round-up percentages was to simplify the building up of costs, e.g. not having to calculate or pay the exact cost of toilet roll for a week's extra time on Site. However, as all this data is now usually in and assembled by IT, round-ups no longer serve their purpose. Indeed, they can cause more hassle than they save, as the *Contractor's* quantity surveyor has to take the downloaded output from the IT system and then remove any items covered by the percentages. Further, I have been party to numerous contractual arguments over whether an item was paid for directly or in a percentage (and hence not paid for directly), with significant sums hanging on the final interpretation.

3. You ask if those resources are allowable and directly reimbursable in one of the eight sections of the full SoCC or SSoCC, whichever you are using. If they are, you build up the cost of those resources as stated in the relevant Schedule: either (a) from first principles, (b) by using a tendered rate, (c) by using a rate in a referenced standard industry publication or (d) as calculated in accordance with a Subcontractor's subcontract.
4. For those items for which have a rate, you multiply the change in each unit of resource due to the compensation event by the rate for that resource.
5. You add the costs for the individual resources, as calculated above, together.

Lastly, to calculate the change in the Prices, you apply the tendered *fee percentage* to this change in Defined Cost. The same principles apply whether it is a forecast or is based on records, except that for a forecast the *Contractor* is entitled to add in an allowance for risk (see clause 63.8).

Before I go into the detail of the full SoCC and the SSoCC, and explain some anomalies (of which there are few, and these are where the games can arise), I draw your attention to clause 52.1, which is as follows.

> All the *Contractor's* costs which are not included in Defined Cost are treated as included in the Fee. Defined Cost includes only amounts calculated using rates and percentages stated in the Contract Data and other amounts at open market or competitively tendered prices with deductions for all discounts, rebates and taxes which can be recovered.

I particularly draw your attention to the word 'other' in the second sentence. This has quite a few ramifications as it effectively gives precedence to the first part of the sentence.

- If a rate (or percentage) is stated in Contract Data part two as part of the *Contractor's* offer, then, regardless of whether it is above or below market rate, that is what is used. I will draw attention to this a number of times in the following text, but the result is that bidders try to get as many high rates and percentages as possible included in the contract by reference from Contract Data part two, especially if they are not part of any tender assessment. This is then a source of 'hidden' profit over and above that provided by the *fee percentage*.
- If the *Contractor* has gone out and received competitive quotes for, say, an Equipment hire schedule at a national level, then these rates are at 'open market or competitively tendered prices', even if they are expensive at today's rates, say in the light of a recession (although the *Contractor* would have been foolish not to renegotiate them) or local conditions.
- If internal Equipment is being used without any competition (which the *Contractor* could do as a means gaining additional group level profit), then the *Project Manager* can market test these rates and only pay the lesser 'market rate' amount.
- Any market testing would have to be on a like-for-like basis (e.g. Equipment is of the same age, with similar maintenance arrangements if it breaks down).

Note also that Clients are entitled to head office discounts based on, for example, the total volume of group purchases.

As stated previously, there are two Schedules of Cost Components: the Schedule of Cost Components (referred to before and hereafter as the full SoCC) and the Short Schedule of Cost Components (referred to as the SSoCC). Ideally, you will have a copy of the contract to open as this will help you to understand what I am going to write about.

2.3.3.2 The (full) Schedule of Cost Components

PREAMBLE

NEC4 is absolutely strict about the full SoCC being used only with options C, D and E. In NEC3, the *Project Manager* and *Contractor* could also use, by agreement, the SSoCC with options C, D and E, but only in the assessment of compensation events. This choice has been removed in NEC4. The contract does not allow for double counting and, even if the cost is incurred in the Works Areas, it has to be incurred to Provide the Works.

> Years ago, when the first part of the Channel Tunnel Rail Link was being constructed through Kent, a number of Contractors under option C contracts were apparently using their large site offices as mini-regional offices, but charging all staff resources to the contract. This is not allowed and doing so deliberately is fraud – a criminal offence.

1. PEOPLE

The two bullet points state the criteria for people who can be included in the calculation of Defined Cost. Note that under the first bullet point, people working from head office or home can be counted as a Defined Cost and hence their cost is directly reimbursable, rather than being covered by the Fee. However, their normal place of work needs to be within the Working Areas (e.g. more than 2.5 days per week for a 5-day working week).

Historically, a common-sense agreement has been reached to avoid – as in some cases – personnel doing over 3 hours of travel each way plus overnight stays to be counted as a Defined Cost when, in this era of instant communication, they can just as well do the work from head office (if nearer to their home) or their home. In order to be strictly in accordance with the contract, this has sometimes meant a room or an area of a room at the *Contractor's* head offices has been designated as part of the Working Areas and only used for work on that contract.

More recently, in the light of the COVID-19 pandemic, many professional staff's normal place of working is now at home, which implies that to be a Defined Cost, their individual homes need to be classified as part of the Working Areas. During the contract, the *Contractor* can propose this under clause 16.3. However, I can imagine an officious *Project Manager* might well refuse on either of the two reasons in clause 16.3: for instance, 'necessary' in the first bullet point implies a need, not a desirability, for working from home; for the second bullet point, if a person's partner is also working from home, then logic dictates that the house is being 'used for work not in the contract'.

So a tendering Contractor, wanting to reduce site establishment costs, staff travelling time, environmental impact and, in the light of COVID-19, danger to its staff, should ideally specify an area of its head office or 'homes that are used by staff to do work in the contract' as part of the Working Areas in Contract Data part two, to avoid the risk of an officious *Project Manager* denying these as additions.

In stark contrast to my officious *Project Manager* above, many Clients are now actively encouraging working from home for the reasons given above. I suggest this is indicated at tender.

Under the full SoCC, the cost of employed people is worked out from first principles, allowing for all the factors listed in People 11, 12 and 13 and using an hourly or daily figure for each category of person determined. This has involved, for example, the *Client*-appointed accountant locking themself in the *Contractor's* accountant's office and calculating/checking these figures alongside the *Contractor's* accountant. While professional rates are usually calculated individually, more often than not the rates for manual staff are categorised (e.g. unskilled, semi-skilled and skilled labourer), and employees are allocated into these groups.

People 14 effectively covers labour-only people, whether manual or professional (e.g. setting-out staff). Just to reinforce a previous point, these amounts can be market tested under clause 52.1. However, these people are only reimbursed when they are working in the Working Areas. So, an agency quantity surveyor working from home a day or two a week does not qualify as Defined Cost, unless their home is made part of the Working Areas.

2. EQUIPMENT

Given the definition of Equipment, this not only includes what is commonly referred to as (construction) plant (e.g. excavators, generators etc.), but also site accommodation and temporary works. In shorthand, the *Contractor* is now reimbursed as follows.

21. For externally hired or rented Equipment, at cost.
22. For internally owned Equipment, at equivalent market rates, which is not necessarily what the contract, as a cost centre, is charged by the internal plant hire company.
23. For Equipment purchased specially for the contract, the average change in price (purchase minus estimated selling) per *assessment interval* of use, with an adjustment made at the end of the contract for the real selling or open-market price, plus on-site time-related costs.
    In reality, Contractors often want to keep specialist Equipment, as it gives them a commercial advantage going forward, so agree to use an independent assessor to decide a market-rate sale value.
    The on-site time-related costs are tendered in Contract Data part two, so the *Client* either needs to be happy with the amounts entered prior to signing the contract or modify the contract to assess this post-contract award.
24. For special Equipment, the rate listed in the Contract Data. This is intended for items of Equipment for which it is difficult to establish a market rate, such as very bespoke barges and offshore equipment. Clients need to be careful here, as Contractors might try to list out as much Equipment as they can, with rates that are above the market rates, in order to generate 'hidden' profit over above that provided by the Fee. In accordance with clause 52.1, they are then paid these amounts. The selection rules need to take account of or outlaw this game. Often Clients do this by listing out the special Equipment they want priced in Contract Data part two, with the Instructions to Tenderers precluding bidders from entering more items.
25. Cost of consumables, such as fuel, spare parts etc. If the *Contractor* will be using Equipment that it has previously used on another site, beware of a trick that is sometimes used. Instead of giving the Equipment a service back at its own yard, the *Contractor* will service the equipment after it has brought it into the Working Areas and then charge the *Client* for, say, new tyres and other worn parts – costs that have actually been incurred on another contract. Consequently, it is debatable that these costs were only incurred to Provide the Works, especially if the *Contractor* also intends to service the Equipment before it is moved off site and tries to charge the *Client* again.
26. Transportation and setting-up costs.
27. Payments for materials used to construct or fabricate temporary works.

3. PLANT AND MATERIALS

These are reimbursed at cost.

4. SUBCONTRACTORS

This section is new to the NEC4 full SoCC (and the SSoCC). In NEC3, Subcontractor costs were part of the definition of Defined Cost. The *Contractor* is reimbursed what it pays its Subcontractors in accordance with its subcontracts.

However, this creates an incentive for Contractors to turn a blind eye, or even actively encourage, high Subcontractor rates and *fee percentages*, particularly if under a price-based subcontract option (which they normally are). Why is this? Because under a price-based option, the subcontract *fee percentage* and rates only come into play when there are compensation events. So, on the presumption that all subcontract compensation events can be passed on to the *Client* under the main contract, high subcontract rates lead to high Defined Cost for compensation events, to which a high subcontract *fee percentage* is applied, to which the *Contractor's fee percentage* is applied. So, even though the *Contractor's fee percentage* may be reasonable, what it is applied to is not, giving the *Contractor* a higher margin.

For this reason, when preparing option C, D and E contracts I add (as an option Z clause) a new clause 26.5, which states that the *Contractor* is to submit Subcontractor rates and percentages used for compensation events to the *Project Manager* for acceptance, with a reason for not accepting them being that they are above market rates or will not provide value for money to the *Client*. I know this is a bit subjective, so any better wording would be gratefully received!

5. CHARGES

Under NEC3, charges could be thought of as site preliminaries, excluding accommodation (which is counted as Equipment). This is not quite true under NEC4. The items listed here are paid at cost, essentially as per NEC3 (except the direct cost of telephone and internet has been added, while 'specialist services' (i.e. a Subcontractor) has been taken out). However, the percentage for Working Areas overheads has disappeared, with the items that were in here generally being charged for under People or Equipment.

From an open-book accounting perspective, this is good. The data produced by the *Contractor's* accounting system can be used more directly as Defined Cost, without having to take out the items that were previously covered in the ECC3 by the percentage for Working Area overheads and then applying the same percentage to People costs.

However, for calculating the cost of a compensation event, particularly when it extends the time on site, it means that there are more minutiae to add in and consider. To avoid having to do this for every compensation event that extends the time, I suggest that, early on in the contract, the Parties' quantity surveyors sit down together and agree a daily and weekly on-site cost for a given level of site resources in terms of People, Equipment and Charges. This is periodically updated when there is an increase or decrease in resources. Regardless, it saves time during the contract.

6. MANUFACTURE AND FABRICATION AND 7. DESIGN

Both of these headings require the *Contractor* to submit hourly rates for various categories of employee for work that is carried outside of the Working Areas. These should be the direct costs of employing these people, calculated in a way akin to that under People, with a manufacturing facility, fabrication yard or design office overhead applied on top to give all-in rates. Under NEC3, this overhead was given as a percentage and applied to the rates, but it always seemed to be tendered at zero, with the rate given covering these costs. Consequently, in NEC4, these additional percentages have been removed as they are simply not necessary.

Note that these rates are only for the *Contractor's* own people and not for Subcontractors' people. On complex projects, I suggest that the *Contractor* is allowed, even encouraged, to submit different rates for different factories or design centres or for different types of work, rather than one set of rates, as this better reflects its true costs (i.e. a designer based in India costs the *Contractor* a lot less than one based in London).

You can see the game here, tender on the basis of the London-based designer, but use one of equivalent grade in India. However, this will make tender evaluation harder for the *Client*. To address this, albeit for an NEC4 Professional Service Contract, for one of my clients we effectively had a duplicated schedule of rates: one for services delivered in the UK and one for services delivered from abroad.

8. INSURANCE

This covers what is not paid out by the insurance company given the level of insurance the *Contractor* should have taken out. To give an example, say that the stated excess/deductible for insurance in Contract Data part one that the *Contractor* should have is £10 000. In reality, its company-wide insurance policy excess is £50 000 and a £100 000 insurance event occurs. Because we are working under a cost-based payment option (C, D or E), the *Client* is liable for the first £10 000 (under a target-cost contract this would come out of the *Contractor's* pain/gain share). In theory, the *Contractor's* head office is self-insuring for the next £40 000 and will pay the site this amount. That's the theory! The next £50 000 up to the £100 000 will come from the insurance company.

Note that the cost of taking out insurances – the insurance premiums – is not a Defined Cost and therefore needs to be included in the *fee percentage*.

2.3.3.3 The Short Schedule of Cost Components

PREAMBLE

NEC4 is explicit about the SoCC being used only with options A and B and only for compensation events. In NEC3, the *Project Manager* and *Contractor* could also use, by agreement, the SSoCC with options C, D and E, but only in the assessment of compensation events. This choice has been removed in NEC4. The contract does not allow for double counting and, even if the cost is incurred in the Works Areas, it has to be incurred to Provide the Works.

Before my comments on the eight sections, a reminder that the big difference for Contractors, compared with NEC3, is that they no longer need to try and contain Subcontractor rates and the single remaining *fee percentage* within their own. This is because they are now paid their *fee percentage* on top of whatever they pay their Subcontractors in accordance with the subcontract terms.

1. PEOPLE

The first two bullet points match the first two in the full SoCC and the last bullet point is equivalent to People 14, so the same comments apply.

A difference between NEC3 and NEC4 is that we now have tendered People Rates (which, having capital initials, is a defined term). Under NEC3, people costs were worked out from scratch, but the contract gave no guidance on how. Option A and B clause 11.2(28) defines the People Rate as being 'the *people rates* unless later changed in accordance with this contract'. As *people rates* is in italics, these are stated in the Contract Data, in this case by the *Contractor*. Consequently, if incorporated into the contract, the *Client* has to pay these rates regardless of whether they are on the high side or not. So the rates need some sort of evaluation pre-contract, which I discuss later on (see Section 2.4.1.1). The 'later changed' bit could come into play if, for example, option X2 (changes in the law) is specified and (in the UK) the level of employers' National Insurance is changed. Additionally, under options A and B clause 63.16, a new rate can be agreed for a category or individual where there is no rate. Ideally, this is by agreement, but if there is no agreement, the *Project Manager* 'assesses the rate'. However, no criteria for this assessment are given with the SSoCC, so I would assume it is 'at open market or competitively tendered prices with deductions for all discounts, rebates and taxes which can be recovered', as per the last half-sentence of clause 52.1, and not as per how people costs are built up under the full SoCC (because it is not part of an option A or B contract).

2. EQUIPMENT

In the SSoCC, the default mechanism for evaluating the cost of Equipment is that the Contractor puts forward a published list of Equipment, such as the Federation of Civil Engineering Contractors' rates, and then applies the 'percentage adjustment for listed Equipment stated in the Contract Data [part two]' to the rates in this list. This percentage has generally been calculated in the following way.

1. Make a selection of Equipment that will be most commonly used on the contract: think of the Pareto principle, i.e. the 20% of construction plant and accommodation that will make up 80% of the Equipment cost for a typical compensation event.
2. Build up the costs for these items from first principles (which may just be their hire rates).
3. Calculate weightings for these items based on the proportions in which they will be used.
4. Multiply the costs by the weightings.
5. Apply the same weighting to the unadjusted prices for the same items in the referenced published list.
6. Work out the average difference between the two sets of values to obtain a percentage adjustment, which is the figure stated in the Contract Data.

This percentage is usually negative, but depends on such factors as site location and access, and state of the economy. As this figure is normally part of a tender assessment, Contractors at tender want it to be as high and negative as possible to look good, but would then be losing money on Equipment costs from compensation events.

For Equipment not listed in the published list, the *Contractor* is invited to give separate rates for 'other Equipment'.

Tendering Contractors play two interlinked games here.

- They reference the most expensive published list, allowing them to tender the largest percentage discount, thereby making them look better than they are by comparison with their not so sly competitors. As a result, wise Clients now tend to specify in Contract Data part two which standard list all tendering Contractors are to tender the percentage adjustment against.

- They refer to a long list of Equipment, which pretty much covers everything they will use, and include, right at the end, an innocuous statement along the lines of 'these rates are to be used instead of rates in the published list'. This then is part of their offer which the *Client* accepts and, with these words, almost certainly overwrites what is in the unamended contract.[16] It turns out that the extremely high negative percentage adjustment for listed Equipment (and therefore attractive to the *Client* at tender) is never applicable, while the rather generous rates tucked away in the *Contractor's* own list are (not so attractive to the *Client*). Potential Clients need to be wise to this trick, which is equivalent to the one played in the full SoCC under Equipment 24. Consequently, it is normal for the *Client* to list what this Equipment might be and to preclude the bidder from adding in rates for other Equipment.

3. PLANT AND MATERIALS

These costs are calculated in the same way as in the full SoCC (i.e. at cost).

4. SUBCONTRACTORS

As per the full SoCC, the *Contractor* is reimbursed what it pays its Subcontractors in accordance with its subcontracts plus having the *fee percentage* applied, so the same comments apply.

5. CHARGES

Compared with NEC3, the percentage for people overheads has disappeared. The words are identical to the full SoCC and so the same comments apply.

6. MANUFACTURE AND FABRICATION AND 7. DESIGN

The big difference compared with NEC3 is that these now only apply to the *Contractor's* own people and not to Subcontractors, be they in a factory supplying Plant and Materials made specially for the *works* or a designer. The words are identical to the full SoCC and so the same comments apply.

8. INSURANCE

Again, the words are identical to the full SoCC and so the same comments apply.

2.3.4 Filling in the other parts of Contract Data part two

THE *FEE PERCENTAGE*

Under the first sentence of clause 52.1, 'All the *Contractor's* costs which are not included in the Defined Cost are treated as included in the Fee'. What does this mean in practice?

- Under options A and B, the tendered *fee percentage* needs to cover profit, head-office overhead and an allowance for anything else not listed in the SSoCC.
- Under options C, D and E, the tendered *fee percentage* needs to cover profit, head-office overhead, including the cost of taking out insurances (under options A and B, the cost of taking out insurances would already be in the Prices), and allowances for some Disallowed Costs (as listed in clause 11.2(26)) and anything else not listed in the full SoCC.

[16] Referring back to clause 52.1, second sentence, even without this innocuous phrase, there is an argument that they should be used. Inclusion of this or an equivalent phrase gold-plates that argument!

- Under option F (management contract), the tendered *fee percentage* needs to cover profit, head-office overhead, including insurances, and *all* internal costs incurred in the Working Areas, except for those where lump sum *prices* have been tendered in Contract Data part two for work that the *Contractor* will do itself.

Typical *fee percentages*, in decent times, are as follows.

- For a medium to large civil engineering or building construction company the typical *fee percentage* is 6–8%. *Contractors* in the original research who had tendered less and subsequently won the contract found they were losing money – or at least not making any – on compensation events. As a result, they were trying to push up Actual Cost (as it was in the second edition; Defined Cost in the third edition onwards), which caused friction at site level. However, medium to major mechanical and electrical companies, which includes tier 1 companies in the process engineering sectors (such as the nuclear, power and chemical sectors) are likely to have a *fee percentage* around 15%.
- For a typical small building company with an annual turnover of approximately £1 million, the off-site overheads may be 15%.
- For a small specialist subcontractor, the off-site overheads may be 25% and, in some cases, their tendered *fee percentages* has been as high as 40%.

This is logical, as the larger the company, the greater the likely economies of scale, and thus the lower the percentage overhead. In addition, the smaller the value of work being conducted on site, the more likely it is that the *Contractor's* management and administrative staff are responsible for a number of sites, which they visit on a weekly basis, doing substantial work outside of the Working Areas. Therefore, these costs cannot be included in People costs when assessing compensation events, so they are counted as off-Working Area costs, which increases the *fee percentage* further.

THE *KEY PERSONS*

The *Contractor* has to state the names, experiences and qualifications of the *key persons* it intends to use on the contract. The same comments that apply to the selection of the *Client's* key people (see Section 2.2.2.3) also apply to the selection of the *Contractor's key persons*.

It is common practice for Clients to insert the job titles of the *key persons* they want to see. The problem that tendering Contractors then have is that the job titles they use do not necessarily match the job titles stated by the *Client*. Instead, Clients should add the words 'or equivalent' and provide a short description of what that person is to be responsible for as well as the minimum experience and qualifications it requires from these *key persons*. It is also common for Clients to want or require a profile or curriculum vitae of the tendered *key persons* as an appendix to Contract Data part two. Sometimes Clients insert the minimum experience and qualifications they expect of these *key persons*. As a *Contractor*, be aware that under clause 24, if you submit the 'A' team at tender, but actually provide the 'B' or 'C' team when you win the contract, then the replacement people can be rejected (or 'not accepted' in NEC language) by the *Project Manager* if they do not have equivalent experience and qualifications as those put forward at tender.

> One major Client decided to challenge a Contractor that consistently put forward the name of a star project manager for large projects but never actually used that person on the Client's contracts. It turned out that not only had the individual retired, but he had also unfortunately suffered early death in his retirement. As you can imagine, putting forward a dead man to be the lead person on a contract did not go down well with the Client.

THE 'STARTER FOR 10' (OR 20 OR 30) EARLY WARNING REGISTER

The format of what is put in here will normally be decided by what the *Client* has decided when putting the tender out (see Section 2.2.4.4) and the extent of information put in will be determined both by this and by the individual Contractor's risk management capabilities.

As stated previously, the early warning process and contents of the pre-contract Early Warning Register in theory have nothing to do with the contractual risk allocation under the contract. Note the 'in theory', however, and this is where it is transgressed: as a Contractor, you might state a 'physical condition' risk that you think is unlikely to occur, but state it because you think you should put something in and/or to get higher marks in the quality assessment part of your bid. You win the contract and almost everybody forgets about this risk as it is fairly unlikely to happen. However, under compensation event clause 60.1(12), if this risk then occurs, the *Project Manager* would be justified in saying 'an experienced contractor' did judge at the Contract Date that it had a greater than 'small chance of occurring' to allow for it, as it is stated in the Contract Data part two as a matter to be put in the Early Warning Register. Consequently, no compensation event!

Having said this, such a decision could be challengeable on the basis of quantum if, for instance, the *Contractor* had allowed in its submission 'up to a maximum of £10 000' for the impact, but the extent of the physical condition was costing £100 000. In this case, the *Contractor* would have reasonable argument for the additional £90 000. This is all provided the original allowance satisfied the criteria in clauses 60.1(12) and 60.2.

SCOPE PROVIDED BY THE *CONTRACTOR* FOR ITS DESIGN

This is equivalent to where the *Contractor's* design proposals are referenced and incorporated into the contract. If not referenced here, they are not part of the contract! As explained previously, the effect of clause 60.1(1), second bullet point, is that the *Client's* Scope has precedence over the *Contractor's*.

- The onus is on the *Contractor* to ensure compliance with the *Client's* Scope. A change to the *Contractor's* Scope to make it comply with the *Client's* is not a compensation event, so the cost falls within the original Prices.
- Assuming a Contractor's proposals are assessed as part of its tender submission, the Contractor obviously wants them to be sufficient to score top marks. However, the more information the Contractor puts in here
  - the more it obviously costs at tender
  - the more the Contractor limits its ability to innovate and find cheaper ways of delivering the work etc. once the contract is won as they will have a contractual obligation to do it that way
  - the greater the likelihood of ambiguity or inconsistency, not just with the *Client's* Scope document, but also within the *Contractor's* Scope, which, in accordance with clause 63.8, will be assessed as a compensation event against the *Contractor's* interests.

In order to create a level playing field, I suggest Clients specify in their Instructions to Tenderers the state of development they expect tenderers to develop the design to (in the 'Scope provided by the *Contractor* for its design'). For instance, RIBA has various stage gates that could be used.

REFERENCING A PROGRAMME

I have two points on this.

- If a programme is referenced here, it becomes the Accepted Programme on signing the contract, and the *Project Manager* will be unable to retain a quarter of the Price for Work Done to Date in accordance with clause 50.5. This implies that tendering Contractors want their programme to be of good quality, for pure short-term commercial reasons (ignoring any scoring of it the Client may be making at tender and the benefits of starting a contract with a decent programme!).
- Because the *Contractor* has to comply with information referenced as Scope but can change the programme, the general thrust of where a Contractor wants to place method-related information is in the programme, with as little as possible, if anything, in the *Contractor's* Scope. This contrasts with where the *Client* may well want that information – if the *Client* has chosen a particular Contractor to a large extent on its methodology or approach, then this needs to be referenced into the contract via the entry in Contract Data part two for the *Contractor's* Scope.
  However, the unamended Contract Data entry states 'The Scope provided by the *Contractor* for its design is in', which does not quite fit the bill as it is not design. Therefore, a Client may want to
  - insert another entry, which reads 'The Scope provided by the *Contractor* for how it is to Provide the Works is'
  - change all instances in the *conditions of contract* of 'Scope provided by the *Contractor* for its design' to just 'Scope provided by the *Contractor*' (via an option Z clause).

  However, I emphasise that you do not want the *Contractor's* detailed programme referenced here as it is bound to change: it is the approach or methodology expressed to a level of detail that the *Contractor* can be held to.

TENDERING OF RATES AND THE PERCENTAGE ADJUSTMENT FOR LISTED EQUIPMENT

The methods for arriving at the various percentages and rates tendered in Contract Data part two have been described in Section 2.3.3.

2.4. Evaluating the tender

2.4.1 Evaluating the effect of the tendered percentages and rates

The principal difference for the *Client*, when assessing Contractors' tenders under the ECC compared with traditional conditions of contract, is in calculating the effect that the various rates and percentages returned in Contract Data part two. For the priced contracts (options A and B), these come into play only when compensation events occur. For the cost-based contracts (options C, D and E), these come into play both for normal work in the original work and for compensation events.

Why should you do this? In an ideal world, there would be no changes and compensation events after the contract is signed, but in reality there always are. Early experience from years ago, when I was doing my PhD research, and early consultancy experience all strongly indicated that the fact that a Client is doing some sort of assessment makes the tendered rates more competitive. There is also evidence that, on contracts where an assessment was not done, the rates and percentages tendered have been excessive. It is now standard practice to do an evaluation of the tendered percentages and rates.

I will start with a tender assessment of the financial factors under an option A or B contract, then build on it for an option C or D contract, and finally consider options E then F. The aim is to arrive at an 'adjusted total of the Prices for tender assessment purposes' or equivalent.

2.4.1.1 Tender assessment under options A and B using the SSoCC

A model tender assessment sheet is given in Appendix 5 of the NEC4 ECC Guidance Notes (volume 3: *Selecting a Supplier*) for use with the priced contracts (options A and B). As with the NEC3 Guidance Notes version, in my opinion, it does not quite match how the SSoCC works, including missing out some key items that I would want to evaluate, namely the *people rates*.

So here is my stab at a more logical and complete tender assessment process for the financial part only.

A Estimate the number of people hours that will be used for compensation events per type of person listed in Contract Data part two. But there is a conundrum: as the potential *Client*, you do not know exactly what type of people or job titles will be used by the selected *Contractor*, and these will vary from tendering Contractor to Contractor. So, you have two alternatives.

(i) Specify the categories of people (e.g. plumber, electrician) or individual persons (by type, e.g. site manager, site quantity surveyor) in Contract Data part two. For the tender assessment, you then make an estimate of the number of hours *each* will be used for compensation events, to come up with a weighted model that takes account of the number of hours you estimate each category/person will be used for.

(ii) Let each tendering Contractor enter the categories that match the actual job titles and categories it uses, together with its rates for those categories. In the tender assessment model, normalise these rates: for example, if there are ten categories/job titles, sum the rate/category column and divide by ten. Then multiply this by your estimate of total people hours for compensation events.

Both of these methods have their pros and cons, although (i) is generally favoured.

B State in Contract Data part two, which published list is to be used (as described in Section 2.3.3.3) in order to avoid Contractor gamesmanship. Estimate a figure for the value of *Contractor* owned or hired Equipment costs due to compensation events. This is then adjusted by each tendering Contractor's percentage adjustment for listed Equipment.

But you also need to consider how you are going to evaluate rates for other Equipment as, if you are not evaluating them at all, they are likely to be high (as a source of 'hidden' profit for the *Contractor*). Alternatively, you can just ask tendering Contractors to identify any Equipment that they will be using but which is not in the published list, strike through or delete the box where a rate would be entered, and agree pre-contract (ideally) or early on in the contract to use market rates (as per clause 52.1).

C Make an estimate of the cost of Plant and Materials that will be bought directly by the *Contractor* for compensation events.

D Make an estimate of the cost of Subcontractors for compensation events. In building and civil engineering projects, this will typically be the largest item by a fair way.

E Make an estimate of Charges that will be incurred directly by the *Contractor* for compensation events.
Items **C**, **D** and **E** will be the same for each tendering Contractor.

F Using a similar method as at **A**, estimate the rates for Defined Cost of manufacture and fabrication outside the Working Areas and design outside the Working Areas, i.e. estimate the number of people hours that will be used for compensation events per type of person listed in Contract Data part two. It may be that these are estimated to be zero, in which case leave out an entry in the Contract Data. As the Guidance Notes state, design can be for temporary works, not just permanent works.

G Sum the totals of **A** to **F**.

H Apply the tendering Contractor's *fee percentage* to **G**.

I Do an arithmetical check to ensure that the sum of prices in the *activity schedule* or lump sums and quantities times rates in the *bill of quantities* is the same as the entry for the 'tendered total of the Prices'. The reason for this is that nowhere in the *conditions of contract* is the term 'tendered total of the Prices' referenced; it is the Prices, which are the sums in the *activity schedule* or *bill of quantities*. So my view is that the tendered total of the Prices contractually counts for nothing! If there is a discrepancy, ask for clarification from the relevant Contractor.

J Sum the amounts in **G** and **H** to give the 'adjusted tendered total of the Prices' or equivalent. Repeat for each tendering Contractor.

As you may have picked up, there are a fair few estimates in here. Clients need to be realistic here about the extent of likely change and where it will happen. For me, the starting point for doing this is the project risk register. They then need to match this to tendering Contractors' capabilities. For instance, an earthmoving contract will have a much higher Equipment component if the *Contractor* is managing the earthmoving itself (whether using its own tracked excavators and lorries etc. or hiring them in) than if it is subcontracting it out (the cost will then become a Subcontractor cost). On the other hand, a refurbishment of an historic building might have a high people cost, as lots of skilled tradespeople will be used.

If you are subject to EU Procurement Law, you pretty much have to tell Contractors upfront, when you issue the tender documents, how you will take these rates and percentages into account in your evaluation. Contractors tend to look fairly hard at this in order to make whatever rates and percentages they tender come out favourably.

Taking the earthmoving example above, let us assume the *Client* has taken the view in its model that the selected *Contractor* will be managing the earthmoving itself and hence gives a high weighting to Equipment. A tendering Contractor might then deliberately decide to subcontract this work out, allowing it to tender a high negative adjustment for Equipment in the published list (because it will not be used for anything significant), but encourage the Subcontractor to give high rates and/or a positive adjustment for listed Equipment in its subcontract tender, on which the *Contractor* gets its *fee percentage* applied when a compensation event occurs. Knowing now that the biggest element of a typical compensation event will be in subcontracts, they could also choose to submit low people rates and counteract this (together with the high negative adjustment to listed Equipment) with a slightly higher *fee percentage*. Further, if it thinks the amount allowed by the *Client* for compensation events is low, it might reduce the tendered Prices to a level that gives zero or negative profit (should there be no change), gambling there will be a greater amount of Defined Cost to which its high *fee percentage* is applied.

Consequently, if you are a private sector Client, you may wish to keep what this calculation is to yourself. But if you are a public sector Client, if anything you want to overestimate the amount of change, so that a high *fee percentage* and rates have a disproportionate effect on the adjusted tendered total of the Prices compared with the unadjusted Prices, thus encouraging a more competitive *fee percentage* and rates.

So, to summarise, the very fact that the *Client* is doing some sort of assessment means that the rates and percentages are likely to be lower. However, the tendering Contractors will look very hard at the tender assessment model to work out how to make themselves look attractive at tender while maximising their entitlements for compensation events post-award. At a high level, the main games played by tendering Contractors are

- spotting where the tender assessment model does not match how it thinks the contract will turn out, in terms of both the total value of compensation events and the cost weightings given to the eight headings in the SSoCC
- sneaking in above-market rates for those items it thinks are underestimated in terms of time in the *Client's* assessment model. The extreme of this is where they are not assessed at all.

To counteract this, Clients need to ensure its model is realistic in its weightings, and either ensure all aspects are in the model, or exclude them to be assessed at market rates either before or after the award of the contract.

2.4.1.2 Tender assessment under options C and D using the full SoCC

I could simply advocate doing a very similar assessment to that above (for options A and B), but taking account of the differences in rates and percentages under the full SoCC – i.e. there are no *people rates* and no percentage adjustment for listed Equipment when the full SoCC is used, but there may be a 'time-related on cost charge' for any Equipment purchased for work on the contract. However, there is a huge difference when using these options: any rates and percentage entered into Contract Data part two by the winning *Contractor* apply to **all** of the Defined Cost, not just when compensation events occur. Consequently, the effects of any games that a Contractor plays at tender are magnified should they win the contract.

Fortunately, due to the simplifications that have occurred between NEC3 and NEC4, there are fewer games to be played. However, two big ones remain.

- As before, the tendering Contractor sneaks in above-market rates for those items that it thinks are underestimated in terms of time in the *Client's* assessment model. Again, the extreme of this is when they are not assessed at all. The not-quite-so-extreme version of this is when the *Client's* assessment model is only for compensation events and does not take account of the costs of work within the original Prices and Scope.
- The tendering Contractor artificially suppresses the Defined Cost component of the target Prices, but inflates the Fee component (i.e. has a large *fee percentage*).

Let me illustrate the second game with a tale of two Contractors tendering for the same contract. Table 2.2 shows the two Contractors' tenders. Which one would you choose? Most people, if they did not know it was a trick question, would go for Contractor B, as its tender is lower, both with and without the adjustment to allow for compensation events.

Table 2.2 Contractor game under a target-cost contract: at tender

	Contractor A (playing it straight)	**Contractor B** (being devious)
A Contractor's true estimate of Defined Cost for work in the Scope	£910 000	£910 000
B Defined Cost stated in tender	£910 000	£800 000
C Tendered *fee percentage*	10%	20%
D Tendered target Prices (B × (1 + **C**))	£1 million	£960 000
E *Client* estimate of base cost for compensation events	£100 000	£100 000
F *Client's* adjustment for different *fee percentages* (**E** × (1 + **C**))	£110 000	£120 000
G Adjusted tendered total of the Prices (**D** + **F**)	£1.11 million	£1.08 million

However, let us now see what happens if, for both Contractors, the Defined Cost did actually come in at £910 000, as both estimated, and the *Client's* estimate of £100 000 in Defined Cost for compensation events came to pass. The outcome is shown in Table 2.3.

So which one would you choose now? In this analysis, Contractor A looks to be preferable. Assuming the *Client* takes some share of any overrun, it will always be in a tendering Contractor's interest to artificially inflate the *fee percentage* and artificially deflate the Defined Cost element of its bid. Therefore, one of the key things to look out for is not just a realistic tendered total of the target Prices, but also realistic tendered Defined Cost and *fee percentage*, which make up that tendered target Price.

> When I was doing my postdoctoral research in commercial arrangements for effective partnering, I interviewed the people who had put together the contracts for the Channel Tunnel Rail Link (CTRL), which were predominantly let under option C. They capped – as in did not allow – Contractors' *fees percentage* at 10%, which is at the top end of what is reasonable. They also asked for detailed programmes with substantial back-up information, in order to ensure that the Defined Cost element of the tendered target Prices was realistic. It was a lot of work for the Contractors and for CTRL's staff, but how many projects of this size (around £2.5 billion in late 1990s prices) come in on time and cost?

Table 2.3 Contractor game under a target-cost contract: how it works out

	Contractor A (playing it straight)	**Contractor B** (being devious)
H Actual out-turn Defined Cost for work in original Scope	£910 000	£910 000
I Base cost of compensation events	£100 000	£100 000
J Total out-turn Defined Cost (**H** + **I**)	£1.010 million	£1.010 million
K Total out-turn Defined Cost × tendered *fee percentage* (**J** × (1 + **C**))	£1.111 million	£1.212 million
L Adjusted target Price after compensation events (**G**)	£1.111 million	£1.080 million
M Overrun vs target (**K** – **L**)	£0	£132 000
N *Contractor's* share of overrun assuming 50/50 split (**M** × 0.5)	£0	£66 000
O Total paid out by *Client* after application pain/gain share (**L** + **N**)	£1.111 million	£1.146 million

So how do you counteract these two games? Let us take the high *fee percentage* game first. For this you have two options.

- You could state the maximum *fee percentage* allowed, as per the CTRL example above. I have done this a number of times with public sector Clients and their lawyers have raised no objection. Like CTRL, we capped it at 10% as, for civil engineering Contractors, this is at the top end of what you would expect.
- You could tell the tendering Contractors what the profit component of their *fee percentage* is and that the remaining elements are to be built up from first principles, to be audited and verified prior to entering into the contract.

> For Connect Plus, when setting up its Renewals Framework for maintaining the M25 London orbital motorway, we chose the second option. We told the tendering Contractors that the allowance for profit was 2% (a bit below the market rate, because we wanted them to be making profit by generating savings) and gave them a template to fill in for the remaining costs. We actually said this is 'our starter template, but feel free to modify it if we have left something out or it does not match how you calculate your head-office overheads'. In our tender scoring model, if, having audited and verified the build-up, we agreed with it exactly, then the tendering Contractor was awarded full marks – a 10 out of 10 – for that part of the tender assessment. If we were not comfortable with it and it needed modifying, but in a minor way (within a preset range), they were scored a 5 out 10. If it needed to be modified by more than this preset range, then not only did the tendering Contractor score no marks, but they were eliminated from the competition altogether. This was because we did not want to be working with an organisation we could not trust.

So, how do you counter the game of tendering high rates for Equipment and for manufacturing and fabrication and design outside the Working Areas? The answer is similar to the above for the *fee percentage* – don't have any!

In theory, target-cost contracts are open-book contracts, where the *Contractor* is paid at cost. So, why not match the reality to the theory and build up those rates at cost, as per the full SoCC.[17] Instead, you might ask for these rates to be stated by the tendering Contractors, but in a document that is clearly stated as not being a contract document and which is not referenced from anywhere else within the contract documents. Indeed, you might extend this to people costs and any other internal costs. Alternatively, you do not ask for them at all. Either way, you then do a thorough audit, either before or immediately after contract award, to verify or work out the rates. If pre-contract and a public sector Client in the UK, I would recommend the competitive with dialogue selection procedure.

[17] If you are to do this

- the time-related on cost charge for Equipment purchased specially for the contract (in Equipment 23) and the entry in the Contract Data need to be deleted – these costs can be accounted for and paid for under other headings in the full SoCC
- the words in Equipment 24 need to be replaced with a new statement of how the cost of already owned 'special Equipment' is calculated, and hence the entry for special Equipment in the Contract Data needs to be deleted
- if the rates for manufacturing and fabrication and design outside the Working Areas are relatively minor parts of the total Defined Cost they can typically be left as they are.

2.4.1.3 Tender assessment under option E using the full SoCC

Under option E (cost reimbursable contract), the Prices are only a 'forecast of the total Defined Cost for the whole of the *works* plus the Fee' (clause 11.2(34)). So, unless you deliberately want to underestimate your project's out-turn cost, there is absolutely no point in using a tendered figure as part of your selection process.

As there is no pain/gain share arrangement, the games that can be played around this do not come into play. Apart from that, my other comments about options C and D also apply to option E, which means that the financial assessment comes down to the assessment of a realistic *fee percentage*.

2.4.1.4 Tender assessment under option F

Similar to option E, option F (management contract) is a cost plus percentage fee contract, with the Prices again being a 'forecast of the total Defined Cost for the whole of the *works* plus the Fee', so the same comment as for option E applies.

However, the presumption under option F is that all the physical work and any design of permanent works is going to subcontracted, with the *Contractor* only managing the *works*. Consequently, Defined Cost is largely what the *Contractor* pays its Subcontractors, with the *Contractor* having the ability to identify and price, as lump sum '*prices*', work that it will do itself. This is also Defined Cost and, as with options C, D and E, the *fee percentage* is applied to all of the Defined Cost (i.e. Subcontractor cost plus identified *prices*). Therefore, the *fee percentage* needs to cover all the *Contractor's* management costs. And, like option E, the financial assessment comes down to the assessment of a realistic *fee percentage*. In fact, I would say even more so than with option E, as it needs to be realistic because, if it is not, the *Contractor* will start to skimp on management resource, which undermines a principal reason for choosing this strategy.

So what's the game for tendering Contractors? See the box below.

> Very early on in my consultancy career – around the turn of the century, possibly a little bit before – for a period of about 6 months, every civil engineering contract in south Wales seemed to be let as an option F contract. For the financial aspects, the tendering Contractors were assessed only on their *fee percentage* (as it was NEC2 at the time, there was only one *fee percentage*). The game that Contractors played was to identify all their management activities as lump sum *prices* – i.e. there were no costs to cover in the *fee percentage*, meaning it could be very low, especially as it was effectively a 'risk-free' contract from the selected *Contractor's* point of view. As quick as this mini-fad came, it passed.

2.4.2 Other aspects to evaluate

Clients may wish to assess other aspects of Contractors' tenders in the qualitative part of their tender assessment processes. These aspects and methods for assessing them include the following.

- Comparison of tender Prices when using activity schedules. One of the complaints from Clients about using an activity schedule – whether under option A or C – is that it does not give the same level of transparency that a bill of quantities does, especially when it is a method-related one. In Section 1.2.2.2, I argued that this is partly illusory, yet there is some truth in this accusation. Here is how some Clients make a comparison of Contractors' tender Prices under options A and C.

1. The *Client* allows the tendering Contractors to break their activity schedules down into activities that reflect their programmes, albeit perhaps limited in numbers or minimum value.
2. The *Client* asks the tendering Contractors to allocate each activity to a heading, of which there are 10–20. For instance: Preliminaries, Enabling works, Earthworks, Foundations etc., as well as a catch all Miscellaneous heading for anything else.
3. When comparing tenders, the *Client* first of all looks along the sums under these main headings. This tells the *Client* whether one tendering Contractor has put more cost into one heading compared with the others, or vice versa.
4. If desired, the *Client* further investigates each of the activities under that heading to see which are high or low.
5. If desired, the *Client* looks at the programming submission to ascertain resources, methodology etc. related to the high or low values.
6. If desired and able to under any relevant competition rules, the *Client* asks questions of the Contractor.

- Analysis of the submitted programme, to ensure its realism, practicality and completeness. This could lead to further questions in order to clarify the Contractor's intentions. (See Section 3.5 for a thorough discussion on the programme and what to look out for in an NEC context, as opposed to a more general context, when evaluating the programme.)
- Adjustment to the total used for comparing tenders (i.e. the sum arrived at through the process outlined above) to take account of the time value of money. For instance, if it is costing the *Client* £5000 per week not to have the asset in use, then the £5000 could be subtracted from a Contractor's tender for each week that Completion is shown to be ahead of the Completion Date. However, I would only recommend this approach if, as a Client, you are asking Contractors to tender the Completion Date in Contract Data part two. You then do this calculation on the different tendered Completion Dates. This is now illustrated in the Appendix 5 of the NEC4 ECC Guidance Notes (volume 3: *Selecting a Supplier*).
- Analysis of cash flow. For contracts tendered under the activity schedule options, this is comparatively easy, as the Prices for each activity that will be completed in a certain month can be summed. For other options it is not so easy. Sometimes, Clients have asked tendering Contractors for an S-curve showing planned expenditure, in the form of cumulative payments by the *Client* plotted against time. This then forms the basis for any earned value analysis.
- An analysis to take account of the time value of money. The sums arrived at in a cash-flow analysis can be discounted back or forward to a set date using a pre-agreed discount percentage.
- Assessment of the *Contractor's* design. This can be based on, for example, architectural merit in the case of a building or the technical solution in the case of process plant. Following any discussions, it is important to confirm what aspects of the *Contractor's* design and what commitments are to be incorporated into the contract by referencing them in Contract Data part two under 'Scope provided by the *Contractor* for its design'. Failure to do this does not tie the *Contractor* to fulfilling its proposals or what was agreed.
- Assessment of the *Contractor's* proposed methodology or approach. See 'Referencing a programme' in Section 2.3.4.

- Assessment of the *key persons* involved. The type of people that it is desirable to have on site was discussed in Section 2.2.2.3. The qualifications and experience of the *key persons* put forward by the *Contractor* can be analysed. In a few cases, Clients have told a Contractor that a certain individual is not appropriate, due to previous experience with that individual. Other Clients have used face-to-face meetings and structured interview evaluation techniques to assess tendering Contractors' project teams and included them as part of the tendering procedure.

REFERENCES

Australian Construction Services (1988) *Strategies for the Reduction of Claims and Disputes in the Construction Industry. A Research Report.* Department of Administrative Services and several other participants, Queensland.

Broome JC (2002) *Procurement Routes for Partnering: A Practical Guide.* Thomas Telford, London.

Broome J (2010) NEC3 Key Dates – more risk on the Contractor? *NEC Users' Group Newsletter.* Available at www.jonbroome.com/resources/articles (accessed March 2021).

Earnshaw N (2020) Key Dates in the NEC3 and NEC4 Engineering and Construction Contract. Online article. NE Consult. Available at https://www.neconsult.co.uk/resources/ (accessed March 2021).

Houghton E and Trebes B (2011) *The Z factor: to what extent have Z clauses impacted on the NEC's success as a modern contract?* Paper presented at NEC User's Annual Seminar, 11 April 2011.

NPWC/NBCC Joint Working Party (1990) *No Dispute: Strategies for Improvement in the Australian Building and Construction Industry.* Presented at National Public Works Conference, Australia.

Patterson R (2007) NEC and risk management. *NEC Users' Group Newsletter* **40**.

Patterson P (2017) NEC: how to require something is done by a certain date (the difference between key dates and sections). *NEC Users' Group Newsletter* **86**.

Patterson R (2020) Helping NEC clients decide the best way to get a first programme in place. *NEC Users' Group Newsletter* **108**: 10. Available at: https://www.neccontract.com/About-NEC/News-and-Media/Helping-NEC-clients-decide-the-best-way-to-get-a-first-programme-in-place (accessed March 2021).

Patterson R and Essex R (2010) Ground conditions and risk allocation: combining the NEC Engineering and Construction Contract (ECC) and the geotechnical baseline report (GBR). *Tunnels and Tunnelling*, Dec.

Patterson R, Davies M and Trebes B (2015) *NEC contracts and the Construction (Design and Management) Regulations 2015.* News article. Available at: https://www.neccontract.com/About-NEC/News-and-Media/NEC-contracts-and-the-Construction-Design-and-Man (accessed March 2021).

Perry JG and Thompson PA (1982) *Target and Cost-reimbursable Construction Contracts.* CIRIA Report R85. CIRIA, London.

Revay SG (1992) Can construction claims be avoided? In *Construction Conflict and Resolution* (Fenn P and Gameson R (eds)). E & FN Spon, London.

Broome, Jon
ISBN 978-0-7277-6205-4
https://doi.org/10.1680/necaug.62054.189

Chapter 3
Good practice following award and the crucial first month(s)

3.1. Introduction

This chapter describes the principal steps necessary after the award of the contract (i.e. the Contract Date), but prior to starting work, whether on design or on Site (i.e. the first *access date*). The principal issues that need to be addressed are

- training – or, more precisely, common understanding – for those involved in the running of the contract
- putting in place a system and protocols for the efficient administration and management of the contract (i.e. how you are going to communicate, both informally and formally) – this includes project organisation, which I have gained a better appreciation of since the previous edition of this book
- payment, especially where the *Contractor* is reimbursed its costs (e.g. options C to F) and how the pain/gain share operates under options C and D
- making the best use of the programme, so that there is always an up-to-date, understood, agreed and accepted programme
- operating the early warning and risk-reduction process
- quickly and amicably agreeing quotations for compensation events.

It was suggested by a number of interviewees that a week or two extra should be programmed into the project programme to allow for these issues to be addressed, especially for those new to the ECC.

Why does the title of this chapter contain the words 'and the crucial first month(s)'? Because a finding of the research, years ago, was that on successfully run contracts the necessary changes in procedures, systems, skills and attitudes had started to be addressed in the pre-contract phases, and were substantially in place within the first month of being on site. This finding fits with much research on project management in general. In my consultancy experience, I have found that contract teams have recovered from a poor start, but it is almost always done by going back and doing what they should have done in the first months, while at the same time managing a contract in full flow. For some teams, this has only been partially successful.

3.2. Training and common understanding

3.2.1 General

The importance of skills and attitude has been stressed in Section 2.2.2.3, with the most important attitude highlighted as a willingness to change. However, people have to know what to change to. They also have to know what they have to do and, preferably, why. Indeed, the 'why' is possibly the most important factor, because it gives people the motivation to change. Hence the need for training. As one project participant, who had not had training on the then NEC, admitted 'If you just pick up the NEC and read it and don't have any training on it, then ... you probably miss the whole point of it' or, as a senior project manager put it, 'It's just plain different. It means that people who have been science'd in other forms have got something else to learn'.

Note that the title includes 'common understanding', because there will be reduced efficacy if the *Contractor's* staff are clued up but the *Client's* staff are not, and vice versa. This could be extended down to principal Subcontractors as well. In fact, training is the means to common understanding.

For me there are three aspects to this common understanding

- participants have a common knowledge and understanding of what the contract says
- participants have a common understanding of how they will operate the contract on a day-to-day, week-to-week and month-to-month basis
- participants have a common understanding about the project.

Obviously, if people learn through a common experience (i.e. they are in the same room or see the same video, eLearning etc.), they are far more likely to have a *common* understanding. So often these events are jointly funded and attended by both the *Client's* and *Contractor's* personnel.

Let us take each of the items in the above list in turn.

3.2.2 A common knowledge and understanding of what the contract says and why

There are quite a few training organisations and individual trainers who offer training on the ECC.

- While the ECC is a legal document, it is also, as many observers have noted, almost a manual for project management. This suggests that a trainer should have some understanding of project management in order to understand why clauses have been written in a certain way. Purely legally based training, in the traditional style of training on contracts, is not completely appropriate. Feedback from those who have had it indicates that it misses the point about the ethos of the contract and any insights on how to operate it.[1]
- To gain the most from the ECC involves cultural change and, to some extent, skills change or, at a minimum, a change in how existing skills are used. This also needs to be put over.
- It is one thing to understand what a part of the contract says, but people also need to know how one part relates to another (e.g. how the programme is used for assessing compensation events). (For an excellent article on this, see Patterson, 2010.)
- As always with training, exercises and case studies help with retention and understanding of information. Participants also lose their fear of the contract when they pick it up and start to use it.
- Lastly, I also think it is important that potentially contentious issues are brought up, and how wording is to be interpreted or a situation avoided is discussed. For instance, under options B and D, the way in which items are re-rated when there is a change in quantities (see Table 1.3 in Section 1.2.2.2 for my views on this).

[1] As an aside, someone once said to me that every lawyer, as part of their training, should spend year trying to operate a contract they have written. Given some of the non-sensical and utterly impracticable rubbish that I have seen written (including by top law firms), I thoroughly agree!

Potential commissioners of training should bear this in mind when selecting the training organisation and the individual trainer. However, if sufficient people have previously had sufficient experience and training of this sort, then why repeat it?

Before I move on, I want to say something about the use of eLearning. But be aware that there is a bit of a vested interest here. I, with Glenn Hide (another well-known NEC name), have published a comprehensive set of NEC eLearning modules through Built Intelligence. Thomas Telford Ltd (the publisher of this book and the NEC contracts) also has eLearning available.

One way of approaching learning is to view it as a progression.

1. *Awareness*. You become aware of something – in this case, what the NEC is about – and can make associative links (e.g. you can pick the right answer in a multiple choice questionnaire, but would not necessarily be able to regurgitate the answer).
2. *Knowledge*. You can go to the right place in the contract to find the definition of compensation events, and even repeat what the contract says.
3. *Understanding*. You know how different parts of the contract relate to each other and, for example, why a clause was written in a certain way and what its implications are.
4. *Competence*. Which almost always has to come with experience and use, but can be enhanced and accelerated by taught learnings, as well as by coaching and mentoring and other reflective practices.

My view is that good eLearning can be better for raising awareness and imparting basic knowledge compared with face-to-face training. This can be because of, for instance

- the bite-sized nature of modules, which, if correctly labelled, allow the user to access the specific training relevant to the issues they are currently encountering
- start and finish multiple-choice questions – the start questions raise the user's awareness of what they do not know, so they look out for the answers
- the user having to at least click through the eLearning
- frequent mini-exercises, which make user do something with the information provided, rather than letting them just click through to the next module. This aids retention and scratches the surface of understanding.

eLearning can also include pointers on effective implementation. And because doing the eLearning and the marks scored on the quizzes can be recorded and contribute to the user's Continuous Professional Development requirements, that accountability factor can enhance motivation.

However, for deeper understanding and to accelerate competence, people need more complex exercises (which are not of the right/wrong type) and relatable real-life war stories (both from the trainer and other participants), and to be able to reflect on these and their own experiences, be asked questions and ask the trainer questions, and be prompted to relate what they have learnt back to what changes they will make in real life.

So, for a project team coming completely knew to the NEC4, or for individuals new to NEC4, the same eLearning might well be part of the answer to get them up to speed. Less time is then needed with an expensive resource, i.e. a trainer.

3.2.3 A common understanding of how the contract will be operated

This is different from 'how the contract operates' as this is what people have to, or it is advantageous that they do, during the contract on a day-to-day, week-to-week and month-to-month basis. If two key people have different and incompatible ideas of what good practice is and these differences are not discussed, project participants will not be on the same page with respect to implementation. For me, the key elements of this are how the two (or more including Subcontractors and consultants) parties, in accordance with the contract and to bring benefit to the project, are going to

- assess and agree how much the *Contractor* is going to be paid on a 4-weekly or monthly basis (this is particularly important when the contract is under one of the cost-based options C to F, especially if one or both of the Parties is new to these options)
- communicate effectively and efficiently
- make best use of the programme, so that there is always an up-to-date, understood, agreed and accepted programme
- operate the early warning and risk-reduction process
- quickly and amicably agree quotations for compensation events.

Unsurprisingly, this replicates the sequence of topics covered in this chapter! (I would add that all the workshops that I have done which have followed this format have led to successful projects. If you want to read about one, see Freeman (2011) – I was that trainer!) While the trainer might be able to provide direction and guidance on this, sufficient time also needs to be built in for the participants to think through and action plan both what they have to do to put in place the systems etc. and what they want to or can commit to given the time constraints everybody operates under. The trainer should also utilise participants' knowledge gained from previous contracts and trainings of what works and what does not. This is moving into the role of facilitator, albeit that the trainer is knowledgeable on the topic on which they are facilitating.

Speaking personally, I really like doing these sorts of events because I learn from participants' previous experiences, both good and bad!

3.2.4 A common understanding about the project

The previous two headings have been specific to the NEC4. Under this heading I include things such as

- the background and context of the project (e.g. whether it is part of a larger project or programme and, if so, how it fits in with that project or programme)
- why there is a project (i.e. the problem it is to solve and the hoped-for benefits)
- an idea of the key stakeholders and their potential influences
- each Party's project-specific objectives
- the main constraints, key interfaces and critical dates
- the principal threats and opportunities
- the key thrust or approach in terms of the programme for how the contract will be delivered
- getting to know each other, both as individuals and in terms of roles and responsibilities, so that informal communication channels are open.

Ideally, some meaningful work will be done to refine and gain clarity with regard to the above issues, because by doing such work, understanding and ownership of the objectives, risks, issues and actions is gained. This is very definitely the realm of the facilitator! (For more on this, see Broome, 2008, 2016.)

All of this helps build a high-performing team, the members of which are predominantly on the same page in terms what the contract is about, how it is going to be achieved and what could stop them from achieving it.

3.2.5 Conclusions

All of the above involve a cost. Or is it an investment? I think it is the latter, as I despair of organisations that are willing to spend a small fortune on lawyers and claims consultants (which sometimes includes me) to sort out the commercial mess that arises when an even larger amount has been incurred in extra costs on Site. When I have come in at the tail end, it is either because lawyers have been mucking around with the contract or because of ignorance of the Parties. Training and common understanding addresses the ignorance factor.

These trainings/workshops are not necessarily one-off events, held only at the start of the contract. Some follow-up training or group consultancy may be valuable as the contract progresses, such as

- 2–3 months into the contract – so that the team can get an independent view on some contractual areas on which there are differences of opinion and to address implementation issues
- when a phase of a contract is changing – as this is when new challenges and risks may arise and new team members come on board.

Furthermore, training and workshops can be combined, either into a single event or a series of events. For the latter, a logical sequence would be something like: Here is what the contract says; Here is some observed good practice; What are your experiences, good and bad?; What are you going to do on this contract?; and Action planning. The first two elements are training, the other three are facilitation.

Whatever you do, the investment needs to be justified. So when deciding what you need to do and for how long, you should consider the following questions.

- What is the size of the contract?
- How straightforward is it (e.g. will there be much change and many compensation events)?
- What is the existing knowledge and experience (positive or negative) of the NEC?
- Is this knowledge and experience one-sided (e.g. *Contractor* knows it well, but *Client* does not, or vice versa)? When NEC was coming into the civil engineering sector, one of my contractor clients always used to get me to do joint training when the *Employer* was unfamiliar with the NEC. Even though their staff had pretty much heard it all before, what my client wanted was for there to be a *common* understanding of what the contract said, which prompted discussion about how the *Contractor* and the *Employer*, as a team, were going to implement it.
- Do the people and organisations know each other from before? If so, was it a positive or negative experience?

The preferred option might be a short internal event or series of events, or it could be a whole series of externally facilitated events aimed at culture change and upgrading skills – or maybe something in between. Whatever you decide, the investment should be justifiable.

3.3. Communications: principles and systems

3.3.1 Eight principles for effective and efficient communications

In the original research, it was an almost universal opinion that more contractual communications occurred during the implementation phase of an NEC/ECC contract because

- many more issues are dealt with as the project progresses, within the timescales of the contract
- contractual communications have to be 'in a form which can be read, copied and recorded' (clause 13.1) (which means that emails and other electronic means of communication are acceptable).

However, opinion was divided on whether overall there was more or less administration from the *starting date* to the settlement of the final account than with other conditions of contract. As opinion was divided, I thought this would be a good area to research and, based on that, developed eight principles for effective and efficient communications under an NEC contract. Generally, those who did not have these in place struggled, while those who did found the NEC a breath of fresh contractual air!

> One *Contractor* commented that, doing what is described, their site management team probably spent half as much time writing contractual letters to the other party as they would have on a traditional contract. Instead they were able to concentrate on programming and working with the *Project Manager* to save time and/or money. Partly as a result of this, on each of the jobs the *Contractor* was involved in, its profit margin steadily increased as the contract progressed. At the other extreme, one *Employer* had almost 30 different pro formas, which they admitted was quite unwieldy. I was also made aware of one *Employer* who temporarily abandoned the use of the ECC because of 'administrative problems'.

I have progressively updated and enhanced these eight principles based on my consultancy and other experiences. The Principles are as follows.

1. Discuss and agree verbally, confirm and summarise 'in a form which can be read, copied and recorded'.
2. Create the right environment for verbal communication.
3. Agree specific delegations and communication paths.
4. Have regular 'wrap-up' meetings.
5. Develop pro formas.
6. Put in place a logging in and out system.
7. Extend the system up, down and across.
8. Set up principles 1 to 7 by the *starting date*.

For principles 5 to 7, this may mean purchasing an NEC cloud-based communication system, which I will discuss in detail the next subsection.

PRINCIPLE 1: DISCUSS AND AGREE VERBALLY, CONFIRM AND SUMMARISE 'IN A FORM WHICH CAN BE READ, COPIED AND RECORDED'

On the majority of projects, most communication is spoken. This has a number of advantages.

- When speaking, you can exchange information faster than when writing it down.
- You can be a lot less defensive in what you say than if you are going to be 'held' to what you have written.
- You can communicate with more than your words (e.g. intonation, hand movements and pointing at things, whether it is on a drawing, out on site, a box on a spreadsheet or a bar on a programme).
- Whether it is through body language or the words the other person says, you can pick up responses that indicate they are unsure, do not understand or disagree with you, and clarify. You cannot do this with written communication, especially if you are not there when the person is reading it.
- By clarifying what the problem is, what you are trying to achieve, bouncing ideas of each other, doing rough sketches etc. you can arrive a better solution quicker.

Having done this, you can then summarise the problem (if an early warning) and the way forward more succinctly than if you are trying to persuade and agree by writing from a standing start.

Summarising, I have given logical reasons to talk first which, although it was not done on some sites, is only common sense. Or, as one interviewee stated, 'notifications should just be merely backing up what you have agreed by talking to each other'. Richard Patterson, a well-known NEC name and one of the reviewers of this book, has an acronym for this which he shouts at you at every opportunity: TALTEO! It means 'talk and listen to each other'.

PRINCIPLE 2: CREATE THE RIGHT ENVIRONMENT FOR VERBAL COMMUNICATION

Communication on a personnel level can be encouraged by such things as project start events/partnering workshops, team-building days and nights out early on (see Section 3.2). The openness 'kick-started' by these events can be reinforced and sustained by the physical environment.

A top tip for getting people to talk in the first place and for making it easy to talk is to have joint offices. By this, I do not mean offices that divide at the entrance lobby, with the *Project Manager's* and *Supervisor's* offices to the left and the *Contractor's* to the right (or vice versa). As seniority seems to be linked to how far down the corridor you are, this often means the most senior people are as far apart from each other as they can be.

Instead, have joint offices, so that offices belonging to equivalent people are opposite one another. For instance, the *Project Manager* is located directly opposite the *Contractor's* site agent, the *Supervisor* opposite the *Contractor's* chief engineer, and likewise for planners and quantity surveyors. Setting-out engineers have often shared offices with the *Supervisor's* surveyors. Quantity surveyors have even shared offices! This means people can just get up and have a chat with their opposite number. On most of these sites a mature open-door policy has operated. By 'mature', I mean an acceptance that some things are confidential and private, but this should be a rare situation: normally the door is open.

On a few sites, this arrangement has led to a great deal of trust and short 'job swops' when somebody's counterpart is busy. By short job swaps, I mean the following.

- The *Supervisor's* surveyor has done the setting-out for an overworked *Contractor's* engineer, who then checks the surveyor's work, not in a policeman role, but to make sure it has been done correctly.
- The *Project Manager's* quantity surveyor has inputted data into the *Contractor's* financial system, rather than just checking through it.

> On one £100 million design and construct road project, where I was called in to do the start-up workshop, early on we had identified the values that the team wanted to work by. But these 'values' were expressed using high-level words, such as 'collaboration', 'trust' or 'one-team approach', which can mean many things to many people. So I asked and got the group to work on what these would mean in practice. The question I actually asked was something like: 'If I was an alien looking in from above on you as a project team, what specifically would I see and hear that would lead me to believe you were doing [insert value]?'
>
> Different groups fed back for different values and there was, as usual, a lot of overlap.
>
> Towards the end of the workshop, we action planned various critical success factors for the project. (In a project environment, a critical success factor is a key strategy for delivering the stated project objectives. Often these are high-level, so they need be grounded in actionable steps.) One of these actions was to design the layout of the yet to be established site office, for which they wanted:
>
> - the layout to be largely open-plan, but with meeting rooms and booths where people could have informal ad hoc mini-meetings
> - people to be located close to the people they would work with, regardless of which organisation they worked for (e.g. not just all quantity surveyors together, but the individual designers to be located close to the site engineers who would construct what they were designing)
> - hot desks for Subcontractor staff to be located throughout the facility, so that they could sit close to the engineers etc. who they needed to interact with
> - a mature open-door policy, the default being that any member of either the *Client's* or *Contractor's* staff could attend any meeting
> - a space for whole-team briefings, even if this meant that people were sitting on desks etc. for 15 minutes or so
> - key project information to be displayed visually (where everybody would pass regularly, so available to everybody) and updated regularly.

PRINCIPLE 3: AGREE SPECIFIC DELEGATIONS AND COMMUNICATION PATHS

Without any delegations, all contractual communications between the *Client's* designers or specialists and Subcontractors have to be routed through the *Project Manager* and *Contractor* and in a form which can be 'read, copied and recorded'.

If we look at effective implementation of business strategy – which coming from a project management background means, for me, a project or programme to deliver that strategy – then it is not enough to

- give individuals a job title/high-level role
- make it clear who they report to and who reports to them (e.g. an organisation tree).

What really makes strategy fly (see Neilson *et al.*, 2011) in terms of turning grandiose words into actions and results are

- clear decision rights – i.e. who actually makes the decision linked to accountability for outcomes
- good information flows – i.e. who does the decision-maker need to consult and liaise with to make a good decision and who needs to be informed so that, in turn, that person can make good decisions for which they are responsible.

So let us consider the top two bullet points above, which are necessary for the next two but are normally as far as people go. You could do this for your own organisation only, and then come together and see how it fits with the other Party's structure and modify accordingly. This implies a fairly transactional relationship: for example, once a month the *Client's* programmer, on behalf of the *Project Manager*, will meet up with the *Contractor's* programmer in order for the programme to be accepted first time.

But the more integrated you become, the more it needs to be co-creation – while respecting each others' contractual rights and obligations – as there is more dependency and interaction. For example, I would expect the organisation of a project team under an ECC option A contract to be vastly different from that of a team working under the Alliance Contract. And, if under the latter, I would not expect the *Client* to impose on the other Partners what the organisation would be. They might have a straw man starting point, but this would evolve both in early discussions and over time.

Going back to an ECC option A contract type arrangement, each individual within an organisation needs to know who their opposite number is. The net result would be two parallel organisational trees with, say, doted lines going across showing who is the equivalent person that each individual communicates with.

However, I think that this solution would not be sufficient. For example, is the *Client's* programmer **responsible** for reviewing the programme, but makes a recommendation to the *Project Manager*, who actually accepts it under the contract? Or do they have **authority** to accept the programme (in which case they will need to be delegated this authority under the contract)? Who needs to be **consulted** with prior to the recommendation or acceptance? And, once accepted, who needs to be **informed**, either of all of it or only the parts that affect the individual in a particular role? For example, on a *Contractor*-designed contract, the designer who reviews drawings submitted for acceptance – whether just responsible for reviewing or having authority to accept – needs to be informed what drawings are going to be submitted and when, so that they can make time to turn the drawings around within the timescales of the contract (i.e. the *period for reply*).

And you could do a similar thing within the *Contractor's* organisation for programming. Indeed, for all professional roles. For example, while different sections might have different section engineers, rather than stating each engineer's name, you state the role of 'section engineer' and then define the specific responsibilities for that section. Likewise, you might have different technical leads for different aspects of the design, so the role is 'technical lead'.

If you look at the paragraph above, you might notice the words highlighted in bold: responsible, authority, consulted and informed, or RACI for short. These form the basis of RACI diagrams, a reasonably well known management tool. Essentially, you have a matrix, with the decisions that need to be made in the first column and then, in subsequent columns, who is responsible (ideally one person), who has authority (definitely one person), who needs to be consulted (usually a number of people) and who needs to be informed (again, usually a number of people).

I have seen this done for every action in an ECC contract – which seems a bit excessive to me – but it absolutely makes sense for the main actions under the contract for any medium-sized or lager project. By main actions, I mean those shown in Table 3.1.

As you can see from Table 3.1, there are many questions to answer. If they are not answered, project participants will be left in a state of uncertainty and confusion. Either this uncertainty and confusion will never be resolved, or it will be resolved slowly. So to hit the ground running, resolve these questions early!

It is more than worthwhile for the *Project Manager* and the *Contractor's* site manager to do a first draft RACI diagram, possibly by preparing individual diagrams and bringing them together to see how they match up. For instance, a *Contractor's* organisation that pushes accountability down to the lowest level might not be compatible with one in which the *Project Manager* is accountable for everything (i.e. is not willing to delegate anything). The *Contractor's* organisation would be very responsive and able to make decisions fast but, without good information flows, might act in an uncoordinated manner. The *Client*-side organisation might be slow at making important decisions, and not make minor ones at all because the *Project Manager* only has time for the major ones. Somewhere in the middle is a happy medium.

Having decided upon the RACI diagram, a number of things need to happen.

- The people concerned need to be 'informed' about what aspects they are responsible and accountable for and who they need to consult with and inform.
- To provide certainly on the *Client* side, under clause 14.2 the *Project Manager* and the *Supervisor* need to notify the *Contractor* of the actions they are delegating (i.e. the accountable ones). I suggest that the *Contactor* does the equivalent, even though this is not explicitly required in the contract. The *Client* side and the *Contractor* both need to make the other aware of who is responsible and accountable.
- The mechanisms for good information flow need to be put in place, both formal (e.g. for the progress of a compensation event notification for a small change, from its section engineer originator via the site manager (who contractually notifies), while keeping the quantity surveyor informed) and informal or verbal. This is partly addressed by office layout etc. (see principle 2) and partly by the cycle of meetings and agenda items that are set up (see principle 4).

The article by Neilson *et al.* (2011) that I previously cited also mentions the importance of knowing information that might not be directly relevant to a role, but nevertheless may need to be taken into account when making decisions – i.e. that the left hand has an idea of what the right hand is doing. For instance, a small technical change done for a good technical reason can have an unexpectedly large effect on time and cost if it is on the critical path.

Table 3.1 Questions on who has authority and who is responsible for principal ECC actions

Client side	*Contractor* side	Example
Who reviews and accepts the programme?	Who prepares and submits the programme?	Does the *Contractor's* programmer consult with its site manager before submitting it? Or does the *Contractor's* site manager submit it? Likewise with respect to acceptance for the *Project Manager* and *Client's* programmer? Do the two programmers consult with each other before any submission and/or response?
Who can early warn?	Who can early warn?	Can a section engineer raise and send an early warning on the system (in which case they are responsible and accountable) or does it have to go through the *Project Manager/Contractor's* site manager (in which case the section engineer would be only responsible)? Can *Supervisor* staff raise an early warning?
Who can instruct an early warning meeting?	Who can instruct an early warning meeting?	Can a section engineer – from either the *Client* or *Contractor* side – instruct an early warning meeting in order to resolve a technical issue out on site immediately, without delaying works?
Who can change the Scope?		As a result of an early warning meeting, can the *Client's* section engineer, as a delegate of the *Project Manager,* instruct a change? If 'Yes', up to what value? What about *Client*-side designers? Or do they do drafts that are reviewed by the *Project Manager*, who then formally instructs the change?
Who can accept design?	Who can submit design?	Does the *Client's* designer have the delegated authority to accept the design directly or do they merely make a recommendation to the *Project Manager*? Can the *Contractor's* designer submit designs for acceptance directly or do they need to go via the *Contractor's* design coordinator?

Table 3.1 Continued

Client side	*Contractor* side	Example
Who informs the *Contractor* of a test's results or a Defect?	Who informs the *Supervisor* of a test or inspection and its results or a Defect?	Do all test results come through the named *Supervisor,* or do they come from the *Client's* section engineer who actually did the test? Likewise, for the *Contractor*, does it all go through the *Contractor's* quality manager or do the section engineers inform their equivalents directly?
Who really assesses the *Contractor's* payment application and who certifies payment?	Who puts together and who submits the payment application?	Are the respective quantity surveyors responsible and accountable for both? Or does the *Project Manager* certify/site manager submit (i.e. they are accountable)?
Who within the *Project Manager's* team can notify a compensation event and/or reply to a *Contractor's* notification of one?	Who within the *Contractor's* team can notify a compensation event?	For *Client* section engineers, this relates to 'Who can change the Scope?'. On the *Contractor's* side, can section engineers, designers, etc. notify directly, or do notifications have to come from the site manager, or from another *key person* identified in the Contract Data?
Who can respond to compensation event quotations?	Who can prepare and/or submit compensation event quotations?	Most of the time the *Contractor's* quantity surveyor will prepare quotations and the *Client's* quantity surveyor will evaluate them (i.e. be responsible). Can they also submit and respond to quotations under the contract (i.e. be accountable too)? If so, up to what value? Or does it all have to go via the *Project Manager*/site manager?
Who can instruct work to stop or not to start?	Who can instruct work to stop or not to start (as an internal question)?	Is it only the *Project Manager*? What if a someone is unhappy with an aspect of health and safety?

Two final points.

- Encourage people to have face-to-face, video conference or telephone conversations, so that matters can be discussed and agreed verbally and confirmed and summarised in a form which can be read, copied and recorded (as principle 1).
- While keeping the relevant *Project Manager* and *Contractor* personnel in the loop, encourage direct informal/non-contractual communications between those who are responsible. For example, they should send drawings or proposals direct to each other and, once agreed for acceptance, copy – and note I do say copy – the formal communication to their counterpart, even though the formal contractual communication route for acceptance is via the *Contractor* and *Project Manager*. This speeds up responses, which might otherwise become stuck in the *Project Manager's* or *Contractor's* inbox.

> On three contracts within the original research sample, technical discussions were held directly between the *Employer's* designers and the works Subcontractors under the *Contractor*. The *Contractor* and *Project Manager* were then informed of the technical decision, with the latter confirming its implementation contractually by using pro formas. However, on one contract the *Employer's* designers and Subcontractors were communicating directly, often without the *Contractor's* knowledge. This sometimes resulted in small savings in direct costs being outweighed by delay and disruption costs when the *Project Manager* rubber stamped the change without consulting the *Contractor*. Control of unnecessary and late change is a crucial aspect of good project management.

PRINCIPLE 4: HAVE REGULAR 'WRAP-UP' MEETINGS

One of my favourite quotes is that 'good communication is a series of small agreements', and we will see this when we look at having early warning meetings and agreeing quotations for compensation events.

If you are having regular informal meetings and conversations, then hopefully you are reaching lots of small agreements, and so formal meetings will be a lot shorter. However, at some point, decisions need to be put to bed and actions confirmed. The NEC4 conditions of contract include only one regular meeting – the early warning meeting – but I suggest you should be having up to six regular 'wrap-up' meetings under an NEC contract, which are as follows.

- Early warning meetings every 1–2 weeks, which under NEC4 are a contractual requirement, with the period 'at no longer interval than the interval stated in the Contract Data'.
- Programming meetings, with the minimum interval being the interval stated in Contract Data part one for the submission of a revised programme (normally monthly or 4-weekly). The meeting should be held before the formal submission for acceptance of the revised programme in order to agree which delays are allocated to compensations events and which delays are the liability of the *Contractor*. Having agreed these, the logic can be incorporated into the submitted programme and the use of resources on the compensation events handed over to the quantity surveyors to agree the costs.
- Regular quality meetings between the *Contractor* and the *Supervisor* in order to plan out and confirm testing and inspection requirements and keep on top of the correction of Defects. These are important throughout the contract, but particularly so when take over

and Completion approach – when it will be necessary to agree, for example, which notified Defects will prevent the *Client* or Others using the *works* and an action plan to resolve them.

- Compensation event 'wrap-up' meetings at least fortnightly and in sync with, but just after, the programming meetings, where any queries about whether or not an event is a compensation event are resolved, the approach to costing up a compensation event are agreed and quotations are finalised and effectively agreed prior to submission.
- Meetings to agree the *Contractor's* applications for payment before they are submitted, along with, under options C to F, the forecast total Defined Cost to Completion.

You could argue that meetings of this sort happen on the best-run projects in any case. I would agree. However, formalising these arrangements makes it far more likely that the meetings will actually take place.

On one contract, partly to ensure that decisions were made on small items, and partly to impose on his construction managers the discipline of agreeing things to a timescale, the *Project Manager* set up fortnightly meetings to clear minor early warnings, technical queries and compensation events, in addition to the normal progress, quality and safety meetings. The *Client* has since written a procedure into the contract to make these 'wrap-up' meetings a contractual obligation.

Lastly, many contracts involve an element of *Contractor* design, so it makes sense to add regular design meetings to the list above. Here, the *Client's* reviewer can hopefully provide, on behalf of the *Project Manager*, constructive and useful input and the design approach is agreed,[2] queries can be resolved verbally and designs tweaked to ensure drawings are accepted first time.

PRINCIPLE 5: DEVELOP PRO FORMAS

On all but one of the sites researched in the original sample, a system of pro formas was used. Once developed, these saved time by removing the need to write letters from scratch. I would suggest that pro formas make extensive use of tick boxes, to reference (as opposed to quote) relevant clauses and direct people to action what has been agreed verbally, as opposed to stick the blame on the other Party!

I know of a lot of variation in the use of pro formas: at one extreme was a site that had one, while at the other extreme was a site that used over 30! I would suggest something in between, such as the following.

1. An early warning notification and action sheet, which could originate from either Party. As well as categorising the early warning (see Section 3.6.5), I suggest the top part is more concerned with defining the issue, and the bottom part the actions that have been decided at the risk-reduction meeting, whatever form that took (e.g. out on the Site, the fortnightly 'wrap-up meeting'). Sometimes, when this sheet comes from the *Contractor*, it includes a tick box to notify of a potential compensation event at the same time.

[2] Note, however, that the *Contractor*'s obligation is to do design (and ultimately the physical work) in accordance with the Scope, not in accordance with the *Client* reviewer's personal whims or foibles. Nevertheless, it can only be a positive if the *Client* reviewer can both give constructive input and identify where the Scope is deficient early on. An early change is better than a late redesign and resulting delay.

2. A *Contractor's* submission sheet for items for acceptance by the *Project Manager*, such as drawings, programmes, applications for payment and compensation event quotations. There could well be separate pro formas for each main type of submission, but you would then need another form for 'any other submission'. For compensation event quotations, it includes a summary of the proposed changes to the Prices and the Completion Date.
3. A *Project Manager's* response to a communication submitted for acceptance, whether it is a drawing, programme, quotation etc. Again, there could be separate sheets for the different types. If it is a response to a quotation, then it includes a summary of the changes to the Prices and the Completion Date, and if it is for a proposed instruction, whether or not to proceed.
4. A *Project Manager's* instruction sheet, with three classifications on it: general instruction, instruction changing the Scope and proposed instructions (which would be an instruction for a quotation for a proposal to change the Scope!). This includes a tick box for instructing quotations.
5. A *Supervisor's* sheet for notifying the *Contractor* that an inspection is to be done, the results of a test or inspection, to record the discovery of a Defect, or an instruction to search. The *Supervisor* generates a separate communication as the Defects Certificate at the *defects date*.
6. A *Contractor's* sheet for notifying the *Supervisor* that an inspection needs to be done by the *Supervisor* (or will be by the *Contractor*), the results of any test or inspection done by the *Contractor*, the discovery of a Defect generally and, for a Defect, when it is proposed that it will be corrected.

These pro formas will cover 80–90% or more of all communications, especially if there is an 'Other' tick box, with a space for the relevant clause number and a little room for explanation. For instance, if the *Contractor* wants to expand its Working Areas, it would use pro forma 2, tick the 'Other' box, reference clause 16.3 and title it 'Proposal to add to Working Areas'. The *Project Manager* would use pro forma 3 in a similar way to communicate whether or not they accept the proposal.

I also suggest each pro forma has the facility to reference previous relevant communications, so that an audit trail can be worked back through if necessary.

A simple system like this avoids the writing of long letters, in which each Party tries to include contractual terms to suggest or refute a later claim. Instead, people have noted that more agreed information was communicated, that it was more focused on the solution and, more often than not, did not need to be referred to again, as the subject had been put to bed and dealt with.

Four further tips.

- Contractually, under clause 13.7, it is only notifications and certificates that need to be communicated separately (e.g. on an individual pro forma or email). For ease of referencing and retrieval, and for responding, I suggest all contractual communications are dealt with individually. This might seem heavy on administration, but if the pro formas are set up correctly, as suggested, it actually allows much more focused communication, as each one is far easier and quicker to fill in than writing a letter.
- I generally recommend the use of cloud-based NEC contract administration systems (see Section 3.3.2), but for smaller projects, emails will suffice. Rather than having to attach

a Word document or Excel spreadsheet pro forma to the email, the email can be set up as a pro forma. This can be done by copying the pro forma text from a Word document into the signature part of Outlook, so that each 'signature' is a different pro forma.

- Develop a clear labelling system for each type of communication under the contract – and make sure is used religiously! The system would specify what should appear in the title block of all communications. For instance, communications under the contract named 'Lower Beeding Flood Alleviation Scheme' would start with the acronym 'LBFAS'. All early warning notifications would then have 'EWN', followed by the number of the early warning. Likewise for compensation event notifications (CENs), compensation event quotations (CEQs), *Project Manager* instructions (PMIs) etc.
- Each Party should have a contract email address (e.g. the acronym for the contract followed by its organisational address), into to which all emails, both from and to the organisation, are copied.

The effect of the above four bullet points is that communications will be easily traceable and retrievable. They do, however, require discipline to operate.

Many organisations will already have a system of pro formas for use on NEC contracts. In fact, they will have numerous different systems for all the contracts they have been involved in. This is a good opportunity to compare and contrast in order to come up with the best system for your contract.

When one Party is new to the NEC, it is especially good to 'sell' the benefits of the NEC and develop their buy-in to its effective and efficient joint operation. This can be done by sitting down with them and going through each pro forma, explaining what it is for and adjusting it in the light of their comments. This is an ideal way to establish some 'common understanding' without calling on an external trainer or facilitator.

PRINCIPLE 6: PUT IN PLACE A LOGGING IN AND OUT SYSTEM

The pro formas need to have continuity, and a procedure is needed to keep track of the various stages of communication. On small contracts, this could be an A4 book held in a central place, such as the office lobby. I suggest the pages of this book are divided into the following columns.

- Communication type and sequential number (e.g. EWN 5, CEN 7).
- Previous relevant communication, if applicable (so that there is an audit trail).
- Date received (which should be the same date as when it went out, but it is when it is received that counts contractually, as per clause 13.2).
- Written description (e.g. 'Quotation for revised brickwork to bridge front' or 'March programme submission for acceptance).
- Date response required by contractually.
- Date response required by to avoid unnecessary time or expense (e.g. ASAP or a specific date).
- Date response received.

Exactly the same columns could be put on a spreadsheet.

It could be the duty of the receptionist or office manager to keep this log up to date, or, if in electronic format, the log could be on a project server or cloud based, with all the relevant people given access to it. The main thing is that it is accessible to all, so that numbers can be sequential.

On one large NEC2 contract, a full time administrator was jointly employed by both Parties to keep track of communications and ensure each organisation responded within the timescales specified in the ECC. The administrator had the authority to chase people internally within both organisations as the deadline approached, which brings us onto the next principle.

PRINCIPLE 7: EXTEND THE SYSTEM UP, DOWN AND ACROSS

By 'up', I mean that, if the *Project Manager* has to consult back to the *Client*, then the *Client* person needs to get back to them at least within the timescales of the contract. The system could be as simple as highlighting this on the communication to the *Client* (having made a phone call) and putting a reminder in the diary/Outlook to chase them up.

Likewise, on large operational sites, such as an airport or hospital, the system should be extended across, as there may be a need to involve and obtain decisions from technical experts who are only loosely affiliated with the contract related to the project.

For a *Contractor*, if the communication system does not extend down to include Subcontractors, the *Contractor* may be putting itself at risk due to a *Contractor's* action or default, such as in the following situations.

- The *Contractor* has to supply a quotation to the *Project Manager* within set timescales, but does not pass the relevant information on in sufficient time to the Subcontractor who is doing the relevant work. By the time the Subcontractor responds, the *Contractor's* quotation has to be in.
- A Subcontractor gives an early warning that is not passed on to the *Project Manager* and the problem results in a compensation event. The Subcontractor is entitled to its full Defined Cost plus Fee, but the *Contractor* might have a reduced assessment.

Even if a specialist Subcontractor, let us say under the Engineering and Construction Short Subcontract, is not communicating much 'in a form which can be read, copied and recorded', they need to be informed of the key communications, because subcontracting does not dilute the main *Contractor's* responsibilities in any way, shape or form under the main contract. It is, therefore, in the *Contractor's* interests to educate Subcontractors and make them conform to the contract and its timescales.

PRINCIPLE 8: SET UP PRINCIPLES 1 TO 7 BY THE *STARTING DATE*

On a number of sites, personnel admitted that, with hindsight, they had not given the above issues enough consideration prior to work commencing on site. They were therefore having to put communication processes in place rapidly and without due consideration, which often resulted in later modification. The site that ended up with 30 pro formas was one where they originally had none. They became tired of writing out long letters, so they started to take time out to write pro formas as and when a new one was needed. However, they never sat back and thought it through as a system of forms, which is why they ended up with so many.

Not having these principles in place and the time spent putting them in place when the contract is in full flow all detracts from the management of the contract, which means that both extra and unnecessary time and cost are being incurred out on site as well as in the offices, as professionals' time is being wasted. In the words of one *Project Manager*, under an ECC contract 'you need to hit the ground with your feet running, because where you get hit hardest is in the early stages'.

3.3.2 Use of a cloud-based software-as-a-service communication system

3.3.2.1 What is a communication system?

I use the term 'communication system' because this is the term used in NEC4. It was introduced in NEC4 clause 13.2, which states the following.

> If the Scope specifies the use of a communication system, a communication has effect when it is communicated through the communication system specified in the Scope.

The communication system could, in theory, be an email system, as described earlier in this chapter. However, the term is generally taken to mean one of the cloud-based software-as-a-service (SaaS)[3] NEC systems that are available on the market.

The advantage of using a SaaS system for a project is that people involved in the project have access to a common, up-to-date repository for information in 'real time'. That is, because the information is not held on individual computers, people can access the latest documents, and there is one version of the current truth. This is how document-control systems operate.

However, cloud-based SaaS systems are normally not just repositories for information: they also allow you to do things with that information. In the case of the NEC-based systems on the market (which number 6+, at the time of writing), there is quite substantial software that the providers like to label as 'workflow' systems, in that they guide you through the process of administrating an NEC contract by, hopefully, following the logic and procedures of the ECC. Or more precisely, some do, and some do so only for the change management aspects of the NEC contract (e.g. early warning clauses and compensation events). In practice this means that if, for instance, the *Contractor* notifies an early warning to the *Project Manager*, the system will: record when it was sent and when it was opened; give a drop-down menu for the allowable contractual responses (e.g. **not** allow a *Project Manager* to 'reject' or 'not accept' an early warning notification, which I mention for wider educational purposes!); automatically generate the right pro forma for the chosen response; etc. This is the carrot: it makes it easy for contract participants to follow the contract and to retrieve information put into the system.

An NEC SaaS system will typically have the following core features

- automatic generation of pro formas in accordance with the contract
- date stamping of both when the pro forma was sent and when it was opened
- if a contractual response is needed, a countdown facility showing the user how long they have to respond before being contractually non-compliant, with 'amber' flagging as the end of the contractual timescale for response approaches and 'red' flagging where the timescale has been exceeded
- a database or central repository in which all communications are stored and can be located via a 'search' facility (the better systems link communications together so that there is a full audit trail)

[3] 'Software as a service' is a term used for software that is not hosted on the user's computer, but on a remote – and hopefully secure – server run by the service provider (i.e. in 'the cloud'). To use it, each user needs to (a) be able access the world wide web with a sufficiently fast connection, (b) have a browser (such as Internet Explorer, Google Chrome, Safari) installed on their computer and (c) have someone pay the service charge (normally monthly) to the service provider on their behalf.

- ability to attach technical and commercial documents to contractual communications and, for some systems, hyperlink to an integrated document management system
- ability to summarise in customisable format the current financial and time status of the contract in terms of both implemented compensation events and those being progressed
- ability to delegate actions, so that not everything has to come through the named *Project Manager*, *Supervisor* or *Contractor's* site manager
- read-only access for those who are involved in the day-to-day running of the contract but are not authorised to send contractual communications, and for senior people sitting above the contract, allowing them to check the status of any particular contract and compliance in terms of administrating as obliged to.

Taking the last point, all of the systems to my knowledge provide some nice visuals that summarise not only how many early warnings, compensation events etc. there have been and how many there are in process, but also highlight outstanding communications – and this is the stick! I know of both *Client* project sponsors and senior *Contractor* personnel who like the fact that they can look into a contract and check its status in 'real time', and can see whether it is being run properly and, if not, exactly who is not doing what they should be. There are far fewer places for the slack or incompetent – or just plain overworked – manager to hide. And if the contract is being administrated as the NEC is written, it generally means the contract participants are doing good project management, which means that the project has a far higher likelihood of being successful.

3.3.2.2 Why use a software-as-a-service system? Tangible and intangible benefits

The benefits of using a SaaS system can be split into tangible cost savings, which are relatively easy to measure, and intangible benefits, which are harder to measure but often just as important.

TANGIBLE COST BENEFITS

Independent academic research (Ming Sun and Tejas Oza, 2008), partly funded by the then DTI (and partly funded by one of the suppliers), identified the purely financial benefits from three projects (Table 3.2). The stated savings ignore other potential savings arising from more general contract administration. On the other hand, cost is not mentioned. While I do not know the exact

Table 3.2 Case studies and tangible benefit calculation results Reproduced from Ming Sun and Tejas Oza, 2008

	Case study 1	Case study 2	Case study 3
Project description	£34 million building refurbishment project	£6 million substation power plant project	£3 million civil engineering project
Project duration	20 months	18 months	18 months
Number of early warnings	78	53	105
Number of compensation events	123	79	106
Value change before instruction	13	6	9
Value change after instruction	110	73	97
Consumable cost savings	£1969	£1294	£2017
Administrative staff cost savings	£11 445	£7497	£11 052
Managerial staff cost savings	£41 954	£27 629	£41 617
Total tangible savings	£55 368	£36 420	£54 685

cost of the system to these projects, commercial intelligence suggests that it is likely to have been in the range of £1000 to £2000 per calendar month for case study 1 and less than £1000 for case studies 2 and 3, indicating a return on investment of at least 2 to 1, or over 100%.

However, two things I should mention.

- The figures in Table 3.2 for the monthly cost are at 'book' prices with no discounts. If you take into account the increased competition and reduced costs of IT since the study, and opportunities for discounts on bulk purchases by an organisation, the actual monthly costs are likely to be significantly lower. This would push up the return on investment.
- My personal view, based on a reasonably detailed look at one contract, would be that the savings generated per communication are on the high side. This would lower the return on investment.

INTANGIBLE BENEFITS

If the tangible benefits are good, then the intangible benefits were found to be off the scale. Four specific benefits were consistently highlighted by users in the research (Ming Sun and Tejas Oza, 2008), which correlate with my discussions with people who have used the particular system (for the NEC3 SaaS supplier for whom I carried out a customer research survey, I interviewed over 80 individuals, some of whom had used other systems as well). These benefits were

- improved contract change management efficiency (i.e. it takes less time, and hence less cost, to administrate)
- improved contract compliance (i.e. people followed the contract more; or in my language, less room for slackness), which leads onto, from a senior management perspective
- reduced project risks (which could be both due to better project management due to contract compliance and/or to less contractual risk due to less subjectivity in the contract), which in turn leads onto
- improved predictability in project outcomes.

Particularly in the IT/software sector, there has been quite a bit of research on 'project capability maturity', which looks at how an organisation and project team run their projects and correlates it with project success. Put simply, the greater the maturity level on a scale of 1 to 5, the better the inputs and processes, and hence the greater the likelihood of a project being a success.

Ming Sun and Tejas Oza's (2008) research highlighted a number of areas in which use of the system increased the capability maturity levels for how the project was managed over and above the increase resulting from use of the NEC family of contracts alone. The case study contracts in Table 3.2 achieved a maturity level between high end (3) and low end (5), as compared with the construction industry benchmark, at the time, 1–2. That is, use of the NEC family of contracts increases the likelihood of a project being a success, and this is further enhanced by use of an NEC cloud-based SaaS system.

To summarise, I am a big fan of these systems, but I introduce two caveats.

- On one series of contracts let under a framework, the increase in Prices was only an average of 2%. This was because of the amount of joint work put into making the Works Information (as it was under NEC3) right in the first place. The regional managing quantity surveyor thought that, with such a minimal amount of change, the system did not offer value for money. In other words, the less the change that actually occurs, the less the return on investment from a SaaS system.

- A danger that I have seen a number of times is that people find it too easy to administrate the contract by these systems and forget to talk to each. Remember principle 1 for effective communications: discuss and agree verbally (or TALTEO), confirm and summarise 'in a form which can be read, copied and recorded'!

3.3.2.3 What's on the market? The known suppliers

At the time of the final review of this this section (December 2020), I am aware of the following systems on the market.[4]

- The MPS Contract Change Management (CCM) system, the original system that started it all.
- The Aconex system, which was second on the market, now owned ultimately by Oracle. It started off as a document management system and added on an NEC module, which I believe you can have on its own.
- The thinkproject Contract Management System, which was previously known as CM Toolkit or CEMAR.
- The Sypro system.
- Asite, which has a number of project management modules and a document control system which can link into its NEC module.
- BuiltIntelligence's FastDraft system – the relatively new kid on the block.

3.3.2.4 Which system should I choose?

The answer is, of course, it depends, and I am not going to answer this question for you. Instead, I urge you to follow good project management practice and define what you want from the system before going out and looking at them and negotiating a deal. In fact, I would urge you to more than look at them – get your hands dirty by playing on them with predefined test scenarios. More will be said on this at the end of this section.

For the time being, here are some questions to help you write that end-user requirement.

THE BASICS

- Is the service provider financially stable? You do not want it to disappear with all the data halfway through your project!
- What is its historical availability in terms of system up-time?
- What level of security and back-up do you insist on? Your IT people should be able to advise you on this, but what does the system provide in terms of anti-virus protection, firewalls and back-ups? One system I know of is backed up on a daily basis, which in turn

---

[4] Let me declare some perceived, but not actual, conflicts of interest.

- For many years I was on a retainer as technical advisor to MPS, which runs the CCM system. I stopped doing this because I did not think I was earning the money – they rarely asked for my advice and, when they did, I was often too busy to give the amount of input desired. I also never received any commission for recommending the system as I thought it would undermine my impartiality as an independent consultant.
- While I was once a shareholder in BuiltIntelligence and continue to have eLearning published through them, I have had only an afternoon's input into their FastDraft system and have never received any financial benefit from sales whatsoever.

is backed up on a weekly basis which in turn is backed up on a monthly basis, with each back-up stored at a different location.

- How is the system charged for in reality? I know that some providers have a stated way of charging (e.g. per type of user), but in reality will do a deal.

JUST NEC OR THE FULL MONTY?

- Is your organisation already using one of the systems mentioned above for doing other things within our organisation? If the answer is 'Yes', and particularly if the project team is also familiar with and needs that functionality (e.g. a document management system), then the most cost-effective solution might be to just add on the NEC module.
- Related to the above point, how well does the NEC module interface with the other modules? At one extreme, the modules might not integrate at all, which undermines the case for using it. At the other extreme, if the modules integrate well and you want this integration, it will save you significant time. If, however, you will rarely use all the capabilities and features of the 'do it all' system, the user may well have to check more boxes and fill in more fields than is necessary, thus wasting time! One supplier takes the opposite view, and markets itself on the fact that, because it has limited capability, it is simple to use and therefore people do actually use it.
- Do you want the other non-NEC modules or will you be happy with the information, graphs etc. that the pure NEC systems can generate?
- If you go for a purer NEC system, how easy is it to import and export data from and to other systems? Can it be automated, or do you have to delete cells etc. every time.

BREADTH OF APPLICATION

- Does the service provider support the full set of NEC contracts you are likely to use?
- Do you want to fill in the Contract Data yourself, which is then carried forward into both the document pushed out for tendering and into the operation of the contract? This undoubtedly saves time, but what are the constraints on doing this if you want to modify the standard Contract Data for good reason? (See Section 2.2.2.11 for my thoughts on how to write option Z clauses, which might well equally apply to other modifications in the Contract Data.)
- How are option Z clauses incorporated into the system? One provider – presumably for a fee – runs a flow-chart check on Z clauses to avoid ambiguity and inconsistency before incorporating them into the logic of the system. This is a good check (again see Section 2.2.2.11).
- Do you only want the system to administer communications related to change?
- Or do you want it to be able to administer all communications under the ECC or equivalent form? Does this include communications to and from the *Supervisor*?
- And if you do want it to do it all, how comprehensively? A number of systems say that they can do it all, but in reality all non-change management communications are titled as something like 'general communication', to which you have to add the particular clause number etc. and the other Party has to self-generate an equivalent response.

USEABILITY VERSUS FUNCTIONALITY, OR USEABILITY AND FUNCTIONALITY?

- As a very general rule, the simpler the system, the easier it is to use, and potentially cheaper it is, as it has less functionality.
- What is the weighting you are going to give between useability (of which visuals are a subset) and functionality?

- From playing on some systems/having them demonstrated, I know some have looked great initially and have good basic functionality, but once you either try to do something a little more advanced or ask the demonstrator how it does X (e.g. link early warnings to compensation events or arrange data in a certain way), then functionality is found to be lacking.
- I know from the feedback of users of one system that, despite it not looking as visually impressive as other systems, with a little bit of practice it is easy to use.
- Nice visuals eat up data! If your site is located in the middle of nowhere, you may have to – even these days – rely either on limited mobile network coverage or non-broadband connections. Even decent broadband can slow down at peak times. This may mean pages with fancy graphics take minutes to download and be unreliable in uploading. This also needs to be considered when setting up a site if it is going to take weeks or a month or so to get a decent connection.

EASE OF NAVIGATION – TOP-DOWN AND BOTTOM-UP

- Do you want to run multiple contracts from the same interface? For instance, some systems allow a project manager of multiple contracts to see all current communications on all their contracts on one page. This is also useful for a *Contractor* with multiple Subcontractors. This compares with at least one system where you have to log in separately for each contract.
- Related to the above point, do you want the system to be able to amalgamate data from projects up to a programme or portfolio level (or, if a *Contractor* with Subcontractors, the main contract level)? What specific data do you want to be able to do this with, and can the system do it?
- How well does it present information in summary form? For instance, as a dashboard, and how flexible is this dashboard? That is, is it defined by the service provider and 'a given', or is the project or individual user allowed to modify it to suit their role and preferences?
- How easy is it to drill down into the detail from any dashboard?
- If a *Contractor* receives a communication from a Subcontractor, how easy it is it to 'postbox' the same or similar communication up to the *Project Manager* (or *Supervisor*), and vice versa for a communication from the *Project Manager* onto the relevant Subcontractor?

EASE OF NAVIGATION – FOLLOWING A PROCESS AND AUDIT TRAIL

- How well does the system link communications together? For instance, do early warnings naturally link with compensation events? One system goes a stage further, linking compensation events to early warnings to the original activities in the Activity Schedule. This then allows a compensation event by compensation event update on the change to the prices of each activity. Very transparent and auditable!
- How easy is it to follow through (possibly in reverse) a sequence of communications?
- Does the system allow you to add notes to contractual communications? For instance, following an instruction from the *Project Manager* to submit a quotation, the *Contractor* starts to make an assumption. Having had a verbal discussion to confirm that this is the basis on which the *Project Manager* wants it to assess the compensation event, does the system have a means of recording this, or does it have to go back to the start of the process?
- Does the system have an easy to use and decent search facility? For one system, someone complained that it only seemed to look for words in the titles of communications.

DELEGATIONS AND ACCESS

- Does the system have a number of categories for accessing it? For instance: read only; input and send (as in the *Project Manager* or *Contractor's* site manager); draft and only send after the notified individual (say the *Project Manager* or *Contractor's* site manager) has approved it; and so on.
- How does the system do delegations? Does the *Project Manager* have to delegate all their actions, or can they delegate actions in individual clauses (e.g. accepting quotations up to a certain value by their quantity surveyor)? Likewise for the *Contractor's* site manager. What about for the *Supervisor* (if it covers their role)?

SETTING UP THE SYSTEM AND LEVEL OF SUPPORT

- How quickly can the supplier commit to setting up the system from the time of award?
- If being used as an enterprise-wide system, can you set up a contract yourself?

> A number of systems allow you to do this yourself and consequently charge a lot less for the 'self-serve' model. Another one, certainly historically, did not, insisting every new contract is set up by the supplier, which was reflected in the price.

- What level of support does the supplier provide and what support do you want?
  - Does someone visit your site at the start of the project and train the team?
  - Does that person re-visit for further training clarification, if necessary, once the team has understood the basics and wants to start using the most advanced features? Do they charge for this?
  - Does the system have the equivalent of Guidance Notes for each page?
  - Can users toggle between the system, the words in the contract and the words in the Guidance Notes that accompany each contract form?
  - Are there online tutorials and, if so, how good are they?
  - What if there is a problem, either with the system or a user not knowing and unable to find how to do something? How do you contact the supplier, and how quickly do they respond?
  - Are users allowed access to a dummy project, where they can play with the system and make mistakes, rather than make mistakes on the 'live' system. They then may have to gain agreement with the other Party for the service provider to go in and change the data on the system, which is a hassle for everybody concerned!

CHECKS AND BALANCES TO PREVENT MISTAKES

- Does the system have gates, and how effective are they in stopping incorrect input of data? By this I mean, can users see what will happen when they tick an 'Are you sure?' or 'Submit box'. If an extra tick box just appears on the screen to confirm, then it is of little benefit.
- Related to the above point, does the system allow communications to be prepared in draft form and saved?

USE OF DATA AND ANALYTICS

Compared with 2012, when the previous edition of this book came out, we now live in an era of 'big data'. Data can increasingly be used to gain insights into which areas are working and which that are not. A contract change management system, particularly for a repeat-order Client with a lot of projects, will progressively build up a treasure trove of data around sources of additional cost and time. In an NEC context, the most likely sources of this data will be early warnings, compensation events and Defects, but you will often need to go back a stage further than the matter or event to identify the source. For instance, is the change to the Scope due to the end user changing their mind, poor design, lack of design etc. Here are some questions to ask.

- Who owns the data: the system supplier or the Party paying for the system?
- How easy is it to download the data for an individual contract, series of contracts etc. so that it can be manipulated on evaluated on external software?
- What are the system's in-built capabilities with respect to data analysis at project, programme and portfolio levels (e.g. how easy is it to cut and dice by project type etc.)? Is data presented is a way that can give insights?

WHAT HAPPENS AT THE END OF THE CONTRACT?

If you do not have an ongoing corporate service agreement with the provider, then your contract will at some point have to be turned off in order to avoid ongoing service charges. While the service provider will presumably keep the data somewhere, how do you access it if you do need it? One solution is to spend some time (days and possibly weeks) downloading all the information, but this loses the automatic linkages between the various communications. On the other hand, one provider – for an additional fee – can provide all the communications in read-only pdf format with hyperlinking, so that it is has the same navigation as the 'live' system.

SUMMARY

Choosing the right system for your project or projects depends on a whole load of factors, including price. However, the ultimate test comes down to what a system can do that is relevant to your needs and wants, and how easy it is to use. If I were choosing a system, I would do the following.

1. Define what I want from the system, both organisationally (in terms of reporting upwards) and operationally (in terms of day-to-day, week-to-week and month-to-month management and administration of the contract).
2. Involve someone who knows the NEC in the above, who will write test scenarios, relevant to point 1, that involve the participants to the contract. By test scenarios I mean, for instance: how does the system handle the whole change management process between the *Project Manager* and *Contractor* from early warning through to an implemented compensation event? How do I access and find historical contractual communications and follow an audit trail? How do I get information out of the system in a certain format?
3. For each potential system supplier, let them give a general demonstration of their system.
4. Have a play with the system on a dummy contract, with at least two user windows open (normally the *Contractor* and *Project Manager*), and try to do the test scenarios developed in point 2 yourself. This will give an indication of how user-friendly the system is.
5. Get the service provider to demonstrate how the system does the scenarios, if at all – this will confirm, or not, its functionality.

If you are working in the public sector, you will need some sort of scoring system to evaluate the system objectively, and this would have to be communicated to the provider. Even if you are in the private sector, without going over the top, using some sort of scoring system would be professional.

Finally, having chosen a system, you will have to name the system in the Scope document (as the 'communication system'). As stated at the start of this section, this is a new requirement to NEC4 (under NEC3, you had to write an option Z clause). However, what happens if the cloud-based system does not cover a particular communication because it is rarely used, such as an action related to termination or the carrying out of a dispute under the adjudication clauses? For this reason, having specified the communication system in the Scope, I suggest writing an option Z clause that says something like: 'If the communication system does not cover a communication under this contract, then …'.

3.4. Operating the payment clauses

This section is primarily written for when options C, D and E, the cost-based options, are being used. Two other issues are covered first though: the general operation of the payment clauses, and how to work the option B and D clauses that govern when a change in quantities becomes a compensation event.

3.4.1 Payment process

Before I start, a quick word about the payment terms. The exact operation of the payment provisions will depend on whether the work is being done in England and Wales and whether it is subject to the provisions of the Housing Grants, Construction and Regeneration Act 1996, as amended by the Local Democracy, Economic Development and Construction Act 2009 (often just called 'the Construction Acts'). Similar provisions also apply in Scotland and Northern Ireland, albeit with some minor changes in detail, and in New Zealand, with substantial modification. In fact, if you are a New Zealand reader, then disregard the rest of this section. If your contract is subject to the UK Construction Acts, option Y(UK)2 should have been specified.

IF THE CONTRACT IS NOT SUBJECT TO THE CONSTRUCTION ACTS

The *Contractor* has to submit to the *Project Manager* applications for the payment of any amounts it considers are due, giving details of how it has arrived at those amounts. Under NEC4, the *Contractor* has to submit an application **before** the assessment date (clause 50.2). If it does not, then it can get a zero or even negative assessment (under clause 50.4). This change is not only completely needless, but also unnecessarily punitive on the *Contractor*! Under NEC3, the application had to be submitted on or before the assessment date (not just before), and it was not mandatory for the *Contractor* to submit an application (although all *Contractors* I know submitted details of the amounts they considered due).

The *Project Manager* has 1 week from the assessment date to certify the payment (clause 51.1), has to consider or take account of the *Contractor's* application (clause 50.2) and has to give details of how the amount due has been assessed (clause 51.1 again). Under NEC4, the *Contractor* is no longer entitled to interest for a late certification in itself, but it will be if the late certification causes late payment, as it is entitled to interest on late payment.

A result of the 1-week certification period, particularly on Option B (priced contract with bill of quantities), is that those acting on the *Project Manager's* behalf have tended to measure completed work as the month progresses, rather than in the period between the assessment date and certification date. This avoids a week of intense activity to agree the monthly measurements.

Unless stated in the Contract Data, the *Contractor* is paid within 3 weeks of the assessment date, which should be a maximum of 2 weeks after the certification. In some Client companies, this may

Figure 3.1 Timescales for payment if option Y(UK)2 applies

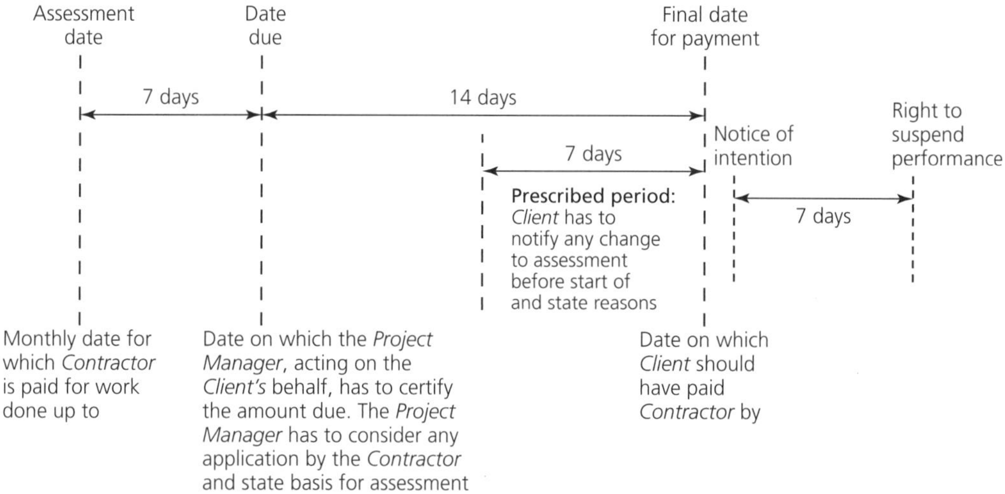

mean speeding up the administrative procedures relating to payment, otherwise the *Contractor* will be entitled to interest on the late payment.

IF THE CONTRACT IS SUBJECT TO THE CONSTRUCTION ACTS

If the Construction Acts apply, then the clauses of option Y(UK)2 will apply. The clauses in this option effectively overlay or add to the core clauses – so read the preceding paragraphs if you have not done so already. Figure 3.1 and the following text illustrate how the payment terms work.

Under the alternative clauses designed to satisfy the requirements of the Construction Acts, the *Contractor* still has to submit to the *Project Manager* an application for payment before the assessment date and the *Project Manager* still has to certify the amount due by the 'due date' – 1 week after the assessment date. The *Client* has to pay the *Contractor* by the 'final date for payment', which is 14 days after the due date. If paid late, interest is paid on the late payment. If the *Client* intends to withhold payment, it has to inform the *Contractor*, with reasons (in what is generally known as a 'pay less notice'), by the start of the 'prescribed period' – 7 days in the ECC – which runs back from the 'final date for payment'. If the *Client* does not follow this procedure and withholds payment without informing the *Contractor*, including the reasons why, or the *Contractor* successfully overturns the *Client's* decision through adjudication and payment is still not made, the *Contractor* can give 7 days notice of its intention to suspend performance. This right ceases as soon as payment is made. In the ECC, the *Contractor* is entitled to the additional time and cost of suspending performance (which is assessed as a compensation event) – I believe this right is enshrined in the Construction Acts. The financial assessment would therefore include any demobilisation and remobilisation costs, as well as the length of time for which work was suspended; and the time assessment would adjust the Completion Date, as well as Key Dates and sectional Completion Dates.

In order to outlaw certain unethical practices, the original Construction Act was amended by the Local Democracy, Economic Development and Construction Act 2009, which came into force in October 2011. The amendments are as follows.

- The *Project Manager* must issue, by the due date, the payment certificate specifying the amount due ('the 'notified sum') and the basis on which the amount was calculated. If this 'payment notice' is not issued within 5 days after the due date, then
  - if the *Contractor's* original application is compliant, it is considered to be the payment notice stating the notified sum
  - if the *Contractor's* original application is not compliant, a default notice needs to be issued.
- The notice of intention to withhold payment is replaced by a requirement to pay the notified sum or, not later than 7 days before the final date for payment, to give notice of intention to pay less than the notified sum, stating the amount considered to be due and the basis on which that sum is calculated.
- If the *Adjudicator's* decision changes an amount notified as due, the sum decided by the *Adjudicator* is to be paid not later than 7 days after the date of the decision.

My last comment is for Contractors: these provisions, albeit with some modified timescales, also have to be in your subcontracts, and applied as such

3.4.2 Payment under options B and D, the bill of quantities options

If you are working under one of the bill of quantities options (B and D) and have not amended the contract to negate my comments in Table 1.3 (in Section 1.2.2.2), I suggest you agree on how you are going to operate the contract. Otherwise, there will potentially be a number of applicable rates for the same item of work, and some rates for work that is not done. Good luck!

3.4.3 Payment under the cost-based options

My first point about using the target-cost contracts is to tell all project participants, from foremen and site engineers upwards on the *Contractor's* side and any *Supervisor's* staff on the *Client's* side, how a target-cost contract works and how it aligns motivations of the Parties to work together to reduce time and cost. When doing this in joint training sessions or workshops, I see the lightbulb going on in people's heads as they realise that 'partnering' or 'collaborative working' is not a soft and fluffy thing resulting in obscure benefits, but will result in tangible savings to each of their employing organisations. It gives people, especially *Clients'* people, the reason for cooperating.

The second – and biggest – point also applies to option E (the cost-reimbursable contract) as well as the target-cost options and, to a lesser extent, option F (the management contract). It is the need to agree a pragmatic approach to assessing and auditing Defined Cost. The extreme of this is that everything is audited every month (i.e. seeing everything right down to the bottom level of the *Contractor's* cost-breakdown structure[5] and matching it up with every individual supplier quotes, orders, delivery notes and invoices and actual payment). From a *Client's* perspective, this is 'pound stupid, penny wise' as the cost of doing this for everything on a monthly basis far outweighs the

[5] A cost-breakdown structure is a bit like a work-breakdown structure used to break the project down into parts, before ordering them in a logical sequence for the programme.

- The top level might be the activities with a cost against them (e.g. 'Construct retaining wall at X location').
- The second level down might correspond with the eight headings in the Schedule of Cost Components (e.g. the value of Plant and Materials in the activity above).
- The third level might further split this down into major elements (e.g. the costs of concrete, re-bar and shuttering).
- The fourth level might be the cost for individual materials (e.g. different types of concrete).

value for money obtained, as the *Client* is paying for both the *Project Manager's* quantity surveyor's time and, ultimately, for the *Contractor's* time in supplying this information.

Unfortunately, I have heard of many Contractors having to supply duplicate monthly files in hard copy of their quotes for orders, invoices, delivery notes etc., as well as all other statements, to the *Project Manager's* quantity surveyor in order to facilitate payment. If nothing is stated in the Scope, there is a strong argument for saying they do not have to. Clause 52.2 states what records the *Contractor* has to keep – accounts of payment of Defined Cost, proof that payments have been made, communications about Subcontractor compensation events and other records as stated in the Scope – while clause 52.4 states that the *Project Manager* has to be able to 'inspect' whatever accounts and records the *Contractor* 'is required to keep'. There is nothing in the contract as written about having to supply any duplicate files.

Furthermore, while the first bullet point of clause 11.2(26) for Disallowed Cost allows the *Project Manager* to disallow cost which 'is not justified by the *Contractor's* accounts and records', we are talking about civil law, so while the burden of proof is on the *Contractor* to 'justify' or demonstrate it has incurred those costs, the standard by which this is judged is not 'beyond reasonable doubt', as with criminal law, but on the 'balance of probability' (i.e. more than 50% likely). Or put another way, if you were to ask a director of virtually any organisation going down the target-cost contract route if they are happy to pay out monies to a *Contractor* on a 51% probability that the cost had been incurred on their organisation's contract, I think you can guess the response. But as written and without additional words in the Scope, that is the contract position.

I was involved as an expert witness in an adjudication on an option C contract. The *Project Manager's* quantity surveyor was insisting on having presented to him, on a monthly basis, pretty much every single quote, order, delivery receipt, Subcontractor application, Subcontractor or supplier invoice etc. and proof that it had been paid. The contract had

- no extra requirements over and above those in clause 52.2 about what records the *Contractor* had to keep
- nothing in it which said the *Contractor* had to present the information to the *Project Manager* (i.e. other than clause 52.4, whereby the *Project Manager* is allowed access to the accounts and records applied, there was nothing in the *conditions of contract* or Scope).

If the quotes, orders, delivery receipts, Subcontractor applications, Subcontractor invoices or supplier invoices did not match up exactly with the payments, then monies were being disallowed.

The problem for the national *Contractor* was its accounts are computerised (who'd have thought these days!), with many of the site costs assembled automatically by the system, fed by data from its national suppliers, be they for construction Equipment or Plant and Materials. So as the *Contractor* said, 'if the quantity surveyor wanted a bit of paper or even a pdf, then all he got was a printout of data'. Not once, despite the *Contractor's* numerous invitations, had the *Project Manager's* quantity surveyor sat down beside the *Contractor's* quantity surveyor with the accounts package open (i.e. inspected the actual accounts and records, as per clause 52.4) or done any sort of audit to ensure that what the *Contractor* got charged and paid as an organisation was modified accordingly to end up as Defined Cost in accordance with the contract (e.g. with a bulk purchase discount or rebate, as per the end of core clause 52.1).

> Further, the *Project Manager's* quantity surveyor was picking and choosing which rates to use: if a local supplier that he had phoned up was cheaper on an item than the *Contractor's* national supplier, he would only pay the cheaper rate. Again, not in accordance with clause 52.1.
>
> To cut a long story short, the *Adjudicator* found predominantly in favour of the *Contractor*.

So, here's what we can learn from all of this.

- The 51% probability that costs charged to the contract as Defined Cost actually are incurred on the contract is not high enough for most Clients.
- The first three items listed in clause 52.2 as the records the *Contractor* has to keep –accounts of payment of Defined Cost, proof that payments have been made and communications about Subcontractor compensation events – are not explicit enough in terms of what the average Client's quantity surveyor wants to see in order to be confident that their client is paying the right amount in accordance with the contract.
- Contractors of any size use computerised systems to assign and assemble Defined Cost. Not understanding how this works leads to ongoing and repetitive arguments about predominantly the same thing every month.
- Seeing and – given the previous bullet point – printing out every single communication related to payment is an awful lot of wasted effort for both Parties, which the *Client* ultimately pays for.

So, if those are the problems, what are the solutions? Here is my take on it from a Client's perspective.

1. Make sure the Client's internal auditor is aware of how the contract works and is party to any discussions on the level of detail needed for monthly assessments and audits. This is especially important if this is the first time that a Client organisation is using an 'open-book' format for a construction project. The last thing either Party wants or needs is for the auditor to find out 6 months or a year into the contract, and want all costs to be retrospectively audited.

2. At tender, ask a question about how a potential Contractor will operate the contract on an 'open-book' basis. I would want to know, for instance, how the winning *Contractor* would, in accordance with requirements stated in clause 52 and the Scope

   - ensure that only costs incurred on this contract are booked as Defined Cost
   - identify potential Disallowed Cost
   - ensure that cost which the *Project Manager* has decided is Disallowed Cost does not reappear as Defined Cost in subsequent applications
   - keep track of and resolve disagreements over Disallowed Cost if, for example, there is insufficient information on which to base a decision.

   And I would want these fine words to be sufficiently precise to be contractually meaningful, so that they can be incorporated into the contract as *Contractor's* Scope. The tender scoring criteria therefore need to give them marks for being expressed in the language of the contract and to a level of detail where the *Project Manager* or their quantity surveyor can hold the

Contractor to account (rather than being vague generalities). The Instructions to Tenderers need to say that the *Client* has the option of including these words as *Contractor's* Scope.

3. With the exception of the *fee percentage*, do **not** have Contractors' tender rates and percentages as part of their offer. As I outlined in Section 2.3.3.2, this can be a source of 'hidden' profit for Contractors and so leads to gamesmanship at tender. So, in the Contract Data part two that you send out, either

   - make it clear that any rates and percentages put forward – apart from the *fee percentage* – will be audited prior to signing the contract, which may result in adjustment prior to entering the contract, as part of any negotiation (i.e. they are proposed rates and percentages, subject to audit)
   - delete the lines requiring these entries altogether, making it clear that these rates will be established within the first few months of the Contract Date, along with an audit of the chosen *Contractor's* financial system and processes.

   For rates, I mean the following (for example).

   - Having got the *Contractor* to put forward, with a build-up, its staff and other people costs, audit them to make sure they have been built up in accordance with People 1 of the full Schedule of Cost Components. Note that an accountant, representing the *Project Manager*, is probably better qualified to do this than a quantity surveyor and, due to data protection legislation, would probably need to visit the *Contractor's* head-office and sit beside the *Contractor's* accountant as they demonstrate the build-up and explain the assumptions behind these people costs (i.e. 'witness' that it is built up in accordance with the contract, but not take anything away).
   - Having got the *Contractor* to identify what, if any, *Contractor*-owned Equipment it will be using, agree – as per Equipment 22 – the rate by reference to open market rates.
   - Where there are long-term supply agreements – whether for Equipment or Plant and Materials – satisfy yourself that the prices to be charged were gained via an 'open market or competitively tendered' process, as per clause 52.1.
   - Check that what the contract is charged, as a cost centre, has any company-level discounts and rebates deducted, again as per clause 52.1. Often the *Contractor's* site-level people are in complete ignorance of such arrangements.

   What I like to do is have these rates put forward and accompanied by a letter – which I have written – signed by the *Contractor's* financial director or equivalent. The wording in this letter confirms that these are a true and fair build-up of costs in accordance with the contract. It also contains a reminder that knowingly making a false statement is misrepresentation in law, which may be a criminal offence and could result in disciplinary proceedings from their professional body resulting in disqualification. For some reason, the rates and discounts seem to be pretty much spot on! As a result of this, many of these rates can be put into an agreed 'contract cost manual', or equivalent document, which is used both to reimburse the *Contractor* the Defined Cost and for the assessment of compensation events.

   For auditing the system, I mean the following (for example).

   - How and whether it actually fulfils the promises made by the *Contractor* in its tender under point 2 above.

- How people hours are tracked, so that you are not paying for staff not working on the contract, even if they are within the Working Areas.
- How and when the rebates and discounts are applied (e.g. by the *Contractor's* site-based quantity surveyor, or before they get to their computer terminal?).
- When hiring Equipment under a long-term arrangement, understanding: the cradle-to-grave system of placing the order, it turning up at the Site and charging starts; how the contract (as a cost centre) is charged (including when the discount is applied); and when the Equipment is off-hired and the charging stops (i.e. when the supplier is notified it is no longer needed, not when it is removed).
- Similarly for purchase of Plant and Materials, including what happens to unused Plant and Materials.

Deficiencies in the system – or just areas where you want more evidence – can be addressed early on in the contract, not months (or even years) after the cost has been incurred or after having had the same arguments every month. Further, errors or uncertainties are viewed as problems in the system which can be addressed, rather than deliberate obfuscation or misrepresentation by the *Contractor*.

4. Be clear about what you require to have presented as part of each payment application, what you must see each payment cycle and what you want to be able to see, if you so choose, every payment cycle. Let's take each of these in turn.

   - *Require to be presented every month.* At a minimum, this will be a breakdown of how the application is built up, probably in the form of an Excel spreadsheet, and ideally in a way that is easily downloadable from the chosen *Contractor's* accounts package. This implies the specific format is developed post-contract award. In order for the *Client* to satisfy its own internal arrangements, this might mean having copies of the 'big-ticket' items, such as Subcontractor applications, major supplier invoices and labour and Equipment time sheets. But I stress, it is in neither Party's interest to generate pages and pages of pdf or reams and reams of paper documents.
   - *Want to see every payment cycle.* This will typically mean the *Client's* quantity surveyor as delegated *Project Manager* sitting down beside the *Contractor's* quantity surveyor, who logs into the system and checks that the figures in the application match the data, documents etc. in the system. For every payment application, I would expect a check on the big-ticket items (e.g. the 20% that make up 80% of the Defined Cost applied for).
   - *Want to be able to see, but will not necessarily check, every payment cycle.* This can be random or periodic checks on specific aspects or because of issues or anomalies that come up. For instance
     - □ one payment cycle, the quantity surveyor could check that the hours booked in the timesheets multiplied by the agreed rates for each person or category of person match up with the amount charged for People
     - □ another time, it could be a similar thing for hired Equipment, and ensuring that payment is made by the next assessment date
     - □ the next time it could be following through on some Plant and Material orders – from receiving the quotation, placing the order, any adjustments for changes, receiving the invoice and ensuring it has been paid within the required timescale
     - □ another time it could be randomly selecting an activity and drilling down through all the detail of that cost

- alternatively, the reason for selecting a particular activity could be because costs have come in much higher than the activity price in the Activity Schedule, or that much more is being claimed than was expected for, say, concrete, so there is a need to explore why that is – it could for a very good technical reason, it could be that the supplier is being inconsistent in its applications or it could be glitch in the system somewhere which can then be fixed.

While I have given the high-level type things that I would check out and the reasons for doing so, get your quantity surveyor to write in the Scope the specific documents or ledgers that they would want to see under the three categories above. Having said that, bear in mind that different contractors use different systems, which generate and present data in different ways, so these should be specific enough to get what your quantity surveyor needs and wants to see, but not so specific that the *Contractor* has to manually create specific documents in particular formats to satisfy the quantity surveyor's personal foibles.

On one target-cost contract with a new client, I was initially brought in to act as the *Project Manager's* advisor or mentor, but ended up being more of a coach to the whole project team. The *Employer's* quantity surveyor (as it was an NEC3 contract) had written a fair amount of text into the Works Information about what information they wanted from the *Contractor* with respect to Defined Cost. However, it seemed fairly generic to me, and it turned out it was largely 'cut and pasted' from a standard industry document on what accounts a contractor should have. The two Parties starting to have regular arguments about what these words meant in practice. The *Project Manager* did not want this to end up as a dispute when overall the Parties were working pretty well together on a tricky contract. My sub-brief from the *Project Manager* was to bang heads together until we had a set of requirements for what the *Employer's* quantity surveyor wanted to see that matched the specific documents and ledgers that the *Contractor* could produce with little or no extra work.

An afternoon of stepping back from the specifics of an individual payment application led to an action plan that largely resolved the differences. This involved rewriting the relevant area of the Works Information to state far more precisely what the *Employer's* quantity surveyor wanted to see and what the *Contractor* could, with relative ease, provide or allow the *Employer's* quantity surveyor to see.

5. Finally, there is still a danger that what is comprehensively entered into the system is fraudulent, so there need to be some checks to ensure that resources booked to the Working Areas – whether people hours or material deliveries – are actually employed and used in the Working Areas. It would be fraudulent if they are not. Most national contractors' senior management I know shudder at the thought of being found to be doing something fraudulent like this: the threat to their company's reputation is simply too large. The most obvious check is to watch out for materials being taken out of the Working Areas, yet I have heard second-hand accounts of a few contracts where there were no checks and this was occurring. As Ronald Reagan said in relation to nuclear disarmament: 'Trust, but verify'!

The opposite of this is that the paperwork or data is incomplete, but people clearly have been working in the Working Areas, the Plant and Materials have been delivered and/or work has been done. For instance, there is a pile of re-bar at the Site, but the delivery note is missing, or an activity could not have been done if a person or machine had not been on site. This involves the relevant people getting out from behind their computer screens and offices and out onto the Site, to see what work has been done etc.

My penultimate point is that some Clients do employ external auditors on their project team to come in and do periodic audits. This is partly in case their own people, who may be working very collaboratively with the *Contractor,* become a little bit too complacent or trusting, and partly because a new set of eyes can look at things from a different perspective, which can lead to suggestions for improvement. This can also include looking in detail at areas where the *Client's* quantity surveyor perhaps does not have expertise and/or there have been issues or niggles.

I can understand this and have no objection to it if the auditors are genuinely impartial and constructive. What I do have issues with, however, is where external auditors who are brought are paid on the basis of a share of any Defined Cost they identify which should not have been allowed in the first place or should be categorised as Disallowed Cost. First, the *Project Manager*, when assessing payments, is meant to act as an impartial administrator and I fail to see how someone who is rewarded for disallowing costs can be impartial: they will naturally err towards a harsh interpretation. Second, if the *Client*, via its *Project Manager*, is only doing this because costs are exceeding its budget, this only ramps up the pressure on the individuals doing the audit.

> I heard of one large *Client* who did this on a large target-cost contract when the electrical and control systems led to a late escalation of the Price for Work Done to Date, such that it exceeded the target Prices when for a long time it had been expected to come in below. The *Contractor* went to adjudication over the amounts that were disallowed. As well as deciding on specific issues, the *Adjudicator* agreed that trawling back through more than 3 years of accounts and payments, when for the majority of this time the *Employer* had been perfectly happy with the Defined Cost paid, was not in the 'spirit of mutual trust and co-operation' required in NEC3's clause 10.1 (now 10.2 in NEC4).

The example above brings me onto my last point, which is aimed more at Contractors and is specific to NEC4: namely the new provision of clause 50.9, whereby the *Contractor* 'notifies the *Project Manager* that Defined Cost for a part of the *works* has been finalised, and makes available for inspection the records necessary to demonstrate that it has been correctly assessed' (which it should have done regardless). This provision was introduced precisely because of the complaints from Contractors of the nature illustrated in the box above.

A 'part of the *works*' could be either a geographical section or an element that may be represented by a Subcontractor account. So the *Contractor* might effectively be saying 'I want to close out this subcontract account (perhaps by making a final assessment under new clause 53), but I don't want to do that only for the *Client's* quantity surveyor on the *Project Manager's* behalf to say I should have not paid the final amount' and then disallow costs (under the second bullet of clause 11.2(26)), which would leave the *Contractor* out of pocket.

The deficiency of clause 50.9 is that it does not allow the *Project Manager* to notify that they want the Defined Cost for a part of the *works* closed out, thus allowing the *Contractor* to 'find' missing costs many months after the work has been completed. It is easy to adjust clause 50.9, via an option Z clause, so that the initial notification can be done by the *Project Manager* as well as the *Contractor*.

SUMMARY

Here is my summary to this subsection.

- Make sure that all project participants of any managerial capacity, from foremen and site engineers upwards, understand how a target-cost contract aligns objectives and the basics of open-book accounting under NEC4.
- Address the issues that arise from open-book accounting early on.
  - For the *Client*, this includes specifying your medium-level requirements in the Scope, asking a tender question on how the *Contractor* will meet these requirements, and ensuring that its response can be incorporated into the Scope. Do not get the *Contractor* o tender rates as part of its tender offer – although you might want it to put forward proposed rates subject to audit – and do not require paper or pdf copies of every order, invoice etc., but do make clear what you expect to be able to see if required.
  - In the first few months of the contract, audit any proposed rates and the system so that any concerns can be addressed early, on rather than having the same ongoing repetitive arguments every payment cycle.
- The monthly check on the *Contractor's* application typically involves the quantity surveyor who represents the *Project Manager* sitting at the *Contractor's* quantity surveyor's computer terminal, checking the big-ticket items and doing some deeper dives. It should not be, and the contract does not require it to be, printouts of data and reams of back-up.

If you are the *Contractor* and the *Client*/*Project Manager* has not initiated the above, then do what you can to resolve issues proactively: it is not in your interest to have payment issues undermining collaboration elsewhere on the contract or to have monies, which you think are booked, disallowed late in the contract. This includes progressively closing out Defined Cost for parts of the *works*, such as Subcontractor accounts, as these parts are completed.

3.5. The Accepted Programme

3.5.1 Overview

This section provides the following.

- A general overview of what seems to be good practice in programming, partly so that you, the reader, know where the authors of the NEC are coming from, but mainly so that you know where I am coming from in terms of the more detailed advice I offer.
- A recap of why the programme is a core feature – or 'the beating heart' – of the NEC.
- A more detailed explanation of terminology used in the ECC related to the Accepted Programme, in order to give you a thorough understanding of the programming requirements of the ECC.
- A number of subsections on different aspects of good practice with regard to how to have a submitted programme rapidly accepted for everyone's benefit. This includes an expanded version of my article 'ECC programme – the six levels of use', which appeared in the *NEC Users' Group Newsletter* (Broome, 2009a), which highlights progressively better practice in its use, with correspondingly better results in terms of contractual outcomes.
- An explanation of why 'acceleration' under the ECC does not mean speeding up work to meet the original contractual timescales, but bringing forward the contractual Completion Date, and the implications of this.
- A conclusion, which brings it all together.

3.5.2 Good practice in programming

The purpose of this section is to provide insights into what is considered 'good practice'[6] in programming so that you, as the reader and a user of the contract, understand where both the NEC Panel and I, as an author, am coming from in terms of the detailed advice I offer.

When I started my PhD, my supervisor and co-instigator of the NEC – Professor John Perry – asked me to review what 'best practice' in project management was with respect to what the NEC was trying to achieve. He was hoping that what I found would justify where the Panel developing the contract was coming from. By and large it did, and central to good project management is programming. In fact, first generation project management – or PM 1.0, as the Association for Project Management call it[7] – was programming!

What came through loud and clear from the literature review was that 'good practice' in programming involves

- regular monitoring, updating and, if necessary, re-planning throughout a project
- as the time for doing the work approaches, the above being done
  - at increasing levels of detail
  - with greater involvement of those who will be doing the work in planning the detail.

In the literature I reviewed at the time (up to 1996), these bullet points were sometimes encapsulated under the principles of what was called 'multi-level planning' (see, e.g., Laufer *et al.* (1993) and Faniran *et al.* (1994)). While the NEC very much encourages the first bullet point above, it does not overtly encourage multi-level planning as there is nothing in the contract which says that the *Contractor* has to supply more detail as the time for doing work approaches. However, this is not to say it cannot accommodate it. For instance, on some contracts in the research sample, a short-term programme, in addition to the main programme, was submitted, either voluntarily by the *Contractor* or as specified in the Works Information. This seemed to result in higher levels of cooperation and satisfaction with programming compared with cases where a short-term programme was not regularly submitted.

Consequently, I suggest you buy into the concept, if your organisation has not already, because

- statements for how the *Contractor* is to do the work and resources for operations occurring in the distant future are likely to be less thoroughly thought through by the *Contractor* and/or scrutinised by the *Project Manager* than those for more immediate operations
- work in the distant future is subject to greater uncertainty and is more likely to change, therefore time spent preparing detailed method statements and resources is more likely to be wasted.

[6] As I have previously stated, I do not like the term 'best practice', for two reasons. First, what is best practice today sometimes becomes just good practice or even accepted practice tomorrow, when the new 'best practice' arrives (and sometimes it just goes out of fashion for valid or invalid reasons). Second, it is contextual: what is 'best' for a £100 million project might be completely inappropriate for a £0.5 million one.

[7] Projecting the Future was an ongoing 'big conversation' about the future of project management, led by the Association for Project Management for its members and other stakeholders. I was on the panel that helped develop the consultation documents. To give historical perspective, we set out the three historical stages in the evolution of project management – PM 1.0 to PM 3.0 – asking what PM 4.0 (which matches up to what society will look like during and after the fourth industrial revolution) will look like. The initial paper is available at: https://www.apm.org.uk/media/36799/projecting_the_future_msgd_digital.pdf (accessed March 2021).

Indeed, without being overprescriptive, the *Client* might like to give direction on this in the Scope, specifying the degree of detail, the format, the frequency of revision (if a separate contractual document) and the planning horizon (i.e. how far ahead the programme is looking), with the idea that the shorter the planning horizon, the greater the detail and the more frequent the revision. This can be done in two ways.

- The short-term programme has nothing contractually to do with the programme submitted to the *Project Manager* for acceptance (and hopefully accepted). It is a separate and unrelated contractual requirement. Note the word 'contractual' though. In reality, it is both informed by and informs the programme submitted for acceptance.
- The short-term programme is part of the submitted programme, but the words in the Scope require a greater level of detail for all operations that will occur within the next planning horizon (say, a month) than is required by clauses 31.2 and 32.1 of the *conditions of contract*.

An advantage of this is that, when a compensation event occurs that affects the timing of the remaining work, only the part of the programme appropriate to the size and impact of the compensation event need be reviewed and, if necessary, updated. Two potential disadvantages are that, if overprescriptive, the *Client* could effectively end up telling the *Contractor* how to programme, and the extra complexity of different levels of programme could result in even more work.

More recently, a number of 'new' ways of running a programme have emerged.

- *Lean project management*, which essentially takes lean manufacturing principles and techniques and adjusts them to project circumstances where operations are 'unique', as opposed to manufacturing hundreds or thousands of the same items. Where this is coming from is to take out uncertainty in each operation so that the duration of each operation is more certain, and hence the whole programme is more certain. If you are progressively taking out frequently occurring everyday 'mini-risks' in operations, a natural byproduct is that the duration of each operation reduces, and hence the overall duration of the project is also reduced.
- *Critical chain project management*, sometimes called *buffered programming*, has a lot of overlap with lean project management in terms of the techniques used. However, the fundamental difference in philosophy is that it aims to complete a project as quickly as possible, as opposed to with maximum certainty (for an article I wrote on this and how it can be operated under the NEC, see Broome, 2009b).
- *Agile project management*, which, as well as having some new approaches, has overlaps with both of the above. It now seems to be the dominant way of delivering IT software projects, but many of the principles, if not techniques, are applied across a spectrum of non-IT projects (indeed, I co-authored a white paper published by the APM which was applicable to any project with agile characteristics; see APM C&P SIG, 2020).

Before I proceed, let me state that there are some useful techniques in all of these methodologies and a few of my contractor clients have embraced them with good results. This even includes some agile techniques on large, multi-year construction projects – which overall are at the opposite end of the agile spectrum – where there are tiered daily scrum meetings.

However, the common theme among all these 'new' techniques is – and this will sound familiar – that those close to doing the work become more involved in the detail of the planning as the time for doing the work approaches.

In the NEC3 edition of this book, I recommended an excellent article titled *Tricks and Tips in Planning and Scheduling* by a highly respected planner called Gary France, who was the Chair of the Planning Engineers Organisation (France, 2006). Unfortunately, all trace of this article seems to have disappeared from the internet. However, one of its themes, which builds on what I have said before, is that a good planner needs to be a good communicator in terms of

- involving those who will do the work in the planning
  - because the planner is not going to have the expertise to plan in detail every particular operation, which could include not understanding what the risks and necessary prerequisites are in terms of completed actions and operations, information, resources etc. for the operation to proceed efficiently
  - to gain an understanding of and 'buy-in' to not just their part of the programme, but how their part affects and is affected by other parts of the programme
- clearly explaining to the right level of detail what different people and organisations have to do and by when
- gaining reliable information on progress in order to keep the programme up to date (monitoring) and realistic, which could include replanning.

3.5.3 The 'beating heart of the NEC'

This somewhat memorable expression was apparently coined by a lawyer who saw the big picture and recognised just how important having an up-to-date, detailed, agreed, and hence accepted, programme is to the effective operation of the NEC family of contracts. Without an Accepted Programme how can

- the *Project Manager*, on behalf of the *Client*, know what they have to provide and by when, except by referring back to the Scope, which may have been prepared months or years before and hence be out of date?
- the *Project Manager*, and hence the *Client*, have an up-to-date knowledge of when the contract is likely to achieve Completion?
- the team be effective in reacting to early warnings to minimise their impact and also proactively take out time and costs?
- the *Contractor* get a compensation event for something that has not been done by the time in the Accepted Programme?
- the two Parties quickly, amicably and fairly evaluate compensation events, not just in terms of time, but also in terms of cost that arises from the use of resources over time?

The answer to all of these questions is 'you cannot', or at least your effectiveness will be much diminished, and as result there will be arguments that will undermine working relationships and the ability to work collaboratively in less contentious areas.

As a result, the original NEC Panel, quite rightly, put stimuli/sanctions/incentives on the *Contractor* to

- provide a first programme that shows the information required and, until it provides the programme, one quarter of the Price for Work Done to Date is retained (see NEC4 clause 50.5)
- submit updated programmes on time and to a standard which is acceptable in accordance with the contract (see clause 31.3 for the acceptance standards) and, if it does not, then the sanction is that the *Project Manager* should assess compensation events until it does

(see clauses 64.1 and 64.2). Despite the *Project Manager* having a duty to be impartial in matters of assessment, a *Project Manager's* assessment will typically be less favourable to the *Contractor*.

These stimuli are still in NEC4, but an additional stimulus for the *Project Manager* has been added. If the *Project Manager* has not responded to the *Contractor's* programme submission within 2 weeks, and still has not responded a week after a reminder notification from the *Contractor*, then the original programme submission is treated as accepted (see the new last paragraph of clause 31.3).

There are also two general sanctions on the *Project Manager*, in the form of compensation events, for

- not responding in time to communications (clause 60.1(6)), which in the case of a programme submitted for acceptance is the 2 weeks stated in clause 31.3
- withholding an acceptance for a reason not stated in the contract (clause 60.1(9), with the reasons again stated in clause 31.3).

However, I am not sure how you would calculate the effect of each of these compensation events as not having the programme accepted does not affect the *Contractor's* delivery of the *works*. So, apart from showing up the *Project Manager*, their effectiveness as stimuli is limited.

3.5.4 Detailed explanation of the terminology related to the Accepted Programme

The main contractual terms, such as *starting date*, *access date*, Key Dates, planned date of Completion and Completion Date, as well as what acceleration means in practice, are described in Section 1.5.4. If you are not familiar with this terminology related to the time and the programme under the NEC, I suggest you read Section 1.5.4 before proceeding.

This section goes down a level of detail in explaining expressions used in clause 31.2 and their consequences in terms of managing the programme. These consequences can be bad if ignored, and hence not managed, or good if managed positively.

3.5.4.1 Clause 31.2

The first five main bullet points of clause 31.2 are not discussed here, so we start with the sixth bullet point.

BULLET 6: PROVISIONS FOR FLOAT

Let us start with two types of float that I think are important. The reason why they are important concerns ownership. By 'ownership', I mean who has use of it when a compensation event occurs.

The first type is **terminal float**. In NEC speak, this is the difference between planned Completion as shown in the Accepted Programme (when the *Contractor* thinks it will complete the *works* by – and, by accepting the programme, so does the *Project Manager*) and the Completion Date (when the *Contractor* contractually has to complete the *works*). When looking at clause 63.5, if planned Completion as shown in the Accepted Programme is put back due to a compensation event, it can be seen that the Completion Date moves back by the same amount of time. For example, if the compensation event is on the critical path and the *Contractor* is delayed by 2 weeks because of the event, the planned Completion will be put back by 2 weeks, and hence, in accordance with

Figure 3.2 Who owns the terminal float?

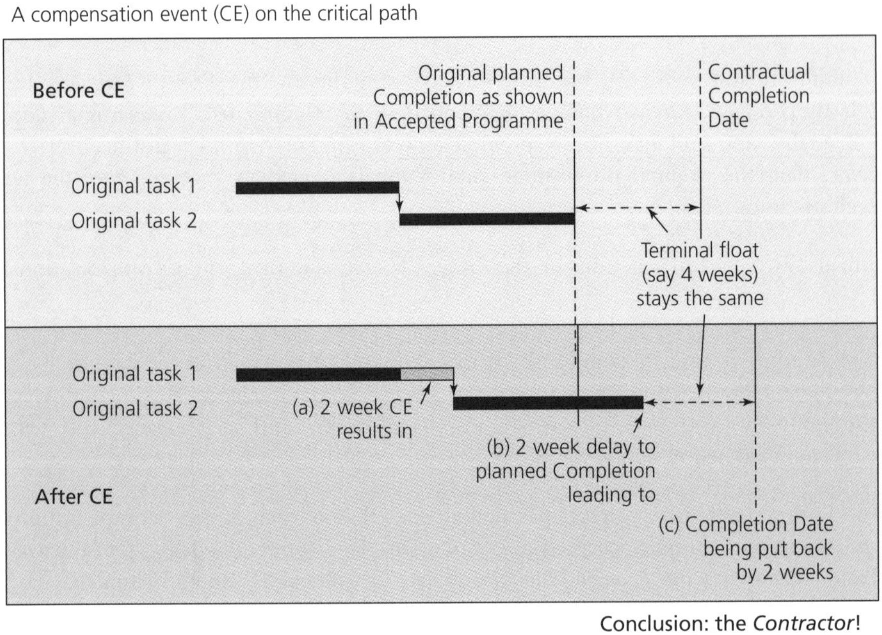

clause 63.5, so will the Completion Date. This is illustrated in Figure 3.2, and applies whether or not the *Contractor* is in delay. In other words, the *Contractor* owns or has use of this terminal float to accommodate its own delays. The converse of this is that the *Client*, via the *Project Manager*, cannot use it up to mitigate the effect of compensation events (unless it wishes to buy it from the *Contractor* for a price agreed by commercial negotiation – see Section 3.5.9 on acceleration).

Consequently, the *Contractor* is free to be realistic about the overall duration of its programme, and show its terminal float.[8] Note that the amount of terminal float shown is likely to alter from programme to programme, as the *Contractor* makes better or worse progress than planned from month to month on its critical operations.

The second type of float is **free float**. Free float is the gap between individual bars shown on a programme, so there cannot be any free float on the critical path. The amount or sum of free float along any path of operations or activities before it becomes critical is called the **total float**. (Note: In some programming software and American literature, total float is called 'slack'.) So, when looking at clause 63.5 again, if the *Contractor* is delayed by 2 weeks due to a compensation event, but there is 2 weeks or more of total float on that path to absorb it, there will be no delay to planned Completion as shown on the Accepted Programme, and hence no delay to the Completion Date. In this scenario, it is the *Client* who has used up the total float on that path, so the *Client* owns it.

[8] This is in contrast to other construction and heavy engineering contracts, where entitlement is based on need. Consequently, no terminal float is ever shown. So, from day 1, and for as long as the illusion can be sustained, the programme the client has is not an accurate representation of the contractor's intentions. The contractor has another programme, which it uses as its working programme.

Figure 3.3 Who owns the free float?

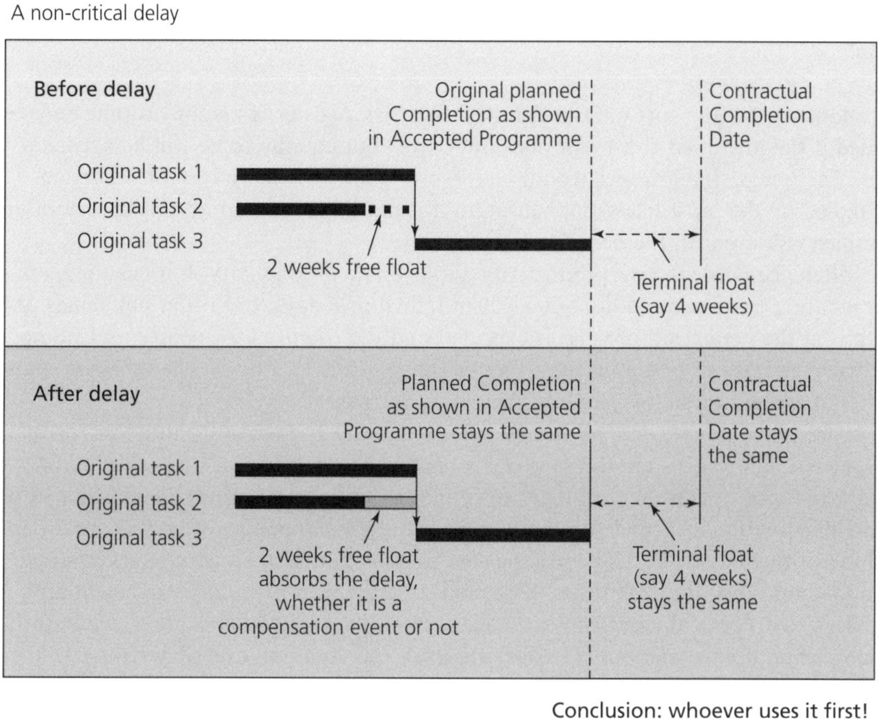

But, if we reverse the situation, whereby we have the same scenario but, instead of it being a compensation event, a *Contractor*'s risk causes the 2-week delay, then the *Contractor* can use up the total float to absorb the delay. This scenario is illustrated in Figure 3.3. If we were then to submit and have accepted a revised programme showing two critical paths, if there was a subsequent delay caused by a compensation event on any of the tasks, we would revert to the previous scenario for a delay on the critical path.

To give another scenario, let us say there is again a 2-week delay due to a compensation event, but this time there is only 1 week of total float available to absorb the delay before it becomes critical. The delay to planned Completion as shown on the Accepted Programme will now be 1 week only – as the first week is absorbed by the total free float along that path – and hence, according to 63.5, the delay to the Completion Date will be 1 week.

Given the above, what is the answer to the question 'Who owns or has use of total float along any path when a compensation event occurs?' The rather nice answer is 'Whoever needs it at the time' – or more memorably, 'Whoever gets to it first'!

Again, if there are no compensation events, the amount of free float, and hence total float, along any path of activities may alter from programme to programme, as the *Contractor* makes better or worse progress than planned from month to month on operations that are not on the critical path.

BULLET 6: PROVISIONS FOR TIME RISK ALLOWANCES

Let us now consider the phrase 'provisions for ... time risk allowances', which is also included in bullet 6 of clause 31.2. A time risk allowance (often abbreviated to TRA) could be one of the following.

- An allowance for the sort of risk identified in a risk register, whereby no time may be needed if the identified risk event does not occur, but equally some will be needed if it does. The time risk allowance would be the probability of the risk event occurring multiplied by the most likely time duration if it did. More is said on this sort of discrete identified risk event in the next section.
- The difference between best productivity and average productivity. For example, on average the *Contractor* can dig out a 600 m trench in 3 days, but if the sun shines, the earth is at the better end of what is expected and the *Contractor's* people and Equipment can work at close to best efficiency, then it can be done in 2 days. The time risk allowance for this average productivity risk is then 3 – 2 = 1 day.
  There are two other ways of saying this that seem make sense to Contractors' people.
  - Time risk allowances are the sum of average productivity 'mini'-risks that occur within an activity or operation, and that are common enough and minor enough not to be identified within a formal risk register. In fact, they happen so often that we do not think of them as risks. These are the sort of risks that lean construction techniques aim to take out. (See my comments in Section 3.5.2 on lean project management and, to a lesser extent, critical chain project management, both of which contain many of the tools and techniques to take out or reduce these all too common everyday risks.)
  - A target programme is a programme with all the time risk allowance taken out –while this is technically possible, it is unlikely in practice (and hence not realistic, and hence not acceptable – see below).

It is the second sort of time risk allowance that the rest of this explanation focuses on.

If we look at the criteria for an Accepted Programme in clause 31.3, as well as showing the information required, one of the criteria is that it has to practical and another is that it has to represent the *Contractor's* plan realistically, both of which could relate to having sensible timescales for operations and activities (i.e. realistic time risk allowances are included). Therefore, a *Project Manager* who is exerting pressure on a *Contractor* to submit an unrealistic programme should not accept such a programme!

So how is time risk allowance shown in the programme? The answer is that, provided nothing is stated in the Scope, it is up to the *Contractor* how it is shown, just as long as it is shown somehow. Methods include

- simply stating the time risk allowance in the accompanying statement of how the *Contractor* is to do the work
- in a few of the early contracts, the number of bars in the programme was doubled, so that for each operation or activity there was one bar showing the best or target time and another showing the time risk allowance (this was a reflection of the sophistication of the programming software available at the time, and I would not recommend it now)
- probably the simplest way is just to add an additional column within the planning software which, depending on the software, either

  - states, as a negative figure, the reduction in the duration of the operation or activity if no time risk allowance were to occur or
  - states, as a positive figure, the additional time risk allowance for that operation over and above everything going really well
- splitting the bar in some way, by hatching or colouring the minimum duration one way and the time risk allowance another (I prefer the idea of hatching, as I have other uses for colours)
- taking the time risk allowance out of the individual activities that make up an operation and putting it at the end of the sequence to give a buffered programme for that operation. This leads onto the practice of critical chain project management, which I mentioned in Section 3.5.2.

The next question is 'Who owns the time risk allowance'? For this, have a look at clause 63.8, whereby an 'assessment of the effect of a compensation event includes risk allowances for cost and time for matters which have a significant chance of occurring and are not compensation events'. Time risk allowances do, on average, occur, and so they have a significant chance of occurring; and the effect of 'are not compensation events' entitles the *Contractor* to include risk allowances in compensation event quotations up to the point at which an additional compensation event (e.g. for weather or physical condition risk) would occur. Three points that come out of clause 63.8 when assessing a compensation event are

- in allowing a duration for extra work added in as a compensation event, the *Contractor* is allowed a realistic timescale, not a best-case timescale
- by implication, for the existing non-compensation event activities or operations, the *Project Manager* has no authority or right to 'grab' the pre-existing time risk allowance to mitigate delay caused by a compensation event
- if the compensation event increases the risk of delay and disruption in existing operations because, for instance, it
  - introduces extra work in an already cluttered area
  - pushes work affected by weather from summer into winter

  then time risk allowance in existing operations could increase.

So, to come back to the question 'Who owns the time risk allowance?', the answer is the *Contractor*.

Some will ask: why do we, as the *Contractor*, have to show 'time risk allowance' when we own it? The following box gives a real life example of how having it clearly shown helped both Parties.

> The contract was for a large tunnel to stop potential flooding in a city centre. At various points on its route, shafts had to be excavated down from street level to meet up with the tunnel. This involved road closures, which had to be applied for and were limited in duration. The *Employer* (it was an NEC2 contract), using the *Contractor's* programme, applied for road closures of these durations, but which had variable start dates, the earliest of which were if the *Contractor* achieved its best productivities in the preceding operations (i.e. without time risk allowance occurring). With the exception of some early problems, by and large the *Contractor* did achieve close to the best productivities, and it was possible to utilise the early road closure start dates. If the *Employer* had not applied for these dates, then the *Contractor* would have had to have stopped work and waited. As this contract was under option C (target contract with activity schedule), both Parties benefited.

If you chose to go down the critical chain/buffered programming path, then the fundamental difference in philosophy from lean project management is that you do each operation as fast as possible and then, as soon as it finishes, start the next one. This means your 'best case' programme is equivalent to a target programme without any time risk allowance. However, to calculate your worst-case scenario, you put all the time risk allowance at the end of the path of operations (and reduce it as it is combined, so statistically you need less). The 'amber' area in between these two extremes represents the natural variability, and reflects the most likely time-scale within which you will finish the contract.

Regardless of whether you are programming conventionally or using the critical chain/buffered programming approach, for the *Contractor's* programme to be accepted, the *Project Manager* needs to understand what the time risk allowance is and for it to be shown. This will directly affect the *Project Manager's* view of whether the programme is realistic (as per clause 31.3 – see Section 3.5.4.2).

To recap, it is the *Contractor* who owns the terminal float and the time risk allowance, while it is 'whoever gets to it first' who owns the total float on any path of operations. So, if you are a *Contractor* constructing your programme, do you want to show your non-critical float as time risk allowance or free float? The answer is as time risk allowance, because you definitely own it – that is, if there is no free float and hence no total float, it means all your activities and operations are on or close to the critical path. So when a compensation event occurs, the logic of a programme dictates that there will always be a delay to planned Completion, and hence a delay to the Completion Date.

So, if you are a *Project Manager* evaluating a programme for acceptance, your first pass check is to see if any bars are overly long, which would indicate an absence of free and hence total float in the programme. You would then need to use your experience and knowledge to drill down and find out why.

BULLET 8: FOR EACH OPERATION, A STATEMENT OF HOW THE *CONTRACTOR* PLANS TO DO THE WORK IDENTIFYING PRINCIPAL EQUIPMENT AND OTHER RESOURCES WHICH WILL BE USED

This is the last part of clause 31.2 that I want to consider here. The five things I wish to raise about this bullet point are as follows.

- This is obviously relevant to establishing whether operation and activity durations are realistic.
- To avoid duplication of documents and work, ensure that the statements for how the *Contractor* is to do the work comply with any 'provisions for health and safety requirements' of the contract or in law. In the UK, under the Construction (Design and Management) (CDM) Regulations 2015, this will mean the Principal Designer will need to be brought into the process for accepting the programme.
- It is fundamental to the *Contractor*, if a compensation event occurs, that, in order to be able to demonstrate the effect on an operation, there is sufficient information in the statement of how it plans to do the work and the resources it will use. If the *Contractor* cannot demonstrate the effect, then not only will it not be able to demonstrate the change in resources and cost for that operation, it will not be able to demonstrate the knock-on delays etc. to other parts of the programme, which is often where the big costs are.

- In the first and second editions of the ECC, you had to submit 'method statements and resources' for each operation. This implied a full method statement and all resources. Clearly, this is impractical for the *Contractor* at tender and during the earlier stages of the contract, when it has not, for example, appointed the specialist Subcontractors who will do the work: the *Contractor* may be able to give an outline of how it expects the work to be done, but this will be an educated guess at best. It would also be a lot of work for the *Project Manager* to evaluate a whole programme containing detailed method statements and resources, and much of the work by both would become redundant as things change during the course of the project.
  Instead, I suggest that people adopt what is established good programming practice. Namely, multi-level programming, where more detail is supplied by the people who are closer to the work as the time for doing a particular operation or task approaches.
  For example
  - when the operation is 6 months away, the planning might comprise a short paragraph on methods
  - when it is 2–3 months away, this might have increased to a list of health and safety risks plus a half-page outline
  - when it is to be done in the following month, there will now be a mini-risk register including responses and a fully detailed method statement.

  Likewise, the description of resources will become progressively more detailed in type and quantity, perhaps even naming the exact Equipment and people in the month before the task is to be done.
- Under option A (priced contract with activity schedule) (and also under NEC3 option C), the *Contractor* 'provides information which shows how each activity on the Activity Schedule relates to operations on each programme submitted for acceptance' (clause 31.4). If you think of operations as being made up of activities in the pricing document, then you will have fewer method statements and resource schedules. How do you show each activity? My suggestions are that you either have mini-bar charts as part of each method statement and/or, if the programme is submitted electronically, you double click on the 'operation' bars, which then spread out to give the 'activity' bars.

BULLET 9: OTHER INFORMATION WHICH THE SCOPE REQUIRES THE *CONTRACTOR* TO SHOW ON A PROGRAMME SUBMITTED FOR ACCEPTANCE

This is the last bullet point in clause 31.2. Requirements that the *Client* might wish to include in the Scope could include the following.

- The programme expressly shows the critical path through the programme. However, progress might not just be limited by the order and sequence in which operations and activities can be done, but by the flow of critical resources.
- If under option C (and if not done as an option Z clause), the *Contractor* shows how the activities in the Activity Schedule relate to operations in the programme.
- The programme includes an initial narrative giving the reasons and justification for the overall approach taken.
- The programme includes a short narrative explaining the key changes since last month's programme submission, in terms of delays and consequences, better than expected progress, effects of large early warnings and compensation events, etc. One *Contractor* did more than this voluntarily: it actively listed out, in tabular format, all the changes in

the programme since the previous submission, with a brief explanation for each. The *Project Manager* commented how helpful this was, as it enabled them to see what had changed without having to search through the programme. I would have thought a combination of a short 'big picture' narrative followed by a detailed commentary would be especially beneficial.

- Earned value information is provided as an add-on submission. I have to confess that, while I like the idea of progress and financial monitoring and forecasting being all tied together in one diagram and a handful of ratios, the practicalities appear different. Namely, the amount of adjustment and argument needed to arrive at and agree the 'true' state of affairs does not seem to justify the benefit. My observation is that a couple of cost consultants sell it as the be-all and end-all to unwary clients and then sell their consultants to try and make it work. Nevertheless, if earned value is your thing, it makes sense to specify the requirements in the Scope.
- On a similar theme (although I have a much more positive view of this), that there is a monthly dashboard, which is aimed at getting over key information, on one page, to those not directly involved in the management of the contract. While the *Client* might know what information is wanted, I suggest the exact information and format is agreed once the *Contractor* is selected, so that it is easy for it to assemble.

IN THE FORM STATED IN THE SCOPE

Lastly, new to NEC4, at the end of clause 31.2, is the requirement to issue the programme 'in the form stated in the Scope'. The form may be a pdf document, but is often the native electronic file format used by the software that the *Contractor* uses to create the programme. So, to open and interrogate the programme, the *Project Manager* will also need access to the software and a licence to use it – and, if they are unfamiliar with the software, training on it as well. These requirements also need to be specified in the Scope.

Alternatively, the *Client* could specify which programming software it wishes the *Contractor* to use. However, this might well mean that the *Contractor* is forced to use software that it is unfamiliar with, and which is incompatible with how it runs its business. Consequently, I would only suggest specifying the programming software if, as a Client, you have a large programme of projects/contracts and you need to integrate different contract programmes into one master programme. An example of where this was the case was in contracts for the London 2012 Olympic Games (where it apparently caused a lot of grief!).

3.5.4.2 Clause 31.3: criteria for acceptance

The four criteria that the *Project Manager* uses when deciding whether not to accept a submitted programme are in clause 31.3.

- *Practicable versus realistic.* One reason is that the programme is not practicable and another is that it is not realistic: but what's the difference? There is no case law on this and the NEC4 Guidance Notes give little direction. Most people's view is that 'practicable' relates more to method and sequence, which could be within a method statement for an individual operation or how the programme fits together, while 'realistic' relates to durations and what sits behind them, i.e. resources and productivities (which are not asked for). However, a conversation with Rob Horne – a well-known, pragmatic and respected NEC lawyer – made me realise a nuance in the wording of the second reason. The contract actually says a reason for the *Project Manager* not to accept a programme is that 'it does not represent the *Contractor's* plans realistically'. So, if the *Contractor* has two programmes – one which it submits to the *Project Manager* for acceptance which, for example, has an approach and sequence with little free and total float in it, and another which it is actually working to – and this becomes apparent, then the *Project Manager* can decide to not accept the submitted one for this reason.

- *Does not show the information required.* This relates to the previous clause 31.2 (discussed above) and to additional information for revised programmes under clause 32.1 (discussed below).
- *Does not comply with the Scope.* This partly overlaps with the last bullet point of clause 31.2, but could also include
  - 'constraints' about how and when the *Contractor* is to do work
  - *how* – as opposed to the *what* of clause 31.2 – the *Contractor* is to present information (e.g. using a particular software package). Or, in NEC4 language, the new last piece of text added onto clause 31.2, namely it is not 'in the form stated in the Scope'.

3.5.4.3 Clause 32.1: Revised programmes

Have a read of clause 32.1. From this, do you think that the Accepted Programme is a living programme to be updated and used throughout the contract, or one to be stuck up on the wall and forgotten about? Fairly evidently, it is the former.

Let us now take each bullet point of clause 32.1, one at a time.

BULLET 1

This states that the revised programme needs to show progress on each operation and its effect on remaining work. Being honest, if not a lot has changed in the preceding 4 weeks or month, I am quite happy for a 'line of progress' to zig-zag down the main programme showing where the *Contractor* is slightly ahead or behind on progress. However, once the shape of the programme, in terms of sequence and dependencies, start to change (whether due to compensation events or not), I would not be happy with this approach, as I do not believe it adequately shows 'effect upon the timing of the remaining work'.

BULLET 2

This requires the *Contractor* to show how it 'plans to deal with any delays', which could have been the subject of early warnings, whether the liability will fall on the *Contractor* or will ultimately result in a compensation event (although for these, the onus is on the *Project Manager* to give direction). But how are all these different classifications shown in a revised programme? Here are my suggestions, which only apply if the event has already caused a delay or is more than likely to in the future.

1. Assign a colour, which can be shown on the programme, to the various classifications below. My classifications (with the colours plucked from thin air) would be
   (i) work in the original Scope, including work that has been delayed by a compensation event (black)
   (ii) non-assigned early warnings, by which I mean those where it has not been agreed whether or not the events are compensation events – this should be a fairly small number of events if regular 'wrap-up' meetings are being held (orange)
   (iii) *Contractor*-owned early warnings, i.e. where it has been agreed they are not compensation events or, if they are, arise from the *Contractor*'s fault and hence have a

zero time assessment (green)

(iv) events that have been accepted as compensation events, but are in the process of being quoted for or having the quotation accepted (purple)

(v) implemented compensation events (red).

2. If the event under classification (ii) to (v) above is worthy of a new row or bar in the programme, then add in a new operation or activity. If not, then add the work onto or into an existing bar.
3. If a new bar is added, in the same way that you show time risk allowance for normal work (category 1(i) above), show time risk allowance as a different hatching. However, there is a nuance here for those instances where agreement has not yet been reached (i.e. category 1(iv) above and, to a lesser extent, category 1(ii)) in that the hatching represents the best and worst estimates of impact, with the former likely to be the *Project Manager's* estimate and the latter the *Contractor's*! This applies to work to be done. Once it has been done, it is what it is.
4. In an ideal world, the regular 'wrap-up' meetings would mean that there are no orange and purple bars on a programme submitted for acceptance as it is always quickly agreed whether an early warning is or is not a compensation event and all quotations are accepted almost immediately. In reality, there will at times be some ongoing discussion over whether an early warning is a compensation event or not (orange), and ongoing discussions over the time effect of compensation events (purple), which, at best, fall across the date when the *Contractor* has to submit the programme. Having these highlighted in a programme will draw attention to areas for action. Consequently, each successive submission of the programme should see orange and purple bars turning to green and red ones, with the amount of hatching reducing or disappearing. If this not happening, then perhaps the Parties should review the way in which they are working together.

BULLET 3

This allows the *Contractor* to reprogramme the work, provided: (a) it is within the constraints of the Scope and (b) it is realistic/practicable to change it, which includes giving the *Project Manager*, acting on behalf of the *Client*, realistic notice of any changes in timescales for things they are to supply. For example, let us say the Scope stated that the *Contractor* could have access to an occupied area at the start of June, but in every Accepted Programme up to April the *Contractor* was happy to wait until November for this access. It may well then not be realistic to suddenly, in the May programme submission, 'demand' access to the area in early June!

> One contract in the original research sample was to prevent tidal erosion in the Thames Estuary. The work consisted of constructing five sheet-pile walls on the estuary bed and placing matting in between them. The *Contractor* that won the bid did so on the basis of doing the *works* with one barge and a piling rig over the whole of the summer. Having won the contract, the *Contractor* was able to gain use of two barges and two piling rigs, so instead decided to do the work over the central 3 months of the summer, as this reduced its land-based costs, as well as risk. There was nothing in the Works Information that said the *Contractor* could not do this, and the *Employer* was quite happy to accommodate this approach.

Those familiar with NEC3 may well have spotted that NEC4 clause 32.1 now only has three bullet points, as opposed to four in NEC3. The one that has disappeared is the requirement to show 'the effects of implemented compensation events'. This was related to compensation events where the

Project Manager has accepted the *Contractor's* quotation, the *Project Manager* has made an assessment (and notified the *Contractor* of it) or a quotation has been treated as accepted due to the *Project Manager's* inactivity (see clause 66.1). Does this mean the *Contractor* does not now have to show 'implemented compensation events'? The answer is 'not directly', but if the compensation event, whether implemented or not, has affected progress on an operation and affects the timing of remaining work, then it is covered by the first bullet point.

A similar argument can be put forward for early warnings. However, early warnings are for matters that 'could', not necessarily 'will', delay Completion or a Key Date. I think a bit of common sense needs to be used here, based on the probability of the matter occurring and, to a lesser extent, its impact. A matter that has a 10% chance of occurring is probably not worth showing in the programme, as otherwise the programme would end up with tens, if not hundreds, of 'what ifs'. In contrast, one with a 90% chance of occurring would normally be shown. However, a matter that has a 10% chance but would have a 3-month impact on a 1-year contract might well be worth at least highlighting in the programme, whereas one that has a 99% chance of occurring, but which would only delay work by a few hours, may not be.

3.5.5 Accepting the first programme

In an ideal world, the programme that is referenced in Contract Data part two of the *Contractor's* tender shows the information required in clause 31.2, is practical and realistic, and complies with the Scope etc., all as per clause 31.3. Consequently, the *Client* has no problem leaving the reference there so that it becomes the Accepted Programme on award of the contract. However, this scenario is unlikely.

> In the original research sample of 29 contracts, 14 required a programme to be submitted with the tender and referenced in Contract Data part two. However, this reference was removed from all 14 winning tenders prior to the execution of the contract, so none of the submitted programmes became the Accepted Programme until they had been further refined after the contract was entered into.

It is far more likely that not only will the *Contractor* want to develop the programme once it has won the contract, but also the *Client* and its *Project Manager* will have some questions and require some clarifications on the programme.

First of all, on knowing that it has won a contract, the *Contractor* will want to develop the programme so that it is both a usable operational programme and an acceptable one, so that it does not incur any contractual sanctions. I suggest that this is a joint effort between the planner who assembled the tender programme, the planner who will be involved in the job, and the *Contractor's* agent. On small jobs, all three might be the same person. At the same time, the people who will actually be supervising and directing the work (e.g. foremen and section engineers) should be further developing the initial short-term programmes and statements of methods and resources for the early operations.

Any comments by the *Client* and/or future *Project Manager* on the tender programme should be considered. Indeed, I suggest the *Contractor* actively engages with the future *Project Manager* – and their planning representative if they have one – to gain their 'buy-in' and understanding of the programme. In addition, if the *Project Manager* is experienced, they may be able to add or enhance the programme and/or remove constraints that will hinder progress. To reiterate previous comments, the time to have these discussions is right at the start of the contract, because

- from the *Contractor's* point of view, to avoid sanctions being applied
- from the *Project Manager's* perspective, the ability to withhold a quarter of the Price for Work Done to Date sanction (clause 50.5) is lost once the first programme has been accepted
- from a successful project point of view, those contracts that did not have these 'fierce conversations' early on became progressively more confrontational and took a long time to settle the final account, while those that did became more collaborative and rapidly agreed the effects of compensation events.

Ideally, these discussions and considerations will be resolved before the contract is entered into so that the revised programme can be referenced in Contract Data part two, although the previously mentioned original research data indicates that this is unlikely. (For further insights on whether to require a programme to be submitted with tender or after the contract has been entered into, see an article by Patterson (2020). The table is especially useful if you want to make this binary choice.)

> On only two of the 29 contracts researched was the sanction to retain money from the first assessment used. Because of the manner in which the parties approached the contract and each other, this did not cause any lasting damage to relations. On three of the contracts, money was not retained when it could have been, either as a result of the *Project Manager* not going through the submitted programme with sufficient rigour or through fear of being seen as adversarial. This caused problems later in the contract in the assessment of compensation events, which did result in friction between the parties.

3.5.6 Six levels of use of the Accepted Programme

The text below is an expansion and adjusted form of my article 'ECC programme – the six levels of use', which appeared in the *NEC Users' Group Newsletter* (Broome, 2009a). It seemed to go down well, as I was invited to give a talk on it at the NEC Users' Group – which also went down well, with one person even thinking I was a planner – as well as being asked to make a video film of it![9]

I wrote the original article as a result of both observing and helping site teams achieve higher levels of success, and trying to categorise the different things they did that led to different results.

I identified six uses of the programme in managing an NEC contract which apply in roughly sequential order; that is, you cannot operate at the later higher levels unless you are already operating at the initial ones. Think of it like a ladder where you cannot move up to a higher rung unless you are already on the rung below. Identifying these uses enables teams to identify the actions that will lead to greater project success.

[9] This 30-minute video can be found either via my website: www.jonbroome.com/resources/videos, or via the NBS website: http://www.thenbs.com/topics/ContractsLaw/videos/NEC3_six-levels-of-use.asp.

USE 1: BASIC CONTRACTUAL COMPLIANCE FOR THE *CLIENT* AND OTHERS INVOLVED IN THE PROJECT

The primary purpose at this level is to avoid compensation events caused by a default by the *Project Manager*, *Supervisor*, *Client* or Others. For this the *Contractor* puts forward a programme to the *Project Manager* for acceptance under clause 31.3. The programme should therefore be in accordance with this clause and contain the information identified in clause 31.2.

The *Project Manager* is evaluating the programme submission to ensure that the order and timing of the work done by themself, the *Client* and Others whom they may be coordinating is as agreed or, if not, is in accordance with the Scope (as per the fourth bullet point of clause 31.2). In other words, making sure that the *Client* or Others fulfil their side of the contractual bargain, so there will be few or no compensation events.

Here, there is an obvious 'What's in it for me?' for the *Project Manager*, as they are professionally representing the *Client*, but what's in it for the *Contractor*? When I have asked my contractor clients if they make more profit from contracts with lots of claims/compensation events, the answer – perhaps going against conventional wisdom – is not. Indeed, one did a formal review and concluded exactly that. The reason being that when they can work efficiently and as planned, they save time and hence money. On the other hand, they never fully recover the time and cost associated with client-originated delay and disruption, and have to spend a lot of management time to get to a position that they can live with. In the meantime, under a priced contract, their cash flow suffers.

The simplest means of identifying what is to be supplied (whether it is access, documentation, drawing acceptances, *Client*-supplied Plant and Materials or Equipment etc.), by when and by whom is to state within the Scope that the *Contractor* is to supply with every programme submission an up-to-date, three-column schedule (i.e. what, when and by whom), and that if something to be supplied is not on this list, but is hidden away in the depths of a programming submission, then it is not a compensation event should the Party fail to supply it. (To absolutely ensure this would apply, you would need to revise the compensation events using an option Z clause.)

USE 2: CONTRACTUAL COMPLIANCE BY THE *CONTRACTOR*

In this use, the *Project Manager* is checking the programme to ensure that the *Contractor* is contractually compliant with its obligations under the contract – for example, the *Contractor* is fully aware of what needs to be done within the constraints of the Scope, that durations of operations are realistic given resource levels, and that the methods are practicable. Under clause 14.1, acceptance by the *Project Manager* of the programme does not change the *Contractor's* responsibility to Provide the Works, so, for this bullet point, it is really a check that the *Contractor* has a realistic, practicable and fully inclusive programme. (The caveat to this statement is in use 1: if the *Project Manager* accepts a programme that shows an action of the *Project Manager*, *Supervisor*, *Client* or an Other and then the party does not do the action, it will be a compensation event.)

If the *Project Manager* discovers that the *Contractor* has missed something out, then presumably the *Contractor* would like to know so that they can plan it in as soon as possible, and possibly tell their bosses that they need to make a provision for it. Likewise, if the *Project Manager* discovers that something has been left out of the Scope, is deficient or needs correcting in some way, then they want to know as soon as possible. That way the item can be added/clarified by a change to the Scope as soon as possible, in order to minimise the delay and disruption effects of a compensation event.

USE 3: EFFICIENT AND RAPID ASSESSMENT OF COMPENSATION EVENTS

For this to take place, the level of detail within the programme needs to be sufficient so that, when a compensation event occurs, the resulting change in method and resources in the affected operation is transparent and easily agreeable. The agreed changes within an operation may affect its duration, which then affects subsequent operations, so the linkages between operations, in terms of resource usage and proceeding/succeeding subsequent operations, need also to be transparent. From this, the additional time and cost can be calculated in order to agree the compensation event quotation before it is formally submitted for acceptance.

Very briefly, the benefits of this are

- for both Parties, it reduces the hassle and management time needed to agree the assessment of compensation events
- both Parties therefore have a rolling final account, albeit with a bit of time lag from what has happened, which means that they can provide realistic forecasts of where the contract is heading, both financially and in terms of time
- if the *Client* wants to take proactive action to change the course of the trend, it is more able to do so
- for the *Contractor* under option A or B, it should get better cash flow, as the sooner compensation event quotations are agreed, the sooner they can be included in the monthly assessment, and hence the sooner they are paid for.

USE 4: MINIMISATION OF CONSEQUENCES BY THE *PROJECT MANAGER*

This is where it starts to get proactive, but not necessarily collaborative, in management. For this use, the consequences of the *Project Manager* not making a decision by a set time or of making a certain decision are transparent from the programme. Hence, as things arise during the contract, the *Project Manager* makes timely and intelligent decisions on behalf of the *Client* that have minimal impact on time and cost, even though they may still be compensation events.

> Very early on in my consultancy career – some 20 + years ago – I was giving training to consultants at which a senior project manager – or Resident Engineer under the old ICE conditions – was present. While doing an exercise on the programme, I still remember him exclaiming something like 'Wow! Numerous times in the past, I have had to make a decision not really knowing the consequences of that decision or when I had to make it by. These [pointing at the NEC's programming provisions] will really help me make better decisions faster'.

USE 5: IT ENABLES JOINT WORKING TO MINIMISE THE CONSEQUENCES OF AN EARLY WARNING

In this use, either Party gives an early warning, which may not necessarily result in a compensation event (i.e. liability for the 'issue' can fall on either Party). For a medium to major event, this will usually involve using the programme as a tool to analyse the 'do-nothing' consequences of the event, and whether the Parties can live with it (e.g. a 2-week delay, when there is 3 weeks of free float to absorb it) or need to collaboratively plan their way out of it in order to minimise the time and cost consequences, be it a compensation event or not.

The difference between this use and

- use 3 is that the Parties are proactively *managing* the situation, as opposed to just rapidly agreeing the contractual consequences (i.e. good *administration*)
- use 4 is that *both* Parties are working together through the early warning and risk reduction process in order to minimise the consequences of that risk, rather than the *Project Manager* doing it on their own.

Consequently

- the 'What's in it for me?' for the Party on whom the liability falls is that a better course of action can be chosen that is more in line with its objectives
- regardless of on whom the liability falls, the chosen actions can be implemented more efficiently as, even if the same course of action is chosen (as opposed to just happens), it can be planned out and done more efficiently.

USE 6: REGARDLESS OF WHETHER OR NOT IT IS A COMPENSATION EVENT, THE PROGRAMME IS THE CENTRAL DOCUMENT AROUND WHICH THE PARTIES COLLABORATE TO REDUCE TIME AND COST

This level of usage involves proactively using the programme to realise opportunities, as opposed to responding to potential threats (e.g. early warnings as under use 5). It requires a high level of understanding of the programme by the *Project Manager*, and the authority of the *Project Manager* to relax constraints, and sometimes technical requirements, stated in the Scope.

> As an example of this, on one site in the original research sample, the *Contractor* was required to put a service main through the perimeter fence of an operational site with high security requirements – by which I mean it had two perimeter security fences. The Works Information stated that the *Contractor* had to erect a temporary security fence, take down part of the permanent fence, place the service through the gap, and then reinstate the permanent fence before dismantling the temporary fence. This was shown as a near week-long operation on the Accepted Programme. They then had to repeat for the second security fence. The *Project Manager* asked whether the *Contractor* could dismantle the permanent fence, put the service main through, and reinstate the fence within a day, thus saving the erection and dismantling of the temporary fence. The *Employer* would supply two security guards for the day. The answer came back that by working a long day with additional men, this could be done. Even though this contract was let under option A (priced contract with activity schedule), using the first edition of the ECC (which had no value engineering clause), it was agreed to split the savings, so both Parties gained.

> Another example of this sort of working is a contract for which I was the project facilitator. It was a new inner ring road in a medium-sized city, and the contract was let under option C (target contract with activity schedule). Due to problems in gaining access to work areas in the Site, planned Completion was going back almost every week and, as the consistent lack of access was a compensation event, so was the contractual Completion Date. The *Contractor* could have sat back and said '*Employer's* problem'. Instead, a couple of months into the construction phase, the *Contractor* organised a programming workshop, not just for its site

team, but also for the *Employer* and the key Subcontractors. Despite there being three critical or near critical paths in three of four areas (i.e. we had to find parallel time savings in three areas), during the workshop the team managed to identify 12 weeks of time savings with an 80% or more chance of being implemented. After developing some of the other ideas further over the following couple of weeks, 20 weeks of time savings were identified. With site costs at £50 000 per week, this translated into a shared saving of approximately £1 million: not bad for a 1-day workshop! This was on a 2-year contract, where the *Contractor* had been involved in the early pre-construction phases as a consultant.

However, uses 5 and 6 do come with a slight health warning, which is that the programme is ultimately the *Contractor's* programme, and that the *Project Manager*, while positively and constructively contributing to its development, cannot 'tell' the *Contractor* what do. More precisely, the *Project Manager* can tell the *Contractor* what it **cannot** do by reference to the criteria for acceptance in clause 31.3, but if they instruct a change, then it is a change to the constraints stated in the Scope and so will be a compensation event.

3.5.7 Practical application of the six levels of use

When helping set up a project team, either as a facilitator, NEC trainer or a combination of the two, I have used this framework to help the team reach consensus on what level of use they aspire to. Following on from this, decisions and tangible actions can be fleshed out further by asking the following sorts of questions.

- What information is actually wanted in the programme, both in terms of additional information not stated in the contract and what is deemed to be superfluous for this project?
- What level of detail does the programme have to show, especially in terms of methods and resources for each operation, given the time to the actual operation starting?
- How is this presented in terms of how the whole programme document fits together? (see Section 3.5.10 as an initial guide.)
- How and when are the Parties going to agree the programme, both initially and on an ongoing basis, prior to formal submission and acceptance? For example: will they issue a draft for a read-through before discussion; will they project it onto a screen or white wall and present on it to the whole project team for discussion (as is sometimes done when lean construction principles and practices are applied); will the *Project Manager* and *Contractor's* programmer sit down together side by side; etc.
- How are 'alterations to the Accepted Programme' (see clause 62.2) due to a compensation event agreed and shown?
- How precisely and when are the Parties going to identify opportunities for further development? On an ad hoc 'wing and a prayer' basis; through a strategic planning workshop of the sort outlined in Section 3.5.6; or through tweaking the early warning process to include the identification and managing in of opportunities, as outlined in Section 3.6.8.3?

Sorting out these issues early on in the project reduces misunderstandings and potential conflict, leading to a greater likelihood of project success. (If you want to see an independent write up of what happens when these principles and practices are applied, together with some of my suggestions on running the early warning and risk-reduction process, see Freeman (2011). I was the trainer on this project, where the Parties decided to operate at level 6.)

3.5.8 Agreeing and updating the programme so it can be accepted promptly

What does all that has previously been said mean for a *Project Manager* or the person who is delegated the *Project Manager's* duties with respect to the programme? For me, it means that

- they need to have a technical appreciation of the work type, how it can be done and how long it takes to do it, or have access to people who do
- particularly early on in the project – prior to award or in the first month – they need to make time to evaluate and understand the programme and, if needed, clarify and challenge the *Contractor* prior to acceptance of the programme; this does, however, need to be an ongoing process.

And, if the *Project Manager's* duties have been delegated, whether contractually (i.e. the person is accountable and can accept or not accept on the *Project Manager's* behalf) or they are just responsible, how is the *Project Manager* and wider *Client* team consulted and informed.

What does all the above mean for a *Contractor*? For me, it means that

- the *Contractor* needs a greater programming resource in terms of a dedicated planner on a medium-sized project or a visiting one supporting its agent on smaller sized contracts
- the quantity surveyors used for putting together compensation event quotations need to understand the programme and be able to build up and forecast costs from first principles; this also applies to any cost consultant representing the *Project Manager*
- the *Contractor* needs to be prepared to spend time explaining and justifying its programme to the *Project Manager*, especially in the early months.

As stated in Section 3.3.1, one of the regular meetings required for the effective operation of an NEC contract is a programming meeting before the formal submission of the programme for acceptance. This is so that the formal submission can be rubber stamped in a matter of days, as opposed to the 2 weeks allowed for under the contract.

At this meeting, the cumulative effect of groupings of compensation events for the past period can be assessed together, as opposed to arguing over the delay and disruption effect of each and every minor compensation event. For more detail on this, see Section 3.7.4.

3.5.9 Acceleration under the NEC

As briefly mentioned in Section 1.5.4, under the NEC 'acceleration' does not mean speeding up work to achieve the original planned Completion as shown in the Accepted Programme or to keep the original Completion Date where it is. From looking at clause 36.2, it means bringing forward the contractual Completion Date and any Key Dates.

A change in NEC4 is that rather than it being solely down to the *Project Manager* to initiate the procedure, as per NEC3, now either the *Project Manager* or the *Contractor* 'can propose to the other' an acceleration. And if the other is 'prepared to consider the proposed change', then the *Project Manager* can instruct the *Contractor* to provide a quotation.

However, there is no reference to any criteria for how this quotation is assessed. Consequently, if the *Contractor* knows a 2-week delay is costing the *Client*, say, £100 000, the *Contractor's* quotation for a 2-week acceleration could comprise costs of, say, £10 000 and profit of, say, £80 000, giving £90 000 in total. I sense a bout of righteous indignation from some who represent the *Client*! Yet this represents the situation in law, where one party to a contract cannot unanimously bring forward the delivery date. That party would have to have the other party's agreement, and the contract would then need to be amended by a deed of variation or a supplementary agreement. Under the NEC, the *Project Manager* effectively has the contractual power to negotiate and agree on the *Client's* behalf, using terms in clause 36, without the need for a deed of variation or supplementary agreement (as would otherwise be required under clause 12.3).

The *Contractor* has the same 3-week timescale as for compensation events in which to prepare the quotation, which seems a long time as it does not need to justify the Defined Cost at all. In reality, I would expect more of a commercial negotiation between the *Project Manager* and the *Contractor* with, once agreement is reached, the *Contractor* submitting the quotation. Consequently, as the quotation only needs to include the proposed changed Prices and earlier Completion Date or Key Dates and a revised programme, with no financial back-up being necessary, the 3 weeks allowed for the *Project Manager* to accept the quotation seems longer than necessary, even taking into account the probable need to consult with the *Client*.

So, imagine you are a *Project Manager* (or *Contractor* with a Subcontractor) on a time-driven contract. When a compensation event occurs that delays the planned Completion and hence the Completion Date, which of the following options would you choose?

- Have one compensation event for the delay (which would be assessed via the rigour of the compensation event quotation procedure) and then propose to the *Contractor* that it submits a quotation for acceleration to bring the Completion Date forwards to where it was previously.
- Having had a discussion with the *Contractor* about practicality, use clause 62.1 to instruct the *Contractor* to submit a quotation through the compensation event procedure to keep planned Completion, and hence the Completion Date, where it currently is.

I hope you agree that the second of these is the better option, as that way the *Contractor* cannot charge what it likes to bring the Completion Date forwards.

For completeness, if the second option above is adopted but planned Completion cannot be kept the same (e.g. it is not practicable or realistic as per clause 31.3 or for reasons stated in clause 63.9), but the *Project Manager* wants the contractual Completion Date to stay the same, then there is an implication under the contract that the *Project Manager* and *Contractor* can do a deal under clause 63.8. In this scenario, through the compensation event process, the *Project Manager* 'buys' the *Contractor's* time risk allowance in exchange for an increase in the cost risk allowance. This would probably involve an element of negotiation, but I believe the contract gives authority for this to happen. However, I am pretty certain that the *Project Manager* could not impose this on the *Contractor*, because it would contradict the express wording of clause 63.5 with respect to how a delay to the Completion Date is calculated.

3.5.10 What does a programme submitted for acceptance actually look like?

First, I hope you can clearly see that the Accepted Programme is not just a large single sheet with lots of bars on it! If I was a programmer, I might produce a model example, but I am not. Regardless, think of it more like this.

1. A single large sheet showing the key contractual dates, as per those needed under the first two bullets of clause 31.2, and 30 to 50 main 'level 1' operations.
2. A bar chart of, ideally, the last month's work, showing the requirements of clause 32.1 in terms of progress, any delays, regardless of who they are attributable to, and colour coded as outlined in Section 3.5.4.3. I say 'ideally the last month' as, ideally, everything has been agreed prior to formal submission for acceptance. In reality, there may well be some unresolved issues, in terms of contractual ownership or quantum of delay for a compensation event, which go back more than a month.
3. A similar look-ahead programme for the next month or three's operations.

For clarity, items 1, 2 and 3 above are different views of the same electronic programme.

4. For each level 1 operation, a separate section showing the current thinking (i.e. it is an evolving document) of the *Contractor* and its supply chain in terms of
   - activities that make up that operation, and their linkages
   - perhaps a short list of risks associated with the operation, which may include relevant early warnings to date
   - a statement of how the *Contractor* will do the work (which may be very brief initially, but in the month before the operation takes place, will be very detailed and complaint with health, safety and environmental legislation, to avoid duplication)
   - a statement of resources, which may be approximate if the operation is some way off, but may name the specific Equipment and people if it is to happen in the next month.
5. A three-column schedule identifying what the *Project Manager*, *Supervisor*, *Client* or Others have to do or supply, and by when.
6. Other information required under the last bullet point of clause 32.1, which, I suggest, at a minimum is a highlighting of the critical path and a short narrative explaining the initial logic and key changes since the last submission.

3.5.11 Further reading on programming under the NEC 3 ECC

Glenn Hide has written two excellent papers (Hide, 2010a, 2010b) that have appeared in the *Proceedings of the Institution of Civil Engineers – Management, Procurement and Law*, much of which correlate with what I have said here, and a little bit of which does not!

3.6. Early warnings process and application

This section consists of the following sections.

- Why early warn?
- The basics of risk management.
- Where risk management has gone wrong.
- Overview of a practical process, including practical pre-contract risk management.
- Notifying early warnings: who notifies, to whom and what is notified?
- Setting up and revising the active Early Warning Register.
- Running an early warning meeting.
- Managing in opportunity.

3.6.1 Why early warn?

As previously stated, the early warning procedure is the missing step between how risk management is normally implemented – identifying what can go wrong and developing strategies to minimise their impact, which often go unimplemented – and monitoring the programme to identify something that has already gone wrong and has affected the *works* and only then taking action.

An example, taken from my postdoctoral research on how using early warnings and the risk reduction process can reduce time and cost impact, is given in the following box.

> On a tunnelling contract, while sinking the shaft, the workers found that they had hit a layer of chalk that contained flint. Initially, they decided to increase the jacking force, and therefore the number of jacks. However, the flints were at the level where the tunnelling machine was meant to start boring, and the machine was not designed to go through flints this size. The Parties had expected a bit of flint, but not people-sized rocks of it! Within 2 days, a decision was made to deepen the shaft and start the drive 3 metres below the original level. This committed the *Employer* to an extra £0.75 million (in the late 1990s). However, both Parties believed that they had probably saved around £3 million compared with starting the tunnel at the originally intended level. Despite the contract being let under Option C (target contract with activity schedule) and the existence of a partnering agreement, all parties thought it unlikely they would have made that decision, and certainly not as rapidly, if the early warning procedure had not been specified in the contract.

It is worthwhile reiterating that the early warning mechanism is intended to identify problems early, in order that actions can be taken to avoid or minimise their impact. This is regardless of who the liability falls on (i.e. whether or not it is a compensation event or an insurance event).

- ***Contractors***:
  - do early warn 'as soon as' you become aware of any matter that could affect time, cost or quality (in non-NEC speak), especially if the matter is likely to become a compensation event, as otherwise your entitlement could be reduced (see clauses 61.5 and 63.7) and, if under options C to F, any additional Defined Cost could be disallowed (see clauses 11.2 (26)/(27) third bullet point)
  - do not just early warn as a 'set up' for compensation events, because otherwise that is all they will be seen as by the *Project Manager*. And, if an early warning is likely to result in a compensation event, you should actively contribute to its resolution, otherwise why should the *Project Manager* help you out when it is your liability?
- ***Project Managers***:
  - do not just see early warnings as a 'set up' for compensation events to be automatically ignored or rebuffed – indeed there is no contractual mechanism to rebuff early warnings. If an early warning is going to result in a compensation event, your *Client* will pay for it, so it is both your professional and contractual obligation to contribute in order to minimise the impact. If it is not going to be a compensation event, then it is not in your *Client's* interest to have a late contract (even if it can recover delay damages) or for the *Contractor* to be losing money as it tries to find less positive ways to recover it.
  - Remember – and this is a common complaint from Contractors – you also have an obligation to early warn.

Whatever organisation you work for, please do not blame the messenger, or stick your head in the sand and wish early warnings away without taking action.

> On one NEC3 contract I was asked to try and resolve issues on, the *Project Manager* and his staff were *administrating* (as opposed to managing) the contract superbly, in terms of implementing compensation events within the contract timescales (i.e. wrapping them up contractually in terms of time and cost). However, not one of the 110 or so design changes that had come through were early warned to the *Contractor* in advance. They just appeared as changes to the Works Information. Likewise, the *Contractor*, who was early warning with the best intentions, was incredibly frustrated at the lack of help from the *Project Manager* and his staff. Many, but by no means all, early warnings were resulting in compensation events, and the *Employer* was having no input into their resolution, except with 'benefit of hindsight' arguments when assessing the compensation event. Just for balance, the *Contractor* was not administrating the contract very well in terms of notifying compensation events and submitting quotations on time. The net result was a big ongoing dispute at the end of the contract.

I emphasise these points upfront as the cultural legacy of only notifying problems in order to generate a claim is perhaps the biggest hindrance to the effective operation of the early warning processes. Indeed, if I am doing a start-up training for the project team, one of the things I get the project sponsor to say, if they are there, is 'Notify early warnings. Not notifying early warnings will be viewed badly'. Or, as one NEC3 *Employer* bemoaned of their somewhat combative *Contractor,* 'How can we help him out if they are not telling us their problems'.

Two quotes from my original research illustrate what project participants thought of the early warning and risk-reduction process when it works well

- 'it's the jewel in the crown of the NEC'
- 'it's the starting point for cooperation, and not giving them would undermine the benefits of using the NEC'.

So, have I convinced you of the importance of early warnings? This need to give early warnings and respond to them positively should be driven into the minds of all site personnel, from the foreman and site engineer upwards.

Having said all of this, sometimes the opposite is true: the *Project Manager* receives so many early warnings for every minor matter that could possibly result in a compensation event that they are swamped. This is due the *Contractor* being petrified of the sanctions for non-notification, namely reduced assessments of compensation events due to the operation of clauses 61.5 and 63.7. After a discussion and reassurance, the threshold for giving early warnings will normally settle down to a common-sense, workable level.

3.6.2 The basics of risk management

First of all, some basic definitions taken from the APM Body of Knowledge, seventh edition (Association for Project Management, 2019).

RISK EVENT

An uncertain event or set of circumstances that would, if it occurred, have an effect on the achievement of one or more objectives.

In other words, a 'risk' is an event that *may* happen and, if it does, *will* impact on objectives. It therefore makes sense that those objectives are defined and agreed early on in any risk management process. By this definition, risk can be both positive and negative, as expressed by the following definitions.

THREAT

A negative risk event; a risk event that, if it occurs, will have a downside/detrimental effect on one or more objectives.

OPPORTUNITY

A positive risk event; a risk event that, if it occurs, will have an upside/beneficial effect on the achievement of one or more objectives.

How to manage in opportunity is discussed in the last section of this chapter.

In simple terms, the combination of estimating uncertainty – the natural variation in the duration of individual operations and activities and the resulting financial impact – and the probability and impact of individual risk events gives the overall project risk or range of potential outcomes for the project.

Way back in Section 1.5.7.3, I said it was helpful to see the Early Warning Register as an 'emerging risks and issues register', so, for completeness, here is my definition of an issue.

ISSUE

An event which *is* happening or, if action is not taken, *will* definitely happen and, without action, *will* have a downside/detrimental effect on one or more objectives.

For example, if the *Contractor* is missing a small detail on some structural steelwork, at some point that detail has to be supplied, otherwise the building or asset cannot be built or will be built unsafely, thus affecting time, cost and/or quality. Do you notice the similarity in my definition with that for a threat? In other words, I am saying an issue is a risk with a 100% probability of happening because it either has already happened, is happening or will definitely happen if we do not do something about it. So why deal with it in a fundamentally different way? Back to the theory of risk management.

There are many ways of dividing up the basic process of risk management into its elements, but essentially the stages are as follows.

1. An **initiation stage:** depending on the stage of the project when risk management is started and/or the iteration of the process, then contributors, stakeholders and participants attend a meeting that includes a mixture of presentations, group work and feedback sessions to gain a common view on
   (a) the current situation and need for a project
   (b) the aims and objectives of the project
   (c) the main constraints
   (d) the basic methodology or options on how to get from (a) to (b) above.
2. A **risk identification stage**, which normally involves some form of brainstorming of ideas, which can be stimulated by looking at generic, sector-specific or project-specific lists, as well as talking to participants from previous similar projects.

3. An **assessment** or **analysis stage**, which consists of assessing each identified risk in terms of probability of happening and potential impact should it occur. This stage can range in sophistication.
   (a) In its basic form, this might be just placing a risk on a 3 × 3 matrix of high, medium and low for both probability and impact.
   (b) In a more sophisticated form, you might use a 5 × 5 matrix, where you define the boundaries between levels of risk probability, such as very low and low risk (say 1 in 100), low and medium risk (say 1 in 25) etc., and likewise for the most likely impact in terms of quantifying the effect on the defined objectives. For instance, on a £1 million contract, you might set the very low to low impact interface at £2500, the low to medium interface at £10 000 etc. You then multiply the probability of a risk by its most likely impact to get an overall risk ranking, based on where it is placed on the 5 × 5 matrix (i.e. you multiply a score of, say, 2 for probability and 3 for impact together to give an overall risk ranking of 6 out of 25).
   (c) Slightly more sophisticated, you multiply the agreed probability (say 3%) by its most likely impact should it occur (say of £20 000) to give a monetary figure (in this case 3% × £20 000 = £600) for its overall severity.

   For both (b) and (c), you would then typically put these values on a spreadsheet in order of decreasing ranking/severity.
   (d) Much more sophisticated is where, as well as having the probability estimates from (c), you make a three-point estimate (minimum, most likely and maximum) for the impact of each risk in terms of cost and/or time and then do a Monte Carlo simulation. This is where a computer does a run through of the project, effectively rolling a dice for each risk event. It does this 1000 or more times to give an S-curve showing the range of outputs for the project, from the best-case scenario (i.e. where no risks occur – the P0, which the project has a 0% chance of coming in below) to the worst-case scenario (i.e. where all the risks occur with maximum impact – the P100, which the project has a 100% chance of coming in below). Various probability (P) values and the variation from the P50 are used to get an idea of the overall project risk.
   (e) Even more sophisticated than above is a programme-based Monte Carlo simulation. Here, the risk events are mapped onto the project programme in terms of when they would occur, together with three-point estimates for the duration of each operation or activity on that programme. The computer again runs a simulation 1000 or more times to get a supposedly more realistic S-curve.

   Substages (a), (b) and (c) are normally referred to as a 'qualitative assessment', whereas (d) and (e) comprise a 'quantitative analysis'.
4. A **response planning stage**, where you work down your list of risk events by overall ranking/severity, developing responses which, in order of preference
   (a) ideally prevent the event happening altogether
   (b) reduce the probability of it occurring
   (c) reduce or mitigate the potential impact.

   For some of these actions, there may well value-for-money considerations. For instance, if I am spending £10 000 to reduce the probability of a risk event with a 20% chance of occurring and a £20 000 cost if it does, then it is not worthwhile doing so. What you are left with is residual risk. You must then decide whether to
   (d) transfer the residual risk, either via a contract to another participating organisation or to an insurer.

   With these actions in place, you then repeat stage 3.

5. **Response implementation and monitoring**, where you actually undertake the actions that arise from your outline responses, and monitor the current situation of each potential or actual risk and the actions you should be taking to avoid, minimise or mitigate it.

Going back to stage 1, you will notice that I mentioned the word 'iteration'. These iterations could be done in formulating the business case, the concept and definition stage, for each delivery option, including the overall procurement strategy, prior to selecting an individual contractor, at the start-up of a contract and on a weekly, fortnightly or monthly basis to varying degrees of sophistication.

3.6.3 Where risk management has gone wrong

Before I start, I want to let you know that I have been around risk management for more than 30 years, from the time when I was an undergraduate and introduced to it by my ex-Professor John Perry. Following my 3 years with a contractor, John then became my PhD tutor. He was, in fact, the original investigator of risk management in construction, publishing the first report on it in 1979 (Thompson and Perry, 1979). I also run practically orientated risk management workshops.

Despite the considerable number of guides now available on the subject, both in the engineering and construction sectors and project management literature, my view is that not a lot has changed in the basic five-stage process of risk management. However

- where risk management has consistently and persistently fallen down in being of worth is that people do not implement the responses
- the focus of attention in terms of its development has been on ever-increasing sophistication – and complexity – in terms of how it is analysed. This is because the risk management profession, which develops the guidance, is generally not composed of risk managers, but of risk analysts.

For instance, around 10 years ago I attended a seminar where the risk expert effectively said that he 'needed' 49 columns in his spreadsheet to do a 'meaningful' risk analysis, and that a rule of thumb is that a project risk log should have around 100 risks in it. That's 4900 entries in the risk log! Is this practical for the average project manager to do on a one-off basis, let alone update on a regular basis?

The answer of course is 'No', which is why there is a profession of risk analysts. Yet often, having constructed and run their Monte Carlo simulation, these analysts move on to the next project, taking with them any insights gleaned.

Let us now consider whether this is 'meaningful'. I often do risk (threat and opportunity) workshops, either at the bid stage or soon after a contract has been signed, where the emphasis is on getting out tangible actions, with action owners and stated dates, to de-risk the project and/or reduce time and cost. As well as some significant de-risking, a 10% reduction in time and cost for a day's worth of team input is a very good, but not uncommon, outcome.

I was asked by one client to run a workshop where the emphasis was on getting out S-curves to justify the investment case, so I brought in an acknowledged risk expert to help me. During the workshop, he made a statement along the lines of 'human beings, even professionals in that field, are notoriously poor at predicting probabilities, so don't bother, I will put them in. What is important is the relative ranking in terms of probability'.[10] Isn't this then a case of potential garbage in (in terms of probabilities) = garbage out?

Furthermore, because most of the time this risk log is put together at the early stages of a project, many of the risks are fairly generic and general.

- The assessment of the impact is an educated guess, but not an estimate, i.e. a 'guestimate'! For instance, a risk with a direct cost of £10 000 can quickly become a £100 000+ one if it is on the critical path.
- The responses which come out are fairly generic, so evaluating their effect on reducing the probability and potential impact is fairly broad brush. Again, guestimates.

Even if you take out the human element of guessing probabilities and 'guestimating' the impact by using 'big data', I still have reservations. Let us say you have to build a bridge across a motorway as part of a larger improvement contract

- statisticians tell me that you need a sample size of 1000 + to have a statistically significant dataset
- that dataset would have to contain data on durations, costs and variations of a 1000 *similar* bridges built in *similar* conditions using *similar* methods to what you are proposing in order to be relevant to your project's analysis – is that likely?
- and if, as a result of using this dataset to focus on the areas of maximum risk and having done further research to ascertain the root cause of this variability, you decide to something new and different to avoid the problems, by definition you do not have data on your new solution – so how does that feed into your revised analysis?

So, for any risk analysis, we have garbage in for the probabilities and guestimates for the potential impacts. Yet, in the case of the workshop I did, the end client was extremely happy with the resulting S-curves, because of the application of pseudo-science.

Lastly, even if all my criticisms are invalid, how do we know if these analyses are flawed or not? As all projects are, to some extent, one-offs, we cannot statistically correlate the results of Monte Carlo analyses with the outcomes for the same projects.

As a result of all these factors – and there are more I could mention – I am deeply sceptical of the relevance of quantitative risk management techniques on all but the largest projects, where a risk analyst can be retained on an ongoing basis to advise a project manager. I know that many contractors' personnel, while not saying so to clients, have similar views on its relevance and worth.[11]

[10] Indeed, having read a number of books on cognitive biases, I can confirm that all the evidence points to human beings being absolutely rubbish at probabilities. Cognitive biases are mental shortcuts that we all frequently use throughout a normal day in order to avoid having to think things through from first principles each and every time we come across a situation. Most of the time they are helpful. The problem is that they are not necessarily logical. I would say that about a quarter of them have some relationship with risk. As an example, the 'recency bias' favours recent events over historic ones, giving greater importance to the most recent event. So, if a risk has only occurred once on 10 similar projects, but the project it did occur on was the most recent one, then we would naturally give it a higher probability than 1 in 10.

[11] If you want to hear a fuller justification for these views, then listen to a webinar titled 'Risk management is dead (or dying)? Long live the management of risk!', which I delivered for the Association of Project Management and which, at the time of writing, can be found on YouTube at https://www.youtube.com/watch?v=AmDlTxjoI3M.

3.6.4 Overview of a practical process

Before reading this, if you are not familiar with the early warning procedures in the contract, I suggest you (re)read clause 15 of the ECC and Section 1.5.7 of this book.

Despite the reservations expressed above, the effective and efficient management of risk is absolutely core to the successful delivery of projects, and hence the project management professional. However, in my view, the theory and guidance needs to be much more grounded in the day-to-day reality of managing a project or, in our case, a contract, with an emphasis on action (which is what I am attempting to do here).

Before I explain how the various parts interact, here is my terminology.

- The ***Client's* risk register** is the register of risks for the whole project or programme (of which the contract is one 'project'), including their assessment, to whatever detail and complexity they have done it. It is the log developed by the risk analyst from which a statistical quantitative analysis can be done if wanted. On small to medium-sized contracts, you may not have such a register, but on larger projects and programmes run by repeat-order clients, you probably will.
- The **contract-level Early Warning Register** is the risk register that initially contains the matters referenced from the documents in Contract Data parts one and two. Under NEC4, the *Project Manager* is required to collate these risks and responses and to issue a contract-level Early Warning Register to the *Contractor* within 1 week of the *starting date*. This could be, say, the first tab of an Excel spreadsheet and would typically have about 14 columns in it, as per a typical risk register for non-quantitative management purposes. Subsequently, the register is updated either periodically, say every 3–6 months, or at changes in the stages of a contract. For example, on a design and build contract, it could be updated: just before start of construction; when the civil engineering groundwork is being finished off and the building work is starting; when the fitting-out works, such as mechanical and electrical, are starting up; and at the start of the commissioning and handover phase. Emerging contract-specific risks from the *Client's* risk register, and more major long-term risks in the active Early Warning Register (see below) will also inform any update.
- The **active Early Warning Register** is the live register on which early warnings – emerging risks and issues – and actions are added and deleted by the *Project Manager*. In project management literature, this is more akin to an 'issues register' and, let us say, could form the second tab of an Excel spreadsheet (the first one being contract-level register, as described above). The register that I supply to contract teams has nine columns in it.

Figure 3.4 gives a visual overview of the process. It should be stressed that this process assumes that a quantitative Monte Carlo analysis is being done for the project, which may not be the case. In which case, there will be no *Client's* risk register.

Here is the process in some detail.

1. I suggest that the risks and other information relevant to the contract (which is different from the project) are extracted from the ***Client's* risk register** and put in the *Client's* 'starter for 10' (or 20 or 30) contract-related risk events, which are referenced from Contract Data part one. This should definitely result in a reduction in the number of risks presented

Figure 3.4 Illustration of how to integrate a quantitative risk register with the two levels of Early Warning Register

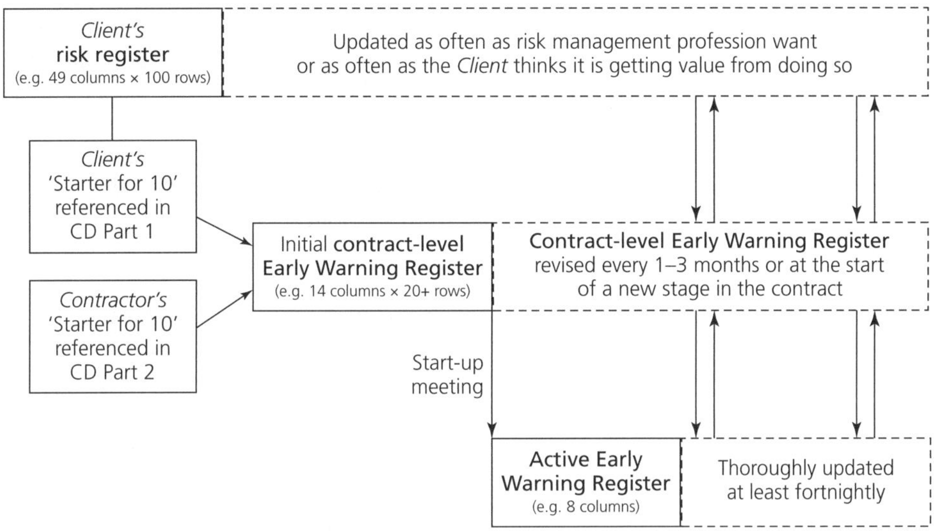

(i.e. fewer rows), and hopefully result in a reduction in the number of columns. What we are looking for here is a useful aide-memoire for the contract team in terms of having their eyes open for emerging risks and a starting point for actions.

2. The tendering Contractor fills in Contract Data part two in the same format as used for (1), and possibly adds to or enhances the *Client's* 'starter for 10' (or 20 or 30), and returns it with its tender. In order to encourage good quality submissions, I suggest Clients consider evaluating this material as part of tendering Contractors' quality submissions.[12]
3. On signing the contract (ideally before), and not later than one week after the *starting date*, (1) and (2) above, plus any losing Contractors' insights, are combined to give the first **contract-level Early Warning Register**. As stated before, let us say this is the first tab of an Excel spreadsheet and would typically have about 14 columns in it, as per a typical risk register for non-quantitative management purposes. The contract team is briefed on the register, critiques it etc., to gain understanding and ownership of it. It may be enhanced by some sort of mini risk and issues workshop early on in the contract, which (while perhaps in a different format from subsequent meetings) would contractually be the first early warning meeting. Generally, though, the risk descriptions and resulting actions will still be relatively generic, albeit more detailed than those in the *Client*'s project or programme risk register. Major *Contractor*-identified risks may find their way back up into this project or programme risk register. Where risks and issues and their corresponding actions are described precisely enough to be associated with a design or construction operation, they are incorporated into the programme document. For instance, where an action is definitely to be done, it can be included in that operation's method statement. Risks that are looking

[12] But remember, just because the *Contractor* has put a response in its pre-contract submission for the Early Warning Register, does not mean it has a contractual obligation to do it. If the *Client* wants the *Contractor* to do it, as a matter of contractual obligation, it needs to be incorporated into the Scope document.

more like certainties (i.e. issues) are entered into the **active Early Warning Register**, say on the second tab of the Excel spreadsheet. For those matters that need to be dealt with early on in the contract, tangible actions are developed as described below.

4. During the contract, only issues (i.e. matters that have happened and which, if not dealt with, will cause delay, increase cost etc.) and emerging risks are given as early warnings and entered into the **active Early Warning Register**. These emerging risks may or may not be in the contract-level Early Warning Register, but the point about 'emerging' is that they are starting to look more likely to occur. If they have been previously identified, participants may refer to the contract-level Early Warning Register for direction on what to do. However, they can now be much more precise about the following.
   (a) What the risk is.
   (b) Where and when it is likely to occur (i.e. on which bar in the programme).
   (c) What its impact is likely to be (by looking at the programme, and on cost).
   (d) What exactly can be done about it in terms of individual and tangible 'tick-offable' actions. (By 'tick-offable', I mean that an action is described in such a way that someone else apart from the person doing it can say 'Yes, it is done'. The question I have in mind is when someone describes an action to me is 'Do I have a clear picture in my head of what the result of that action looks like?' If not, I ask them 'How will I know you have done that action? Describe the output.')
   (e) Who is best placed to do each of these actions (i.e. which person, not organisation).
   (f) By when each of these individual actions are to be done by (a specific date, not 'end of Summer').

   Major emerging risks and some issues may rise up into the contract-level Early Warning Register and then into the *Client's* risk register (at the project or programme level).
5. As per the ECC provisions, as risks do or do not happen and actions are done, they are taken out the active Early Warning Register. Once again, for major risks, this may result in modification to the contract-level Early Warning Register or even the *Client's* risk register, if it exists.

3.6.5 Notifying early warnings: who notifies, to whom and what is notified?

Before the contract starts operating, there are number of things to consider at the contract mobilisation phase relating to who early warns, to whom and what is early warned. 'Isn't this defined by the contract?' I hear you say (in my imagination). The answer is 'Yes, it is'. If in doubt, see clause 15 for a quick reminder of what the contract says. However, there are a few things that sit beneath this.

WHO NOTIFIES?

- Contractually, the *Supervisor* cannot notify early warnings, yet it makes obvious sense for them to do so. If this is the expectation, they need to be told so. But who do they report their early warnings to? Presumably to the *Project Manager*, but in what form – on an internal pro forma, by email or verbally? If the former two, should they be copied to the *Contractor*? Or does the *Project Manager* delegate the action of notifying early warnings to the *Supervisor*, so that the *Supervisor* can notify as well?
- Legally, quite a few people can represent the *Contractor*, including some who have not been put forward as *key people* in Contract Data part two. But who specifically can formally notify the *Project Manager* of an early warning? All of these people? Or does it have to come through the site manager or the quantity surveyor, who often handles the more administrative-type things? And what about the rest of the *Contractor's* site team, such as

engineers and foremen? Do they put something in writing (on an internal pro forma) to the site agent/quantity surveyor, or just tell him or her, who then does the paperwork or sends the email? Whatever route is chosen, the administrative route needs to be made clear and the people involved need to be told of the consequences of not following it (i.e. reduced payments for any resulting compensation event and, under the cost-based options, the extra cost is Disallowed Cost).
- For designers under the Professional Service Contract and Subcontractors under the full Engineering and Construction Subcontract, I would expect formal notifications in a form that can be 'read, copied and recorded', whereas for the Short Subcontract, while notifications have to be 'in writing', some Subcontractors are not capable of doing this. Nevertheless, they do need to be told that they have to early warn and how they should do it (even if it is verbally, with the *Contractor* then doing any paperwork) and what the consequences of not doing so are (i.e. reduced payments for any resulting compensation event).

TO WHOM ARE NOTIFICATIONS SUBMITTED?

In terms of who notifications are submitted to, my question is this: Is the *Project Manager* the right person to maintain and update the Early Warning Register? Sometimes, and for good reasons, the *Project Manager* will be the right person. However, some Contractors have put forward the following arguments for them to be the one who maintains and updates the Early Warning Register and take primary responsibility for following up on the recorded actions.

- Once the project enters the detailed design and construction phase and the Parties are in contract for this phase, then it is the *Contractor* who is usually taking the majority of risk, both in terms of ability to manage it and commercial consequence.
- As a result, the *Contractor* is at the centre of the risk management process, in that, while some actions will fall on the *Project Manager*, the *Client* and its consultants, the majority will fall on the *Contractor* and its Subcontractors, suppliers and consultants. Having separate Early Warning Registers as required under the long/normal forms for each of these subcontracts, which are then merged into a main contract Early Warning Register, would create a confusing plethora of Early Warning Registers. It makes far more sense to have one central register, with the Party at the centre of the process managing and updating it (i.e. the *Contractor*).
- This reason is not stated to the face of the *Project Manager*: the *Contractor* does not believe that the *Project Manager* has the competence and/or the time to manage the early warning process properly.

The drawback to the arrangement is that the *Contractor* may not have the influence within the *Client* organisation to compel people to do the actions that fall upon them and their consultants, whereas the *Project Manager* may have.

Contractually, the *Project Manager* could delegate their authority to the *Contractor* under the contract. However, if not already stated in the Scope, whether the *Client's* or the *Contractor's*, this would be a change in the Scope and hence a compensation event. (Furthermore, if the *Contractor* did not want to do it, it would have a strong argument for saying that this is a *Project Manager's* obligation as defined in the *conditions of contract* (which can only be altered by agreement between the Parties), and so has precedence over the Scope – i.e. you can't force the *Contractor* to do it!)

Lastly, whoever the individual – as opposed to organisation – who runs the Early Warning Register and process is, that person needs to have the authority and personality to ensure that the recorded actions are done. This implies a relatively senior person involved in the contract (and not a introverted risk analyst tucked away in a remote room).

WHAT IS NOTIFIED?

- Are general issues notified as early warnings? My definition of an 'issue' is 'an event which *is* happening or, if action is not taken, *will* definitely happen and, without action, *will* have a downside/detrimental effect on one or more objectives'. So, in early warning terms, an issue could be a matter for which no actions are planned to overcome it, where planned actions are not happening or, as per the new last bullet point of clause 15.3, where actions are being taken but are proving to be ineffective.
- A large subset of 'issues' are what are traditionally called Technical Queries (TQs) and Requests for Information (RFIs), so how are they going to be dealt with? Contractually, they should be notified as early warnings as failure to resolve them may lead to delay and extra cost, but the danger is twofold.
  - At one extreme, the sheer number of TQs and RFIs can overwhelm the early warning process, drowning out the ones that matter. This suggests that some prioritisation or categorisation system for early warnings needs to be put in place.
  - At the other extreme, on one site I know of, when the ECC was coming in, the Subcontractor was quite happily giving TQs and RFIs on old forms – and being constructive in their resolution – but **not** giving them on pro formas titled 'Early Warning'. Unfortunately, the main contract was not going well, so the main *Contractor* started to look at how they could reduce payments to their Subcontractors. Because the Subcontractor had been using old pro formas, the *Contractor* took the position that these informal 'early warnings' did not count.[13] If you go down this route, it does imply that some sort of monitoring process and conversion of the important and/or urgent ones into formal early warnings notifications is needed.

  My preference is for the former, as this protects the *Contractor*. See below for my simple prioritisation system.

3.6.6 Setting up and revising the active Early Warning Register

The active Early Warning Register that I give in Excel format and explain to new contract teams is shown in Table 3.3. Before I explain it, let me stress it is my starting point, so take it and adapt it to how you want to run your contract. On this point, a general initial comment from people who see it for the first time is that it leaves out this or you could add in that. My answer is 'yes', and that is how we end up with 49-column risk registers. It is balancing what is sufficiently useful in terms of having another column against the extra burden of filling it in. The fewer columns the better, but the register does have to be fit for your purpose.

So here are my comments on it.

- First of all, in my mind this contract started in late April/early May, so notice that it is on revision 3 already. After each full early warning meeting (by which I mean the one

[13] One of my reviewers had a similar situation arise in an adjudication where he was representing the Subcontractor. The *Adjudicator* took a dim view of the *Contractor's* actions, concluding that an early warning did not have to be on a piece of paper or email titled 'Early Warning Notification'. On the other hand, the legal view of one of my other reviewers was that TQs and RFIs have to be framed as early warning notifications to be considered as such.

Table 3.3 Model format, with example entries, for active Early Warning Register

Early Warning Register			Revision No. = 3		Date Revised = 23/05			
EW No.	Description + consequence	Date notified	By whom (PM, C, SubC)	Priority (1 to 5)	Actions	Who	By when	When action actually done
2	Soft spots at Chainage 250–300. Some extra cost and delay if not planned out. Maybe a CE depending on extent	06/05/20	C	2	~~Do brief investigation with backactor to establish extent and whether a CE or not. Call out others once uncovered~~	~~Edward Engineer~~	~~13/05~~	12/05
					Include digging out in Method Statement as part of planned operation	Edward Engineer	31/05	
5	RFI on rebar for concrete base of Bridge No. 5: bar bending schedule appears to be incomplete. Needs to be resolved soon to avoid delay given ordering time	10/05/20	Concrete SubC	Was 3, now 4	~~Meeting to confirm and resolve tech issues~~	~~Derek Designer Barry Barbender~~	~~13/05~~	15/05
					Revise schedules and/or drawings and forward to Paula PM	*Derek Designer*	*22/05*	
					Issue any revised schedules as change to Scope and hence CE.	*Paula PM, Tim QS, Barry Barbender*	*23/05*	
7	Security/safety fence has been pulled down near school. H&S issue!	18/05/20	PM	5	~~Reinstate initially~~	~~Syd Site-Agent~~	~~18/05~~	18/05
					Visit first and last thing each day + before school finishes. Reinstate as necessary until permanent solution implemented.	Syd Site-Agent	Ongoing	
					Review and implement sturdier solution	Syd Site-Agent	25/05	

C, Contractor; CE, compensation event; PM, Project Manager; QS, quantity surveyor; SubC, Subcontractor specified to happen at the intervals set in Contract Data part one), it is updated and saved in a read-only format. Then, either it is sent out to each participant, or a link is sent to the relevant people, with the register being placed on a project server to which all have access.

- Notice that when going down the first 'EW No.' column, it jumps numbers. This is because some early warnings have passed and/or all the actions have been done. Consequently, those early warnings and actions have been either removed or hidden. This is in accordance with the contract and enables contract participants to focus on current risks/issues.
- Going down the 'Description + consequence' column, note how I have merged rows to avoid having overly cramped descriptions. I have also included consequence so that people are aware of what may happen if nothing is done.
- The 'Date notified' and 'By whom' columns should, ideally, not be relevant because the Parties will notify early warnings and constructively resolve them as a matter of course. Consequently, there should be no need to look back and use the contractual sanctions in clauses 61.5 and 63.7 of the main contract and clause 11.2 (26)/(27) of options C to F to punish the *Contractor* for not giving early warnings when it should have done. Note, however, that Subcontractors are included as a party who can early warn (on the basis that it is the *Contractor* who is running the Early Warning Register, as per my suggestion in Section 3.6.5).
- For the 'Priority' column, as a starting point, I suggest a scale of 1 to 5. In the actual spreadsheet that I use these are colour coded, so that the highest priority risks are identified more readily (red for priority levels 5 and 4, amber for level 3 and green for levels 2 and 1). Below is my categorisation system for this prioritisation.
  5. **VERY HIGH PRIORITY.** Actions need to be agreed and taken NOW, as is or very soon will be delaying Completion or a Key Date or increasing the Prices/Defined Cost.
  4. **HIGH PRIORITY.** Needs to be resolved at or soon after the next scheduled early warning meeting in order to avoid delaying Completion or a Key Date or increasing the Prices/Defined Cost.
  3. **PRIORITY.** A plan of actions need to be put in place or confirmed at the next scheduled early warning meeting in order to avoid delaying Completion or a Key Date or increasing the Prices/Defined Cost.
     (This is where the more immediate TQs or RFIs may be placed, but they must be noted as TQs or RFIs, so the *Project Manager* can just 'postbox' them onto the relevant people.)
  2. **LOW PRIORITY.** Consideration can be put off for up to a month. Needs to be resolved within 2 months.
  1. **VERY LOW PRIORITY.** Consideration can be put off for 2 months or more.
     There are number of things to note about this system.
     - It is based more on time until actions have to be taken, rather than a combination of probability and impact, which would horrify risk management purists!
     - It could include matters that affect health and safety or the environment, as ultimately these will impact on time, cost and performance.
     - I suggest that colour coding should be used both in the explanation and the active Early Warning Register.
     - If it would work for you, you could simplify it to three categorisations.
- Note that the 'Actions' column has a separate row for each action for a particular early warning. As stated previously, where traditional risk management has failed is its lack of

emphasis on actions, and part of the reason for this is that all the actions for an individual risk tend to be crammed into one box on the spreadsheet. Consequently, they are very brief and hence unspecific, including not specifying who does what action by when.
From my experience, I cannot emphasise the next point enough, and will therefore say it in a number of ways.
  - If you want actions that are done, you need to make them specific or 'tick-offable', so that there is no argument about whether they have been done or not.
  - That means there is clear physical evidence of an action, i.e. something physical happens that you can see (e.g. a meeting; or something is produced, e.g. revised drawings). Take the last but one action for EW No. 7. Initially I wrote 'keep a watch on ...', but how would I know if someone was 'keeping a watch on'? But if they visit first and last thing plus before the school finishes for the day, then, if I desired, I could physically check to see if they are doing the action.

  In order to make these actions specific and tangible, and with a stated owner, I think each action should have a separate box in the spreadsheet.
- The 'Who' column shows the specific person (ideally) or persons who will do each action. This person should be the one best placed to do the action (e.g. the designer, engineer or foreman who is responsible for that operation or activity). This differs from most risk management literature and practice in that
  - actions, not risks, are assigned to owners
  - the owner is an individual, not an organisation (although you could have more column to show the organisation).

  All this is in the interests of making the action specific, and an individual more accountable for doing it, thus increasing the likelihood that action will be taken.
- Notice that the 'By when' column contains specific dates – not, as I sometimes see, 'end of May' or even 'late spring'. People use this vagueness to get out of doing the action: I have witnessed a discussion over whether late spring meant late May or late June!
- The last column shows the date when an action was actually completed. This provides both an indicator of what actions are outstanding and a record of how quickly issues are being resolved.
- Finally, note that, in the 'Actions', 'Who' and 'By when' columns, late actions are in italics (red in the actual spreadsheet) to draw people's attention to them.

As one of my reviewers pointed out, the *Project Manager* and *Contractor* have a contractual obligation to operate the early warnings procedures as described in clause 15.1 of the *conditions of contract*, so none of these more detailed suggestions are contractual obligations. If you want to make them contractual obligations (perhaps after modifying them to suit your circumstances), a good way would be to describe them in the Scope document, which makes them enforceable.

3.6.7 Running an early warning meeting

Before I start, let me state that part of how I earn my living is through facilitating project-based workshops, which often have very little to do with the NEC and may have anything from five up to, in one instance, almost 100 participants. Consequently, I would like to think I know a thing or two about running a successful meeting, which is not to say I know everything or always run them perfectly. Far from it, as some of what I am presenting are tips I have picked up from professionals. On the flip side, I have also discovered what does not work, and tried to address this. Having said all of this, I hope what I am saying does not come across as teaching people to suck eggs.

While I think that regular early warning meetings should be scheduled into the project plan, and under NEC4 you are required to have a regular early warning meeting, that does not mean that all early warnings are or should be resolved at that meeting. Take the following examples.

- On one site in my postdoctoral research, the *Project Manager* would wander into the site manager's office opposite his (or vice versa), have a discussion about an issue and resolve it, before realising that they had in fact just had a risk-reduction meeting (as they were called in NEC1 to 3). Consequently, they recorded the early warning and outcome of the meeting.
- Frequently, engineers and foreman might have an informal meeting out on site while looking at an individual problem, which they resolve there and then. Again, they recognise this as an early warning meeting, even though no early warning has as yet been formally issued. The formal communications as required by the contract rapidly follow.
- On some contracts, early warnings of major events have resulted in special meetings, the sole purpose of which is to resolve that individual early warning, with relevant technical experts from outside the project team attending.

Clause 15.3 states what has to happen at an early warning meeting. However, as with all good meetings, preparation is the key, so, first of all, who should be at the meeting?

The obvious answer is those who can contribute to resolving an early warning, which does not necessarily mean they need to be at the whole early warning meeting if parts of it are not relevant to them. For instance, participants at one mid-project workshop I did identified an 'Area for Improvement' as being shorter meetings. They had fortnightly day-long meetings with 17 people there – of which only five were from the *Contractor*, so consultants were sitting around charging fees for the majority of the time. Given that five participants is apparently the optimum size for making good decisions,[14] this seems a bit extravagant.

I would also suggest that the *Project Manager* and *Contractor's* main person sit down together beforehand to review the Early Warning Register in terms of risks/issues that have or have not happened, actions that have been done, and any reprioritisation of risks/issues remaining, in order to identify which matters should be addressed at the meeting and in what order.

Here's a tip I got from a project manager working under the JCT form of contract. He said he took to every meeting the overall project programme, the subprogramme relevant to the meeting's purpose (e.g. design programme) and a risk/issue log that had the relevant actions in it. He started every meeting with a review of the actions that should have been done since the last meeting, and was not afraid to tell people off if they had not done the action. In addition, any outstanding actions would appear in red in the revised risk/issue register. People soon got the message that the register was not an abstract thing to be filed away and forgotten about – they needed to look at and act on it.

If people are attending for a specific early warning matter, then it is the best use of their time if these are tackled first, especially if they can be resolved relatively quickly. Alternatively, the

[14] As an ex-academic, I should be able to reference the research that confirms this, but I can't find it! However, from my experience as a facilitator, where I break large groups down into smaller groups, this number corresponds with what I have observed in practice.

individual can go to a quiet place nearby – or wait to be invited, perhaps via a video link – so they can work on other stuff until their input is required.

Once these are out of the way, it makes sense to work your way down the list, dealing with the early warnings with the highest priority first. Given my prioritisation system – which may not be yours – items with a priority level of 4 or 5 will need to be resolved within the meeting. For those with a lower priority, it may be more effective and efficient just to action plan what needs to happen to resolve them. EW No. 5 in my example Early Warning Register (Table 3.3) would be an example of this.

For many, if not most, early warnings, the solution and resulting actions are relatively obvious. However, for more complex issues, I think the direction the contract gives in clause 15.3 (about how to resolve an early warning) lacks at least a couple of stages. For instance, if you look at formal decision-making processes, both 'hard' (e.g. risk management, value management and other business-level methodologies) as well as 'soft' (e.g. neurolinguistic programming, solution-focused coaching and cognitive behavioural therapy)

- all to my knowledge have what could be described as an 'information and understanding' phase, where a common base of information and understanding is established of what the problem or issue is (sometimes this is called a 'discovery' phase) and normally involves an exchange of perspectives
- many, but by no means all, have what could be described as an 'outcome-setting' phase, in terms of agreeing what is trying to be achieved.

How does this apply when resolving an early warning? Here is my method.

1. In this information phase, share perspectives on what the problem is and the 'do nothing' consequences of not resolving it. For instance, if there is a 2-week delay on some goods arriving on site which will be incorporated into the *works*, first of all establish if it matters. If there is 3 weeks of free float before the supply of goods becomes critical, then the nature of the solutions will be to make sure that the delay stays at only 2 weeks and does not become 3 weeks or more. If, on the other hand, the problem is on the critical path, the solution is more likely to be an active 'throw money at it' type, as the direct costs of solving it are likely to be outweighed by the indirect costs of time-related costs and damages. Related to this, on a project management point, I personally think everybody from foremen and site-based engineers upwards should be made aware of the time-related costs of a day's delay to the *Contractor*, the additional contract damages if Completion exceeds the Completion Date, and any other costs to the *Client*, whether they be tangible (i.e. a monetary figure can be put against them) or intangible (e.g. reputational). This way, sensible informed decisions can be made about the trade-off in terms of the direct costs of solving a matter versus the consequences of not solving it.
   Exchanging information on what the problem is may also include establishing the causes of the problem. Note I say 'cause*s*' (plural), as complex problems often have more than one cause, yet think how often we find ourselves arguing about something being caused by this or that, when in reality the problem has arisen from a combination of factors.[15]

[15] If you want some tools to establish cause and effect, look either at UK value management literature and/or lean six sigma literature. These tend to be good for 'hard' technical problems. My favourite is a prerequisite or core reality tree, which can either be constructed using Post-it Notes or software.

The discussion then becomes more educated around which was the primary cause and/or to what extent each cause has contributed to the problem.

1a. Once people have established a common base of understanding, more often than not participants will be able determine whether or not the early warning is likely to be a compensation event. If it is not, the *Contractor* takes the lead on solving the problem, with the *Project Manager/Client's* help (as it is not in their interests if the contract runs late). If it is a compensation event, the *Project Manager* takes the lead with the *Contractor's* help and input. To take the simple example above, if the *Client* is meant to supply the goods to the *Contractor*, then, if nothing is done, it will be a compensation event under clause 60.1(3) or (5). If something is done, then either it will be a compensation event under 60.1(1), as the *Project Manager* will be changing the Scope, or the *Client* will be incurring costs so that the goods are supplied on time. Either way, the *Client* pays. On the other hand, if it is the *Contractor* who is meant to supply the goods, it will bear the Defined Cost.

2. Agree what you want to achieve. For instance, is it least cost, least time, best quality etc. For our late supply of goods example above, it could be minimum direct cost if the event is off the critical path (and provided it stays there) or, if on the critical path, it could be least overall cost to the *Client*, including the indirect time-related costs of the *Contractor* (for which the *Client* will pay if a compensation event) as well its own.

> These two steps might seem obvious, but as a facilitator I have been called in a number of times to help teams resolve issues that have remained unresolved for months. In essence, I have got participants to be quiet and just 'listen to understand' the other party's perspective, allowing clarification questions only – no challenges or 'yes, but…'. We might then have done a little work to map out the causes and effects.
>
> I have then asked the participants to state what each other's ideal and acceptable outcomes are, and again merge these into a joint statement which both parties can sign up to.
>
> I kid you not, the solution, in terms of how to get from the now defined problem to the now agreed desired outcome has not even taken an hour.

Virtually all decision-making processes have some sort of idea identification stage, followed by a rationalisation phase where the ideas generated are developed and reduced down into the ones that could work, before selecting the ones that are to be implemented as the chosen solution. The problem I find with engineers and other technical people is not so much that they cannot be creative in terms of generating ideas, but that they have trouble switching off their critical minds. Consequently, someone mentions a potential solution and people immediately critique it for why it will not work. Another solution is identified, which is again critiqued and dismissed until, often fairly soon, a workable solution is found. The problems with this approach are that: (a) people often do not mention a solution unless they think it will work, (b) solutions that have a germ of good idea in them are too easily dismissed and (c) the first solution that will work is that the one that is implemented, thereby cutting out all the other, as yet unthought of, approaches. How does this relate to the what the contract says?

3. The first bullet point of clause 15.3 says 'making and considering proposals for how the effects of each matter in the Early Warning Register can be avoided or reduced'. In other words, generate ideas and then rationalise them. Here is a simple way to get more ideas out than by a normal discussion.
   (a) Get meeting participants to generate, on their own and in silence, three to five ideas of how the problem could be solved, writing each idea up on a separate Post-it Note or equivalent.
   (b) Put the ideas on the table or on a flip chart, take out obvious duplication and group similar ideas close together.
   (c) For each idea: (i) get more explanation from its originator if it is not obvious what is meant, (ii) explore what is good about the idea, and then (iii) explore what is bad about the idea or why it will not work. Do not be afraid to develop ideas as a result of these conversations, but also throw away ideas that everybody agrees will not work.
   (d) Arrange, group, sequence and develop the remaining ideas into a workable scenario or scenarios. Some of the ideas will be mutually reinforcing (i.e. on their own they may not work, but in combination with others they will) while some scenarios will be mutually exclusive.
   (e) Select the best overall solution or scenario given what you are trying to achieve or, in the words of the contract, 'solutions that will bring advantage to all those who will be affected' (clause 15.3, second bullet point).
4. Action plan what are the specific and tangible 'tick-offable' actions needed to implement the chosen solution, including who is to do what action and by when. I have already laboured the point a number of times about how there needs to be clear physical evidence of an action – i.e. something physical happens that I can see (e.g. a meeting, or something is produced, say a drawing). Having mentioned it again in the preceding sentence, I will not mention it any more!
5. Record these actions, who is to do them and by when in the Early Warning Register.

Let us take a simple school example, where there is a strong will for the school to be open in time for a new term. A 2-week delay has occurred – our late supply of goods to be supplied by the *Client* example – and the current Completion Date is slap bang on when the school has to open, but the *Contractor* has 1 week of terminal float in the Accepted Programme which it owns (i.e. the current Accepted Programme shows planned Completion 1 week ahead of both the contractual Completion Date and required school opening date). A method for resolving this issue would be as follows.

1. In our information phase, everyone agrees that the delay is (i) on the critical path, (ii) that the 'do-nothing' consequence will push planned Completion beyond the opening day by 1 week, (iii) that it is caused by the *Client* and, therefore, (iv) it is almost certainly going to be a compensation event. Consequently, (1a) the resolution of the early warning will be led by the *Project Manager* with the *Contractor's* input and help. The *Project Manager* (on behalf of the *Client*) ideally will have a contingency plan for this, and therefore knows approximately how much a week's delay is valued at if they have to make alternative arrangements for the children. (I say 'valued at' on the grounds that, as well as the direct quantifiable cost, I would imagine there would be a lot of political embarrassment, which would be factored into the equation.)
2. The minimum outcome, therefore, is to find a solution that means the school opens on time for less than this cost or value. A stretch outcome might be to recover all the time just

in case there are further delays, regardless of whether any such delays are compensation events or not.

3. The team identifies a number of potential 'part' solutions, which could include: find an alternative supplier; find out why the supplier will be late[16] and potentially address this with the supplier (who, if under the Supply Contract, could be called to an early warning meeting); resequence the programme to reduce the impact; bring in extra resource on operations after the delivery to speed them up; and/or partially demobilise the Site, so that some of the school can be used. After discussion, a number of potential scenarios are identified.
4. The potential scenarios are then action planned to test their viability and cost. Included in this action plan is another early warning meeting to be held in, say, 3 days, to decide which scenario to proceed with. The chosen one would be assessed as a compensation event.
   At the meeting, the last thing that the *Project Manager*, or person in charge of keeping the Early Warning Register up to date, does is recap all the actions, who owns them and when they are to be done by, before going off and revising the register.
5. The *Project Manager* or person in charge of keeping the register then updates the Early Warning Register and circulates it to participants. Alternatively, and ideally, the update of the Early Warning Register is done live, while being projected onto a wall or displayed on a large monitor screen which everybody can see.

3.6.8 Managing in opportunity

3.6.8.1 General

If you go back to Section 3.6.2 and look at the Association for Project Management's definition of 'risk event', it can be either a threat or an opportunity. Therefore, risk management can and should include managing in opportunities. This can take a number of forms.

- Managing in time and cost savings, which can be used to partially or wholly balance out or negate unlinked threats (i.e. some downside threats are almost certain to occur on a project and have a negative impact). By managing in opportunities, you give yourself some more 'fat', or contingency, to absorb downside threats.
- Managing in what I call 'value enhancements' or, more colloquially, 'bells and whistles'. As savings are managed in and threats are avoided or minimised, the *Client* can use the unneeded contingency to add in additional features, thereby getting a bigger bang for their buck – by which I mean more benefit from the same budget.

Yet, the focus of risk management is, and always has been, on managing out threat, as opposed to managing in opportunity. This is despite leading thinkers in risk management broadening it to include opportunities. Table 3.4 highlights the parallels between the risk management process (as outlined in Section 3.6.2) and the formal value engineering process, which ideally is used by designers to identify and develop best-value/optimal designs. Do you spot the similarities?

[16] With regard to addressing why the supplier is late, some real-life examples are: the supplier had not appreciated the intricacies of welding for some bespoke steelwork; the supplier had won a larger order and put the smaller order to the back of the queue (the solution, through gritted teeth, was to pay the supplier extra to bring the order forward again); on another, the solution was to split delivery, so that rather than being delivered in one go, the order was delivered in batches as and when the individual parts became available.

Table 3.4 High-level comparison of risk management and value engineering methodologies

Risk management *Manage* ***threat*** *out*	Value engineering *Manage* ***opportunity*** *in*
1. Initiation phase 2. Identification 3. Assessment 4. Response planning 5. Response implementation	1. Information phase 2. Speculation 3. Evaluation 4. Development (including scenario building) 5. Recommendation

So why do we not focus more of our efforts on managing opportunity in. A view is that it comes down to one psychological and one commercial trait.

- **Psychological:** most of us – some figures suggest 80% – predominantly have a 'moving away from' motivational pattern, in that we are primarily motivated to avoid pain (i.e. avoid threat and poor project performance), as opposed to a 'moving towards' motivation, where we receive pleasure (e.g. seek opportunity and superior project results). In reality, our motivational patterns are a little more complex, in that people usually seek the positive only after they are confident that they have got the downside covered. For instance, in Maslow's hierarchy of needs, people only start to pursue their higher (self) actualisation *wants*, if and when their lower physiological and safety *needs* have been met. So people need to clear the downside risks from their mind before they can focus on the upside wants.
- **Commercial:** partly as a reflection of this, contracts are typically set up in a way that causes us to focus on avoiding the downside (i.e. with an emphasis on damages as opposed to bonuses), which means that participants do not have the motivation or, for that matter, margin/resource to seek out superior results. See my comments in Section 1.2.3.4 on use of the secondary options for damages versus bonuses.

With regard to the last point, contracts and commercial arrangements can be set up to encourage a greater focus on managing in opportunities and achieving superior performance. For instance, you can use the target-cost contracts as written, or the price-based options A and B, which now, under NEC4, have value engineering clause 63.12, which encourages the *Contractor* to seek out opportunities for savings (as any forecast savings that result from a change to the *Client's* Scope – but not the *Contractor's*, for some reason – suggested by the *Contractor* are shared by the Parties).

The rest of this section explores a number of ways of managing in opportunity.

3.6.8.2 Saving time and cost: headline event

This section comes from my experience of running project-based workshops that are wholly aimed at, or have a significant portion devoted to, identifying opportunities. The earliest stage at which I have run one was when a Contractor was tendering for an early contractor involvement (ECI) project – in this instance, there was a plan view of a medium-sized town with a whole host of proposed measures to reduce the likelihood and impact of flooding. Most commonly these are at a start-up workshop, immediately following contract award (such as that discussed in Section 3.2.4), where it is not uncommon for us to identify 10% of implementable cost savings on the *Client's* budget – a very good result.

In all these workshops, some design refinements are identified. However, the majority of savings come from either using a completely different design solution or from rearranging the programme sequence in some way. For instance, on one £33 million contract to construct part of an inner city road, the biggest saving came from having a second temporary bridge across a canal, to provide early access to an area. On another road scheme, building a temporary haul road across a field from an existing road allowed the construction of a permanent bridge, which was otherwise on the critical path, to start 9 months earlier. This change increased the direct cost, but it saved huge amounts of time-related cost.

Workshops of this type produce other benefits, in addition to the opportunities identified: contract participants have a bigger picture of the project, which means they can more effectively contribute to its planning. And they seem to open up a dialogue with the planner about how the work is done, leading to additional ideas being generated after the workshop – which leads on to the next section.

3.6.8.3 Saving time and cost: tweaks to the early warning process

Before I explain this process, let me outline three significant deficiencies in *only* identifying opportunities as the project progresses.

- Either you miss opportunities completely or, by the time you identify them, it is too late to implement them. As a generalisation, the bigger the opportunity in terms of time and cost savings or ability to add value, the sooner it needs to be identified to be implemented.
- The later you identify opportunities, the more likely it is that changing how things are done will lead to some previous work becoming redundant.
- An oft-heard complaint with regard to value engineering suggestions made as the project progresses is that disruption and delay costs of implementing good suggestions at the last minute equal or outweigh the savings made.

The first two points suggest that a headline event to get ideas out and on the table is a good way to go. All points suggests that, to ensure you do actually gain time and cost savings, you need to have in place a structured way of identifying, evaluating and managing in these opportunities.

When reading this, you might be saying (or you might not): 'This is good stuff, but what has it got to do with NEC?' The answer is threefold.

- Under NEC4 options A and B, clause 63.12, any forecast savings that result from changes to the *Client's* Scope (but not the *Contractor*'s Scope) that have been suggested by the *Contractor* are shared by the Parties. While this clause references the '*value engineering percentage*', implying design changes, it could apply to relaxations of or changes to constraints on how the *Contractor* is to Provide the Works.
- Under the target-costs options C and D, clause 63.13, the Prices are not reduced if the *Contractor* suggests changes to the *Client's* Scope that reduce cost. Likewise, under core clause 60.1(1) (second bullet point), if the *Contractor* suggests changes to its own Scope, the Prices are not reduced. (As this is a core clause, it also applies under options A and B, but as the *Client* would not get any share of savings it has no motivation to go along with the suggestion.) For both these clauses, the actual savings are shared under the pain/gain mechanism. The contract therefore encourages identification of opportunities, be they purely related to design or to the relaxing of a constraint in the Scope, or a combination of both.

- A criticism of the early warning process is that it only directs people to identify threats and manage them out. It does not encourage the identification and managing in of opportunities.

Pertinent to this last point, what if the wording of clause 15.1 were changed (with new wording underlined) to allow early warnings for 'any matter which could

- increase the total of the Prices,
- delay or bring forward Completion,
- delay or bring forward the meeting of a Key Date or
- impair or enhance the performance of the *works* in use'

with the follow-on sentence allowing the *Contractor* to 'give an early warning ... of any other matter which could increase or decrease the *Contractor's* or the *Client's* total cost'?

Steve Goodwin, in a somewhat esoteric article – which mentions global meltdowns, air-raid sirens, Buddhism and the effects of meditation, and more – goes further, suggesting that the words 'early warning' subconsciously cause people to react negatively (Goodwin, 2015). Or, in my words, subconsciously triggers the 'fight or flight' response, which is not a good start to what is meant to be a collaborative process! But back to the point I was making.

What if, in clause 15.3, the first bullet point is changed to read something like 'those who attend co-operate in making and considering proposals for how the effects of each threat in the Early Warning Register can be avoided or reduced and how each opportunity can be managed in'?

If we take my suggested format for the Early Warning Register (Table 3.3 in Section 3.6.6), we could either have a separate register for opportunities (e.g. as a third tab in the Excel spreadsheet) or, in the second column, have threats in, say, red and opportunities in, say, green.

To reinforce the opening comments of this section, having a prioritisation system that takes account of the 'time to implement' is quite important, as a negative of progressive identification of opportunities is that the delay and disruption costs of adding in opportunities at the last moment can equal or outweigh any savings. However, running a register in this way absolutely fits in with running a programme at the sixth and highest level of programming as identified in Section 3.5.6, and I would say, based on my experience as a facilitator, that many of the opportunities suggested relate more to programming innovations than to traditional design changes.

3.6.8.4 Adding in value enhancements to get the biggest bang for the same buck

To almost repeat myself, by 'value enhancements' I mean the *Client* can add in additional 'bells and whistles' to gain a bigger bang for their buck, i.e. more benefit from their budget. To give examples of this from various projects I have been associated with

- on a new road project, a footbridge was added which the *Client* originally thought was unaffordable
- in a school, due to savings being made, the doors were enhanced to be solid wood and the windows upgraded to have much better whole-life costs
- in another school, the originally specified IT in the classrooms was upgraded

- on a project to prevent flooding, significant environmental enhancements were added in
- on a large refurbishment of seven blocks of flats, significant savings that were identified early on allowed for the construction of a modest multi-purpose community facility
- on a road, the wearing course was changed to one which both produced less road noise and had a longer life.

Particularly in the public sector, many large projects are given a fixed grant. If the client's costs come in below this budget, the client has to repay the difference, while if they come in above, the client has to find the additional monies from elsewhere within their own finances. So the client's aim will be to spend exactly their budget, but to get the most from it. The problem is that they will not know with absolute certainty what savings they will have to reinvest in value-adding 'bells and whistles' until the contract is complete, by which time it is too late.

This section outlines a process to give Clients a greater likelihood of having these value-enhancing 'bells and whistles', while bringing their projects in slap bang on budget. It is a simple process, but its implementation calls for high levels of professionalism.

However, there are quite a few prerequisites that need to be in place before this process can be enacted.

- A contract strategy that encourages savings to be identified and allows the *Client* to gain a share of those savings needs to have been chosen. The most common contract strategy which does this is the target-cost contract format.
  Assuming that this is the case, it is important to realise that, at any time as the contract or project progresses, the amount that the *Client* has to invest will be the difference between the current target Prices and its overall budget (i.e. the as yet unused contingency) plus their share of any underrun (or minus their share of overrun).
- It needs to be clear whether the *Contractor* is ahead or behind of the curve in terms of progress made and monies spent with respect to coming in below the target Prices and by how much, or whether it will overspend. This suggests that some metrics, such as those used in earned value analysis, are used.
- However, the target Prices are a moving feast, in that the Prices are adjusted due to compensation events. Consequently, the assessment of compensation events needs to be up to date (see Section 3.7).
- Some monies need to be held back for negative risk/threats that might occur, whether this is within the target Prices or outside in the *Client's* contingency to be spent if and when a risk occurs, be it as a compensation event which adds to the target Prices or spent elsewhere on the project. To do this, a contract-level Early Warning Register which has monetary sums in it (of the sort outlined in Section 3.6.4) will need to be in place and periodically, but actively, updated.
  At any time, the *Client* and the *Contractor* need to be making informed decisions on whether and how much of this contingency can be committed to 'bells and whistles', or held back in case a threat does occur. This will partly depend on the nature of the project. For instance, in civil engineering, much of the risk is in the ground, and hence is at the early stages of a project. In other sectors, such as process plant, much of the risk is at the tail end, in commissioning and making all the component parts work together harmoniously. In the latter, you have to be much braver or foolish to release this contingency. Building projects probably sit somewhere in the middle of this spectrum.

If these prerequisites are in place, the structured method given below can be followed.

A. Generate a menu of 'bells and whistles' that are needed, wanted or desirable. This could be by a combination of brainstorming, asking stakeholders and end users etc.

B. Prioritise these into 'needs', 'wants' and 'nice to haves', or use some other equivalent scale.

C. Look at the construction programme, and from this derive when would be the optimum time to build in these features. This may involve some rejigging of your programme to keep your options open. As a small example, in a school building where you want top-quality solid wood doors that will last, rather than sealing off completed rooms with the end doors you may keep your options open by restricting access in another way. You can then decide much nearer the end of the contract what specification of doors to install.

D. Starting with those features highest up the list in terms of desirability
   (a) work backwards from the optimum time to implement in terms of construction to give reasonable dates to start doing detailed design and reasonable times to start applying serious thought in order to give the designers a good design brief
   (b) add approximate costings, both to construct/implement and to design.

   From this, you should have menu of 'bells and whistles', knowing how much they will approximately cost and when you have to start thinking about them, designing them and constructing/implementing them. This list sits above the 'opportunities register' as identified in the previous section, similar to the way in which I suggest the contract-level Early Warning Register sits above the active Early Warning Register for negative risks or threats (see Figure 3.4 in Section 3.6.4).

E. As the contract or project progresses, compare the cost and timing of adding in these value enhancements with a realistic assessment of how much money is available to spend. Having the three commitment points – thinking about it, detailed design and constructing it – allows three 'get-out' points should downside risks or threats occur.

3.6.9 Conclusions on early warnings

Stripped back to the basics, all the early warning clauses say is

- identify and tell the other Party as soon as you become aware of any issues or emerging risks
- cooperate to avoid or minimise the impacts of those issues or emerging risks

regardless of where the contractual liability will fall (i.e. whether they are a compensation events or not).

This section has given a lot of suggestions on how to get the best from the words in the contract and I hope you agree that most, if not all, of them are not rocket science. It does take discipline and a bit of time to implement them, but they should

- increase the chances of the contract coming in within both Parties' time, cost and quality objectives (which includes 'managing in' opportunities)
- save you time overall as, by proactively addressing early warnings and so avoiding or minimising their impacts, you won't have to spend your precious time firefighting.

3.7. Agreeing compensation events quickly, amicably and in accordance with the contract

3.7.1 Background and why do this

In the minds of the original NEC Panel, their utopian ideal would be that all compensation events are pre-assessed, giving Clients almost 'real-time' knowledge of the current Prices and Completion Date, and Contractors certainty over their income and the motivation to work efficiently in order to beat their accepted quotation. This contrasts with the traditional situation in other standard forms where, while direct costs may be paid on account, the final account is settled many months, if not years, after practical or substantial completion, giving both Parties little knowledge over the financial details or likely completion and hence little ability to influence them. What is more, the post-delivery war of claims, rebuttals and counter-claims costs both Parties money, which, looking in on the construction industry from outside, you have to say is a completely non-value-adding activity.

Even if the compensation events are assessed retrospectively, if they are wrapped up promptly, within the timescales of the contract, then at least the Parties would know where the contract is heading and could then do something about it. For example, if cost is overrunning, some of the original work could be taken out to keep to the *Client's* total budget. Alternatively, if the project is on a fixed budget and looking good, 'bells and whistles' could be added in towards the end of the contract, especially under target-cost contracts (see Section 3.6.8.4).

When I started doing my original research, way back in 1993, many commentators and practitioners derided the desire for compensation events to be assessed within limited timescales as the project progressed as being unrealistic. This is still derided today, albeit to a lesser extent, yet I know it can be and has been achieved.

A question I sometimes ask resistant Contractor personnel is something like: 'How long does it take you to put together a tender?' The answer is normally something like: 'Well, we are given 4 to 8 weeks, but we do the majority of work, having got subcontractor quotes in, in 2 weeks'. I then ask two questions.

- 'Why, for work not yet done, can you not put together a mini-tender/quotation for work that you definitely have in the 3 weeks required by the contract?' The answer is normally a scratch of the head. Without a little bit of probing and prompting, they admit that it is more a question of finding the time to do it when they have other duties to do.
- 'For work already done, how come you think you need months, even years, to put together and be paid for claims?' The answer is, to a large extent, that they have to collect and collate records to substantiate, negotiate and agree claims. That is, they do it retrospectively, a long time after the time and costs have been incurred, often based on missing, incomplete or conflicting records, and when the people involved have often disappeared from the scene.

This comes down to perhaps having greater resources available during the implementation phases, prioritising time and doing it efficiently and collaboratively (because it takes two to tango). Doing it efficiently and collaboratively is what this chapter is about.

So what happened in practice? Going back to my original research sample of 29 projects.

- On three of the projects, the final account was agreed on the date of Completion.
- On 23 of the projects, the final account was agreed within 3 months, with a significant minority agreed within 1 month. Just to emphasise this, one of my contractor clients in the civil engineering sector came to expect their final account to be agreed within 1 month of contract Completion. Likewise, I have been told that the benchmark Key Performance Indicator for one fairly major government agency is that all project teams should have the final account agreed within 1 month. If not, it is below the expected performance, and the worst performance is measured at 3 months or more. Do you think teams achieve this by putting the agreement of compensation events off until the end of the contract? Of course not, they do it as they go along. As a result, in my original research all of these Contractors had better than normal cash flow, as they were largely paid for compensation events as the contract progressed. (The majority of my original research contracts were under the priced contracts (option A and, to a lesser extent, option B), as opposed to cost-based contracts.)
- On three of the 29 contracts, the final account had not been settled within 6 months after Completion, following which time I lost touch. It is not surprising that these were the contracts where participants put the contract in the drawer and carried on as they always had done. More specifically, when interviewed, each of the *Project Managers* made comments to the effect that, with hindsight, they wish they had gone through the *Contractor's* initial programme in much greater detail and not accepted it. In each case, when compensation events started to happen several months into the contract and needed evaluating, the *Project Manager* did then go through the contract and did not accept parts of the programme which they previously had. This led to a charge from the *Contractor* of the *Project Manager* not agreeing to the programme as it did not now suit them. Consequently, the assessment of the compensation events was in dispute, as the baseline from which changes are assessed – the Accepted Programme – was in dispute. As a result, not only were the effects of compensation events not agreed as the contract progressed, but the contract became progressively more confrontational in all respects.

One of the projects that was agreed on the date of Completion was a tunnelling project at Heathrow to connect Terminal 4 with the central terminals, after which some rather advanced (for the time) baggage handling kit would be installed. It had the following major compensation events.

1. The final chamber was increased in volume by six times. Consequently, it was constructed by putting sheet piles around an area and digging down, as opposed to being tunnelled out from within.
2. Due to (1), construction had to start at the opposite end to that planned.
3. Due to (2) and operational constraints, a shaft had to come down on top of the tunnel, rather than to its side.
4. Construction had to be stopped for 17 weeks due to the collapse of the Heathrow Express Tunnel, which was using the same tunnelling methodology.

With the obvious exception of the last one, all of these compensation events were pre-agreed within a cycle time of around a week from asking for the quotation. Indeed, for the last event, as the *Employer* wanted to ensure that resources did not go elsewhere, the weekly costs of standing were also agreed within the week.

> As a result of these compensation events, the contract Prices went up from £6.5 million to £13.5 million (in mid-to-late-1990s money, so roughly double it for the 2020s equivalent), with the timescales increasing by approximately 9 months (from 18 months to 27 months), yet the final account was settled on the date of Completion. This not only saved both Parties a fortune in claims consultants etc., but both thought that, by being able to manage the works, the *Employer* easily saved upwards of a million pounds in construction costs. The *Contractor* also achieved a good profit margin.

Have I convinced you that assessing compensation events as the contract progresses is a 'good thing' for both Parties? Ideally before the work is done? I hope so.

3.7.2 Prerequisites for prompt agreement

Before we go on, the following inputs are needed to agree compensation events efficiently and collaboratively.

- The right attitude or mindset in both Parties. Warren Buffett, one of the world's top investors, once said, when determining the right figure to pay for a company, that he would 'rather be approximately right, than precisely wrong'. There is a tendency for professionals to argue over minutiae, especially in relation to forecasts, but this is pointless as both Parties' forecasts will be precisely wrong when compared with the actual costs. This especially applies to the evaluation of risk. Even doing it on 'agreed' records (i.e. where those out on site have agreed them) is an approximation, as you can rarely absolutely divorce the effects of the compensation event from the original work. In the case of a contract with many compensation events, provided neither Party is looking to unduly take advantage of the other, it will even out. As one *Contractor's* director said to me on a contract with 2700 compensation events, 'some you lose, some you gain, but if neither Party is looking to screw the other, statistically it is going to work out about right'. To do this, as one Project Manager said to me, NEC 'requires people on both sides who can take a view on an issue and then move on'. That also implies that they are empowered to take a view (see Section 3.3.1).
- A collaborative attitude and good 'soft skills' to match. This is linked to the above, but represents a different aspect of mindset. It includes the ability to negotiate constructively while maintaining, even enhancing, the relationship.
- A sufficiently detailed, understood, agreed and up-to-date Accepted Programme. This is the base from which compensation events are evaluated (see my comments at the end Section 3.7.1).
- Programming input, as and when needed.
- An effective working early warning process. This will prevent retrospective 'benefit of hindsight' arguments over what the *Contractor* 'should' have done in the circumstances. Instead, the *Contractor* actively contributes to the management of the events and the *Project Manager* gives prompt direction on how a matter is to be overcome.
- Quantity surveyors – or those who will assess the financial impact – who can read a programme, understand both the operation of the compensation event clauses and the Schedule of Cost Components, and build up costs from first principles.

3.7.3 Changes to the original NEC3, the April 2013 version and NEC4

Under all editions of the NEC Engineering and Construction Contract, clause 63.1 is the key clause regarding how the change to the Prices is assessed when a compensation event occurs.

To reinforce the NEC Panel's desire to have compensation events pre-assessed wherever possible, an extra paragraph was added onto the clause when NEC3 came out. This gave the date from when the *Project Manager* gives, or should have done under the contract, an instruction for a quotation as the date which divides the retrospective part of a quotation from the part that is done on a forecast. That is, even if the *Contractor* has done the work by the time the quotation is submitted, the work done from the time of instruction should be a forecast.

In truth, this was largely a reaction to Project Managers wanting the best of both worlds in terms of choosing a quotation based on a forecast or records, whereby they would do something like the following.

1. The *Project Manager* instructs the *Contractor* to submit a forecast quotation, which would quite naturally have some risk in it.
2. The *Project Manager* would disagree with some element of the quotation (and I am not saying that this disagreement was necessarily unjustified).
3. The work needed to be done, so the *Project Manager* would instruct the *Contractor* to proceed with the work as a compensation event, but records would be taken of hours, resources etc. used.
4. The records would be priced up and compared with the forecast quotation.
   If the records were better than the original forecast quotation, then the *Contractor* would make its *fee percentage*, but not additional profit from working efficiently. If, on the other hand, the forecast was better than the records-based quotation, the *Project Manager* would take this, and the *Contractor* would lose money on the additional work.

Obviously, this is not exactly fair and, unsurprisingly, the *Project Manager* only needs do this a couple of times before the *Contractor* starts to refuse to submit forecast quotations. Neither what the *Project Manager* did nor the *Contractor's* reaction to it were in accordance with the second edition contract (or are for the third and fourth). Furthermore, the knock-on effect of behaving in this way is that cooperation, both on compensation events and more generally, starts to evaporate.

When the April 2013 'upgrade' to the original NEC3 came out, these additional words had been adjusted to make them more precise, and the same words had been introduced into some other clauses to make the contract more coherent.

When NEC4 came out, the wording had been refined further and a 'dividing date' introduced, which is used to separate out retrospective assessments from forecasts. To quote from clause 63.1 of the contract: 'For a compensation event that arises from the *Project Manager* or the *Supervisor* giving an instruction or notification, issuing a certificate or changing an earlier decision, the dividing date is the date of that communication'. For other compensation events – which the *Contractor* should notify – the dividing date is 'the date of the notification'.

3.7.4 Agreeing the time and cost effect of minor compensation events and updating the Accepted Programme accordingly

In an ideal world, the time and cost effects of every compensation event would be agreed prior to the work proceeding. This was the original, somewhat idealistic, intention of the authors of the ECC. However, it has usually proved impracticable for small to medium compensation events. There are two reasons for this.

- The work would have to stop while the compensation event was evaluated and agreed, leading to lots of small delays in addition to those arising directly from the compensation event. Completion(s) – the '(s)' is in case there are any sectional Completions – and hence the Completion Date(s), as well as any Key Dates, would be affected adversely, and the *Client* would be paying for these time-related delays.
- Under the second edition, 'If the programme for the remaining work is affected by the compensation event, the *Contractor* includes a revised programme in his quotation showing the effect' (clause 62.2). As virtually all compensation events affect remaining work in some way, this implied that a new programme – not necessarily a fully revised programme for acceptance – was contractually required for almost every compensation event, imposing a ridiculously heavy administrative load on those attempting to operate the contract. Indeed, I can say hand on heart that this was the one clause in every NEC2 contract that I came across that was ignored.
  In the third edition, the wording of clause 62.2 was changed so that the last sentence read 'If the programme for remaining work is altered by the compensation event, the *Contractor* includes the alterations to the Accepted Programme in his quotation' and this has stayed the same in NEC4. I read this to mean that you could include a simple statement saying that operations X and Y will now start 2 days later. And you could argue that the revised last paragraph of clause 63.5 of the January 2019 version of NEC4 further reinforces this view.

In reality, on many contracts within the research sample and many since, good practice for evaluating the minor compensation events has evolved to something like the following.

1. It is agreed that the events are compensation events at the time they occur.
2. The work proceeds so as not to delay the progress of the *works*.
3. The additional people and Equipment hours are recorded, agreed and signed off out on the Site, together with any additional Plant and Materials used.
4. The direct Defined Cost plus the Fee are assigned to the additional resources used, agreed on a face-to-face basis (e.g. at a weekly financial meeting) before formal submission, and accepted easily within the timescales of the contract.
5. Each month, *before* a new programme is assembled by the *Contractor* and submitted for acceptance, the *Project Manager* and the *Contractor* (or their planners) meet to agree the cumulative effect of minor compensation events, in terms of disruption to other work and overall delay to Completion.
6. These effects – and the effects of delays for which the *Contractor* is liable under the contract – are then shown on the revised programme which, having had input to, the *Project Manager* usually accepts on the first submission.
7. The *Project Manager* and the *Contractor* (or their quantity surveyors) agree the time-related Defined Cost associated with the delay to planned Completion due to the cumulative delay and disruption over the last month.

For instance, a series of small compensation events along a path of operations adds up to a generous 1.5 days of additional time. By the time disruption is taken into account, it is agreed that the overall delay is 2 days – the alternative is to argue over what the disruption is for each and every individual minor compensation event. The *Contractor* is paid the direct costs of this extra half day's disruption. However, this path of operations are not on the critical path, so there are no site overhead/preliminaries associated with this cost.

On another path of operations, let us say that, by the time disruption effects are taken into account, it adds up to 1 day's delay for that month, and this is on the critical path. Consequently, planned Completion goes back 1 day due to compensation events – it may go back further due to *Contractor*-owned delays – so in the monthly 'wrap-up' compensation event, the Completion Date is put back 1 day and the *Contractor* is paid for a day's additional preliminaries costs within the Working Areas.

Most of the time, people who have done this have done it outside of the contract, based on a gentleman's agreement – and it has worked. If you want to do this in accordance with the contract, then the *Project Manager*, when instructing a quotation for each of the individual small-value compensation events, should state something like the following as an assumption under clause 61.6.

> Provide a quotation for the direct costs only with the *Project Manager's* assumption of no delay and disruption. Any delay and disruption will be assessed at the end of the month as an additional compensation event under clause 60.1(17) together with other compensation events in which this same assumption is stated.

This allows the *Contractor* to price the direct and immediately foreseeable costs, whether on records or as a forecast. At the end of the month, the 'wrap-up' compensation event for all the disruption and any resulting delay is done under clause 60.1(17), where the *Project Manager* notifies a correction to an assumption about a compensation event (although, strictly speaking, it is assumption*s* about compensation event*s*, i.e. both in the plural).

This procedure does not provide the *Client* and *Contractor* absolute time and cost certainty at any point in the contract, as the original authors of NEC would have liked. It is, however, a significant improvement on the traditional end-of-contract 'claim' for delay and disruption, which provides absolutely no basis for the effective project management of the contract. Instead, you have a rolling final account, a reasonably up-to-date estimate of when Completion will be achieved and, because the Completion Date is constantly visible, both Parties know where they stand and what they are aiming for. It also avoids wasting both Parties' management time with excessive programming input and repetitive niggling arguments about the extent of disruption on each and every compensation event.

Agreement on the effect of compensation events using this method is also easier as the people involved in the assessment

- are using a more up-to-date programme which has more information in it and has been accepted (ideally not more than a month previously)
- are based on site and the work is relatively recent (unlike under traditional forms of contract, where claims are developed by claims consultants etc. who have become involved months if not years after the event happened and are trying to reconstruct reality at second hand, usually based on partially complete records)
- are discussing disagreements in timescales of days or, at worst, a week or two, rather than weeks and months, as can happen on traditional forms of contract, when assessment is delayed until at the end of the contract.

Another issue that has caused problems, particularly on building sites, is that Subcontractors, many of which are small organisations, may simply lack the sophistication and motivation to price compensation events using the Schedule of Cost Components. They are, however, quite happy to do additional work using tendered bill rates because, to echo Warren Buffet, these are 'approximately right'. Indeed, if the Subcontractor is under the Short Subcontract (ECSS) or the *Contractor's* own form of contract, there is no Schedule of Cost Components in the contract (although under the ECSS the Subcontractor can be required to break its costs down under main headings that match the ECC's Schedule of Cost Components). Pragmatism is needed here – if the *Project Manager* or delegate are happy with the rate and the compensation event is of a fairly minor value with little or no impact on the overall programme, then the work involved in evaluating the change in Defined Cost can be done using rates in a bill of quantities rather than using the Schedule of Cost Components. If the *Contractor* has prepared its own bill of quantities for use with the Subcontractors, then this can be used in order to save time. This is completely allowable under NEC4 core clause 63.2 because the rates or lump sums do not have to be ones already in existence under the main contract – provided the rates or lump sums are agreed, they can be invented or made up by those charged with agreeing the value of the compensation event.

If there are time effects, then the direct costs can be assessed using a bill of quantities, while the time-related costs are worked out in accordance with the ECC's Schedule of Cost Components.

3.7.5 A methodology for agreeing the cost effects of medium to major compensation events in advance

This section puts the case for the prior assessment of medium to major compensation events, and gives a methodology for agreeing the cost effects of these in a rational and constructive manner. This methodology should reduce – and has done so – the often contentious atmosphere engendered when agreeing costs and time.

Under traditional conditions of contract, contractors have generally liked to have additional work assessed retrospectively using records. This is because

- they are risk averse
- daywork rates normally more than cover the cost of labour and construction plant.

In the original research, one of the distinguishing features of Contractors who achieved final profit margins that were higher than they had expected at tender was their willingness to price work in advance of it proceeding, so that they could make profit over and above that in the Fee. A contributing factor to this improved profit was better than expected cash flow: under options A and B, the sooner a quotation is agreed and the work is done, the sooner the *Contractor* is paid.

Clients gain three benefits.

- In order to gain early agreement, the *Contractor* submits more realistic Prices and time for delay. This results in less management time being wasted going through and agreeing the quotation.
- Once the changes in Prices and the Completion Date are agreed, the *Contractor* is no longer motivated to make the most of the *Client's* misfortune.
- There is greater price and time certainty.

The extent to which these benefits can be realised depends on how willing the *Contractor* and *Project Manager* are to assess and agree compensation events where the work has not yet been done, which can include prioritising it over the assessment and agreement of work done (e.g. minor compensation events).

Indeed, a minor complaint from the *Project Manager* on the Heathrow tunnelling contract mentioned in a previous example (see Section 3.7.1) was that, such was the focus on pre-assessing large events that could be managed, the administration of minor historical events was being left open and not wrapped up. He therefore instigated a fortnightly 'wrap-up' meeting for the programmers and quantity surveyors, in which they were ordered to put to bed the effects of minor compensation events as discussed in Section 3.3.1.

The following is an attempt to formalise an approach that evolved informally on some projects for the assessment of compensation events. I suggest that the Parties agree, in order, the following details.

1. The subclause number of the compensation event, its root cause and the 'do-nothing' consequences.
2. The desired outcome or range of outcomes, and any constraints on how the outcome is to be achieved.
3. The risks, what can be done to avoid or reduce them and approximate estimates of their probability and impact, from which decisions can be made on who takes the risks, which may result in a *Project Manager's* assumption being stated.
4. The required change in the *Contractor's* method of working, from which the changes in the resources directly involved in the activity can be determined.
5. The change, if any, to planned Completion(s) and hence the change to the contract Completion Date(s) and any Key Dates.
6. The change in Defined Cost related to each of the previous three points.
7. Communicate and iterate.

You will notice some similarities and overlapping themes between this process and, for instance, the process for agreeing what to do with early warnings.

I am aware that some might say that pre-agreeing the time and cost effects of major compensation events is a little idealistic. It is, however, based on observation of real contracts where agreement was reached rapidly, and it has been critiqued by and improved with suggestions from practitioners. It also often involves a significant change in attitude from both Parties administering the contract, namely that

- the *Contractor* is entitled to fair recompense and additional time when a compensation event occurs, provided it can justify it with sufficient rigour
- the *Contractor* does not try to take advantage of the situation when a compensation event occurs (indeed, this point is both contract law as a 'duty to mitigate' (and be paid reasonable costs for doing so) and effectively stated in clause 63.9 of the ECC, which states that the assessment assumes 'the *Contractor* reacts competently and promptly to the event'; see also my comments on mindset at the start of Section 3.7.2).

STAGE 1: AGREE THE SUBCLAUSE NUMBER OF THE COMPENSATION EVENT, ITS ROOT CAUSE AND 'DO-NOTHING' CONSEQUENCES

Taking each one of these in turn.

- For the subclause number of the compensation event, it is not only good contract administration to know under which heading a compensation event has occurred, but also the starting point for reviewing and improving your – as in the *Client's* – capability and competence in delivering projects under contract. I have seen final accounts with monetary figures against each of the 21 core compensation event headings, which gives an insight into what were the causes of the changes to the Prices and Completion Date(s).
- Going back a stage further, if you note where the compensation event came from, you are further nailing the cause. For instance, taking clause 60.1(1) changes to the Scope, does the change result from: a change in the business environment or law; *Client*-generated changes (e.g. the business requirement changes); end-user changes (which may come from the operational side of the business)[17]; from the design organisation (e.g. poor quality specifications); etc.?
- With regard to understanding the 'do-nothing' immediate consequences, as per the first stage of the risk-reduction meeting. For instance, if it is a 2-week delay to a delivery, is that delay on the critical path (in which case solve it fast), or is it off the critical path, where there is 2 weeks of free float (in which case make sure there is not any further delay). One of my reviewers suggested that you 'traffic light' compensation events with respect to their relationship to the critical path: red is on it; amber is close to it and likely to get on it, while green is no effect.

STAGE 2: AGREE THE DESIRED OUTCOME AND ANY CONSTRAINTS

The Scope specifies the *works* and describes any constraints on how the *Contractor* is to provide it. A *Project Manager's* instruction changing the Scope is almost certainly the most common compensation event heading.

A number of Contractors interviewed complained that, under traditional conditions of contract, too often the client or its representative gives inadequate descriptions or instructions of what is required from the contractor for variations: they just want the problem solved, leaving the contractor to guess at what is required and how to get there. It was thought that this is because they themselves did not know what they wanted and hoped that the contractor would come up with something suitable. This then results in argument when it does not conform to the representative's idea/concept of what is suitable and/or arguments of whether the contractor should have been able to work more efficiently while they tried to figure this out. Unfortunately, this was also a consistent theme of complaints from the *Contractor* about the *Project Manager* under the ECC.

John Rayner, a fellow NEC consultant, tells the tale of a meeting he sat in on where the *Project Manager* was blaming the *Contractor* for not pricing up a compensation event in a way that matched what he wanted. The *Contractor's* site manager responded by saying that they had never been given clear direction on what the *Project Manager* wanted and so were left to guess

[17] One *Client* I know got fed up with their operations people not engaging early on in a project's development and then, close to handover, coming in and saying 'I don't like this, change that'. Consequently, the *Client* then adjusted the budgeting, so that late changes came out of the operating expenditure budget, rather than the corporate capital budget.

when preparing the quotation, to which the *Project Manager* replied 'You should know!' So the *Contractor's* site manager asked the *Project Manager* for a clear description of what he wanted, to which he replied 'I don't know'. The *Contractor's* agent then rather gleefully asked: 'So we are somehow meant to mind-read what is in your mind, when you haven't even thought of it?', to which the *Project Manager* replied, with a smile on his face, 'Yes!'

I hope the above tale illustrates the absurdity of asking for an accurate quotation based on inaccurate details. If the *Project Manager* wants an accurate quotation before the work proceeds, then it is a matter of common sense that the *Contractor* needs a precise description of what is wanted on which to base that quotation. If the *Project Manager* cannot do this because they do not know, then it should be thrashed out, perhaps with the input of consultants, specialists and relevant others.

Do note that specifying the desired outcome is not the same as specifying the solution. Defining the problem is the start point, specifying the outcome is the end point, and a solution is a way to get from the problem state to the outcome state. Under the ECC, the *Project Manager* can ask for quotations with a least-time outcome or least-cost outcome. They could also ask for quotations to reach a desired outcome through a range of technical solutions. If the Parties are really working in a spirit of mutual trust and cooperation, the *Project Manager* could even give the *Contractor* the cost that the quotation must not exceed and ask for the best-value outcome that can be achieved for the money. Constraints restrict the solutions that can be applied to achieve an outcome, and the *Project Manager* should state these up front in order to avoid redundant work by the *Contractor*.

STAGE 3: DISCUSS RISKS AND STATE ASSUMPTIONS

Under clause 63.8, 'The assessment of the effect of a compensation event includes risk allowances for cost and time for matters which have a significant chance of occurring and are not compensation events'. (I have underlined the last part as it is slight tweak to the previous edition's wording.) In other words, the *Contractor* is allowed to include for its risks in any compensation event quotations. These can be both average productivity risks within bars on a Gantt chart and risks of the sort that appear in a risk register as beloved by the risk profession (which might well relate to those in the contract-level Early Warning Register – see Section 3.6.4).

The idea of the *Contractor* taking additional risk once the contract is let is anathema to many contractors. Similarly, the consideration of risk from a contractor's viewpoint is a new experience to many who represent clients. In both my academic and subsequent consultancy experience, I have found this an area which people struggle with. Indeed, in a survey done by GVE Commercial Solutions, this was identified as one of the two most problematic areas when it comes to assessing compensation events, with almost all of those surveyed (98%) agreeing that 'more discussion on risk pricing is required by NEC parties and participants prior to assessing compensation event quotations'.

However, on some contracts, risk is discussed openly, which in many instances leads to the *Project Manager* stating assumptions under clause 61.6. The *Contractor* would then price for and accept construction risks in compensation event quotations with the *Project Manager's* knowledge and approval. The *Contractor* then used its expertise to minimise the impact of these risks on the construction programme and costs, which it otherwise would not be motivated to do.

I outlined the generic process of risk management in Section 3.6.2. In the original research, various levels of sophistication were observed in the approach to the pricing of risk and its management. As a consultant, I also hear about the full spectrum of approaches to risk, which include denying that it exists and therefore the *Project Manager*, in direct ignorance or in contradiction of clause 63.8, strikes out any pricing of it within a *Contractor's* quotation. Equally, some Subcontractors are not aware they can include for it.

Next, Contractors would try to add a standard percentage to the Defined Cost of all forecasts for compensation events. No specific risks would be identified for a specific situation and, consequently, no specific actions to reduce the probability of occurrence or impact would be taken. Within the research sample, no agreement was ever reached on the level at which this percentage should be set.

Others would try to add a percentage for risk, which depended on the nature of the compensation event. For instance, 'We are outside and working in the ground so 10% should be added to the base Defined Cost of doing the work' versus 'We are inside in an uncluttered area, so a couple of per cent'. Here risk is at least identified, even if it is in a rather vague way. Agreement on the level of the appropriate percentage for the risks was not always possible and, more often than not, the work had been done before agreement had been reached.

On a few sites, the *Project Manager* and *Contractor* would identify and discuss risks involved in responding to the compensation event, agree how they were to be managed and who they were allocated to. This does not need to be done in a big meeting – just ask yourselves something like

- What are the top five risks related to this compensation event?
- What is the best way to manage them?
- Based on the answers to the first two questions, who is each risk best allocated to in terms of financial liability if it happens?

If the *Client* is best placed to have liability for certain risks – either because it is better able to manage them and/or bear them – this can then be stated as an assumption under clause 61.6 when instructing the *Contractor* to submit a quotation (which would often be worked out on a face-to-face basis). The *Contractor*, while developing the quotation, would sometimes identify other risks and seek clarification from the *Project Manager*. By identifying, analysing and developing management strategies for risks, an approximate figure could be agreed for inclusion in the quotation. For instance

- if the Parties agree that the *Contractor* can take proactive steps to avoid a risk or minimise its probability, then this can be priced up in the normal way
- if the Parties agree that a risk could occur (but might not), then they price up the consequences that would fall on the *Contractor* – this is then divided by the approximate probability of the risk occurring to give a monetary sum to be added into the quotation.

The *Project Manager* can then relate this figure to the *Client*'s desire for certainty versus potential for least cost or time, and take a view on whether the *Client* would wish to take the risk or allocate it to the *Contractor*.

To give an example, let us say that there is a compensation event which includes digging a new hole and putting in a new concrete base. Both Parties know the bottom of the hole is around the groundwater level and there is therefore a risk of flooding, especially if it were to rain at the wrong time, which would prevent subsequent work proceeding. The obvious remedy to mitigate the effect is to have a pump on standby.

- Does the *Client* pay, as part of the quotation, for the pump to be hired on standby, even though it may not be needed? If there is groundwater, then there is no delay. Whether the operation is on the critical path may affect the decision.
- Do the Parties decide *not* to have the pump on standby, with the risk transferred to the *Contractor*? If this is the case, no assumption probably needs to be stated, but the *Contractor* is entitled to contingency (probability × impact, including cost of reactively hiring the pump plus the cost of delay).
- Do the Parties decide *not* to have the pump on standby, but with the *Client* holding the risk? In which case, the *Project Manager* probably would need to state as an assumption that groundwater would not be found at the base level and therefore the *Contractor* is not to include the need for a pump in its quotation. If it turns out there is groundwater, then it is an additional compensation event under clause 60.1(17).

Which option is chosen will partially depend on the real-life circumstances, but also on the *Client's* desire for certainty.

> A word of warning: risk assessment is an imprecise science. On one site the quantity surveyor representing the *Employer* (as it was a first edition NEC contract) wanted the management strategy for mitigating the effect of a risk, should it occur, to be priced with the same rigour as the main body of the compensation event (and they would discuss it in detail), only for it to be divided by an arbitrary figure for the probability of it occurring. This caused considerable frustration to the *Contractor*, who was not being reimbursed for the cost of preparing quotations, while the consultant quantity surveyor was.

The participants may have these upfront discussions, but when the *Contractor* goes away and develops the detail, it may identify other possibilities. In which case, the *Contractor* should not proceed in isolation, but have that further discussion by telephone, video conference or wandering into the *Project Manager's* office and agree the basis for assessment, including whether to add in another *Project Manager's* assumption.[18] At the very least, keep the other party informed.

For the above process to work, with agreement to be reached prior to the work proceeding, it does depend on the *Contractor* being realistic in its risk allowances and for those representing the *Client* to see risk from the *Contractor's* viewpoint. A open and honest conversation goes some way towards making this happen, in which case, all can benefit.

[18] I know that at least three of the cloud-based NEC contract administration systems mentioned in Section 3.3.2 now have the facility for a *Project Manager's* assumption to be added into a quotation while it is being prepared, rather than requiring the quotation procedure to be started again. And this can be done a number of times. But you would only do this following a verbal discussion and agreement.

On one NEC1 contract, a compensation event occurred that was likely to cause disruption to work specified in the original Works Information. The *Contractor* offered to take the risk of disruption occurring, and, if it did occur, keeping the progress of the *works* on programme. After discussion, the figure agreed was £1000. The *Project Manager* compared this with the *Employer's* objectives and accepted this figure in the quotation. The *Contractor* was then motivated to manage the disruption, which he did, rather than take advantage of the *Employer's* misfortune, which would be the typical situation for claims management on traditional construction contracts.

On a contract for a water treatment works, a big settlement tank had, for some reason, to be moved. The decision was made soon after the *Contractor* had started on site, and there were no borehole logs for the new location. The *Contractor* was paid, as a compensation event, to put a large back-actor into doing some deep trial pits. This found that the ground was substantially the same as the original location. The assumption for pricing was therefore that the *Contractor* would encounter similar ground conditions as per the original location, thus avoiding the need for the *Contractor* to add in additional contingency for the unknown risk.

A technique that can be used with assumptions is to state tolerances. For instance, let us assume, using the example in the above box, that the depth of the foundations for the settlement tank was 5 metres and that a layer of rock was thought to start at 4 metres depth. Consequently, the *Contractor* would have to use a pecker to break out the last metre of rock. The assumption could be that the *Contractor* was to price on the basis of rock being found at a depth of 4 metres, but this could potentially lead to a clause 60.1(17) compensation event if hard material was found at 3.95 metres – that is, as no range has been specified, it is highly likely there will be another compensation event. Instead, the Parties could agree, as a *Project Manager's* assumption, that the *Contractor* prices the quotation on the basis of finding hard material at 4 ± 0.25 metres. If rock was found at less than 3.75 metres depth, and so the *Contractor* had to peck out more than expected, it would be an additional compensation event, but only for the additional rock higher than 3.75 metres depth. If rock was encountered at a depth below 4.25 metres, then the *Client* would get the benefit of a negative compensation event, which would reduce the Prices.

On a tunnelling project, the decision was made for the tunnel to strike out at a level 3 metres lower than originally planned due to encountering large flint boulders at the original level of the shaft. An assumption was made that the *Contractor* would only be affected for the first 100 metres of the bore, whereupon the expected conditions would be encountered. If this was not the case, it would be another compensation event.

Alternatively, we can ignore risk and pretend it is not there. As a result, the *Contractor* makes its own assumptions – which may or may not be stated – and includes an allowance for the possible cost and time effects in its quotation. Unsurprisingly, the *Project Manager*, on being presented with this, disagrees in some way, shape or form. Because the *Contractor's* assumptions are not stated, it is likely the *Project Manager* thinks the rates, time or quantity of resources are excessive, as any risk is hidden in the assumptions. The result is argument and a requotation. If the

Contractor had stated its assumptions, then hopefully the Parties would have had the intelligent conversation they should have had right at the start, following which the *Contractor* would requote on the now agreed and stated assumptions. Both routes involve unnecessary professional time and hassle.

As one *Client* quantity surveyor stated: 'there are a given set of circumstances which can be made into assumptions when requesting a quotation. These make it very difficult for the *Contractor* when there are other things which we may have not stated as assumptions, because he then has to make his own assumptions. Now that could inflate a quotation ... We have to reach some agreement as to what is and is not included.'

STAGE 4: AGREE THE CHANGE IN THE *CONTRACTOR'S* METHOD OF WORKING AND RESOURCES

Having gone through the previous stages, the most efficient and effective way of implementing the compensation event – the solution – can be formulated. As the *Client* is paying for the compensation event and there is a lack of competition, it is entitled to have a say, via the *Project Manager*, in how the work will be done. However, it should be remembered that the *Contractor* has been employed because of its expertise in construction, and this can and should be used to develop a more economic and timely solution.

Under clause 62.1, the ECC does allow the *Project Manager* to instruct quotations based on different ways of dealing with a compensation event, but only after discussions with the *Contractor*. Provided it can be done in a friendly way, the logical time to do this is at the same time as 'making and considering proposals' for how to deal with the 'matter' – that is, at the relevant early warning meeting, where ballpark estimates can be discussed, before detailed pricing of the better options is progressed as compensation event quotations. The problem with giving ballpark estimates is that, once an amount is mentioned, the anchoring cognitive bias tends to kick in and people become emotional if the detailed quotation then exceeds this amount. This occurred on a couple of contracts in the original research sample, and I have heard of similar tales since, so I suggest Contractors' people give a best case and a worst case estimate, which Project Managers can in turn pass on to the Client if needs be. This helps manage expectations!

On one contract, which predominantly involved earth moving, the *Employer* (as it was under a first edition contract) failed to gain planning permission for a temporary dumping area in time. The *Project Manager* asked for three quotations based on the following ways of dealing with the event.

- The contract is terminated (although this is not a compensation event, the ECC has more extensive termination clauses compared with other contracts, which makes termination a viable option without becoming embroiled in legal disputes, especially if done collaboratively).
- The earth moving continues through the winter months.
- The *Contractor* stops work over the winter and remobilises at the start of the next earth-moving season.

After some negotiation the third option was chosen as the most favourable. In the words of the *Employer's* contracts officer: 'it was the first time, in my experience, that we have really been able to make a proper project management decision based on good information'. Asked what

> would have happened under traditional conditions of contract, his reply was: 'we would have probably carried on with the work regardless – i.e. over the winter – and had a bun fight at the end of the contract'. This sentiment was echoed by others within the research sample in the light of their experiences. For the more major compensation events, both those representing the *Contractor* and the *Employer* thought that the process of pre-assessment and constructive negotiation ultimately saved the *Employer* money, as what was desired was more defined by the *Project Manager* and more efficiently implemented by the *Contractor*.

A number of Contractors commented that their expertise is traditionally underutilised in the management of variations, and that they liked being able to contribute to solving problems. An ideal way of encouraging this is to pay for time spent preparing and considering quotations. In previous editions of NEC, under options A and B of the ECC, the cost of preparing quotations was explicitly excluded from being a Defined Cost. Many complained that this is unfair if the *Contractor* needs to prepare a number of speculative quotations based on different ways of dealing with an event. As a result, in NEC4, a new compensation event (under subclause 60.1(20)) was added. If a quotation for a proposed instruction is not accepted, it is now a compensation event. This directly relates to the wording in new subsection 65 of the contract (Proposed instructions). I have two and a half comments on this.

- The new compensation event does not directly cover the situation where, under clause 62.1, following discussion with the *Contractor*, the *Project Manager* instructs quotations for alternative ways of dealing with the compensation events (i.e. different methods to achieve the same result). I suggest that 60.1(20) is interpreted fairly broadly, so that it does apply to these circumstances.
  Alternatively, I see nothing wrong contractually (and think it is better project management) in having one compensation event – a change to the Scope – for, say, 'Feasibility and option evaluation for dealing with [the problem]' followed by another compensation event, and hence quotation, for the chosen option.
- While it may well be a compensation event, the *Contractor* still has to demonstrate additional forecast Defined Cost to get a change to the Prices. If the same salaried staff members, who are allocated full time to the contract, will be working extra hours to prepare multiple quotations, but will not be paid anything extra, then there is no change to Defined Cost.

This point can become quite contentious, not necessarily in the circumstances described for a one-off compensation event, but when a larger than expected number of compensation events causes the *Contractor* to bring in extra staff – which does change Defined Cost – in order to keep on top of its assessments. While the *Contractor* gets a little bit extra due to the *fee percentage*, the amount does not cover the change in its salary costs. Some Project Managers have accepted the morality of this argument and instructed, as a change to the Scope, the *Contractor* to bring in an additional quantity surveyor, or other management resource, for a set time. As it is a change to the Scope, it is a compensation event (which does not arise from the *Contractor's* fault, unless it is under-resourced in the first place: another source of discussion!).

Once the most economic, timely or risk-free method of working has been agreed, then the changes from the original method of working can be compared with what is given in the Accepted Programme. From this, the change in the total quantity (amount × duration) of people, Equipment, and Plant and Materials that are directly involved in the activity can be calculated, provided these resources are in the relevant Schedule of Cost Components. This emphasises the need for the *Project Manager* to study the method statements and resources for each operation that the *Contractor* is required to supply in a submitted programme. Any disruption and consequent loss of productivity arising from the compensation event that indirectly affects other activities should also be taken into account.

STAGE 5: AGREE THE CHANGE TO PLANNED COMPLETION(S) AND HENCE THE COMPLETION DATE(S)

If the compensation event is on the critical path, then planned Completion of the contract may be delayed, and hence the Completion Date will be put back by the same time. Likewise for Key Dates and section Completion Dates (which is why I have used included '(s)' in the heading) – see Section 3.5.4 for an explanation of this. The indirect resources which are to be kept on site longer (traditionally referred to as preliminaries) need to be identified.

At this point, it is worth exploring the new words added onto the end of clause 63.5 (previously 63.3 in NEC3) of the January 2019 amended version of the NEC4 ECC. The original June 2017 version of the NEC4 ECC included different – and shockingly bad – wording, which was replaced by some quite good wording. The clause now states that in the time assessment of a compensation event

> The assessment takes into account
>
> - any delay caused by the compensation event already in the Accepted Programme and
> - events which have happened between the date of the Accepted Programme and the dividing date.

Taking the first bullet point first, this stops the *Contractor* from double counting any delay in the programme and, which is more likely, the *Project Manager* from arguing that, as the delay is already in the programme, there is no delay to planned Completion and hence no change to the contract Completion Date.

The second bullet effectively says the latest Accepted Programme is the line in the sand, but, from that line, it is updated to take account of delays due to 'events', whether compensation events or *Contractor*-owned delays, up to the dividing date. This reinforces the stated requirement in clause 62.2 to show the effect of the compensation event on remaining work. The discussion over ownership of float in Section 3.5.4 is very relevant here, as the free float is owned by whoever gets to it first, while the terminal float is owned by the *Contractor*. In practice, this means the people responsible for agreeing the programme need to sit down and update the programme to see, on a day-by-day basis, when and where float contracts or expands. Central to this analysis would be the reasons for the change – that is, whether it is due to a compensation event (in which case the *Client* has got to it first) or a *Contractor's* risk. (For the NEC Board's perspective on this, see NEC, 2019a.)

STAGE 6: AGREE THE CHANGE IN DEFINED COST RELATED TO EACH OF THE PREVIOUS THREE HEADINGS

With the exception of risk, where the figure will be less clear cut, if the previous stages have been followed, then it should merely be a question of putting costs to the changed resources listed in the relevant Schedule of Cost Components and adding them up to gain the total change in Defined Cost, to which the *fee percentage* is applied.

Please note that I am not saying ignore costs up to this point. Rather, I am saying costs come from the use of resources and risk: if you understand and manage these effectively, the agreement on cost is faster and the amount of cost is reduced.

If the users have a contract cost manual or, on smaller sites, have pre-agreed the cost of individual resources then it is a simple and quick matter of arithmetic (see Section 3.7.6). In a similar vein, rather than calculating the time cost in terms of site-based preliminaries from scratch for each compensation event, it makes sense for the quantity surveyors to agree a daily and weekly financial amount for a given level of preliminary resource. This agreed figure is then used for all compensation events that delay Completion (unless the preliminaries change, in which case they should agree a new daily or weekly rate based on those resources).

STAGE 7: COMMUNICATE AND ITERATE

A feature in the majority of projects in the original research sample was that the *Contractor* was given initial parameters and assumptions to prepare the compensation event quotation, and there would then be little or no consultation with the *Project Manager* or the quantity surveyor representing him until the quotation was submitted. At this point, the *Project Manager* would see the final figure for the first time, think it was high (and bear in mind that most people are only used to seeing direct costs during the construction phase, with delay and disruption costs being covered in the claim that is determined after construction) and instruct the *Contractor* to resubmit the quotation. Consequently, much of the detailed work was redundant because, for instance, productivity and risks had not been agreed. Having been instructed to resubmit the quotation, it was only then that the two Parties would discuss and agree these principles, and, having done so, the next formal quotation submission by the *Contractor* was agreed informally, prior to formal submission.

> One frustrated *Contractor* in the research sample made the point that: 'I have better things to do than churn out quotations which, on the basis of past experience, are going to being chucked into the bin and redone!'

It is clear to me that initial and, if necessary, interim discussions before the first formal submission of the quotation would have been beneficial in almost all circumstances. This is really an extension of the collaborative working principle that runs through the ECC.

This did happen on some contracts, with both the *Employer's* and *Contractor's* quantity surveyors sitting in the same office and working out quotations together. They would reach agreement on each of the previously mentioned headings before considering the next. That is not to say there was no iteration – for instance, a risk may emerge when considering the method of doing the work. On a tunnelling project, the two planners representing the *Employer* and *Contractor* would agree the programme effects, from which flow changes in costs, before passing the financial assessment and detail onto their quantity surveyors to agree. When this was done, it meant that the acceptance of the final quotation was almost a formality by the time it was submitted, resulting in less time, effort and acrimony in agreeing the quotation, and a greater likelihood that both Parties can achieve the benefits outlined.

3.7.6 Other tips

These are some other tips for quickly and amicably agreeing the time and cost effects of compensation events in accordance with the contract.

- On smaller sites with fewer workers, where the contract is under option A or B (the priced contracts), it is suggested that the Defined Cost of the individual resources is agreed either prior to entering the contract or in the post-award phase. On larger sites, a quotation manual can be agreed that subsumes the Schedule of Cost Components. This records the agreed Defined Cost for different types of Equipment and categories of workers etc. It should also be agreed which workers fit into which category. When a compensation event occurs, having agreed the units of time worked for Equipment and People, it is then simply a matter of multiplying time by cost per unit of time. The manual can be progressively added to if and when new resources are identified and priced.
- Prepare and use a preformatted spreadsheet which, I suggest, has the first tab as the summary sheet, and the next eight tabs match the headings of the relevant Schedule of Cost Components (although you might wish to combine the later ones onto one tab) and an additional one for risk calculations.
- Where the compensation event involves additional design by the *Contractor*, it may be helpful to split the compensation event into two parts to conform to the ECC's timescales for submission of quotations. The first part is for the additional design, and the second part for the additional physical work. Problems will still be encountered with the second part, when there is no original drawing to compare the revised design with. This, however, is one of the problems with design and build under any form of contract.
- On some occasions there may be insufficient knowledge of the scope of the compensation event, or the degree of uncertainty is too large, for the *Project Manager* to state assumptions that will not be changed, yet the work still needs to proceed. Hence it is impracticable and wasteful for the *Contractor* to pre-assess the compensation event (e.g. when an unexpected area of unsuitable material is discovered in the ground). The physical works may proceed and be evaluated from records as one compensation event, until sufficient knowledge is gained or the degree of uncertainty is reduced for the *Contractor* to pre-assess the remaining part as a separate compensation event. This might go against the strict adherence to using records before the 'dividing date' and forecasts afterwards, as per clause 63.1.
- The general pattern encountered in the research was that the more realistic the initial quotation, the more likely it was to be accepted first time. Under options A and B, this gives the *Contractor* a cash flow advantage – as the sooner the quotation is accepted and the work done, the sooner the *Contractor* is paid. Where the *Project Manager* found the initial quotation was 20–30% over what was expected, they would generally evaluate the quotation in detail and, in the process of negotiation and clarification, any profit outside the *fee percentage* would be taken out to arrive at a sensible amount. The *Contractor* would then be instructed to submit the revised quotation, which had in effect already been agreed. On the few sites where the *Contractor* consistently submitted unjustifiably high quotations, the *Project Manager* would eventually lose patience and start to impose their own assessments on the *Contractor*, to the overall detriment of the *Contractor*. The recommendation is that it is not in either Party's interests for the *Contractor* to consistently submit quotations for compensation events that it knows to be excessive.

In the original research, I was made aware of three large compensation events where there was a significant difference between the change in Defined Cost worked out using the Schedule of Cost Components and the real change in the *Contractor's* cost. (There were other cases, but these were either due to the *Contractor* being unable to justify the change in its resources due to a poor

Accepted Programme or to a lack of realism in the various percentages tendered.) This was to the *Contractor's* advantage in one of the cases, and to the *Employer's* in the other two, predominantly due to the deficiencies in the Schedule of Cost Components in the first and second editions of the NEC (which were much reduced in NEC3, and are even more so in NEC4). In these cases, the injured Party protested sufficiently strongly that the other Party listened and was eventually satisfied of the authenticity of the injured Party's case. An agreement was then reached that satisfied both Parties.

3.7.7 Summary

The following scenario applies whether the event is pre-assessed or not.

Think about agreeing a quotation from a practical viewpoint. Whether on an ad hoc basis or a weekly basis, the two quantity surveyors sit down at the *Contractor's* quantity surveyor's computer and work up a quotation using the suggested process and a preformatted spreadsheet that matches the structure of the relevant Schedule of Cost Components and has a section for risk. They use pre-agreed rates logged in another spreadsheet. At the end of agreeing an individual quotation, the *Contractor's* quantity surveyor summarises the change to the Prices and any change to the Completion Date on a preformatted 'front sheet' email, attaches the spreadsheet (with any agreed supporting information) and sends it off as the formal quotation to the *Project Manager's* quantity surveyor. Ping! The *Project Manager's* quantity surveyor has received an email on their laptop which says exactly what has been agreed. They send back an email accepting the quotation. Ping! The *Contractor's* quantity surveyor receives it and the compensation event is 'implemented'. Apart from adding the assessment into the pricing document (e.g. as a new or amended activity in the Activity Schedule), the compensation event can be put to bed.

For this to work, it does mean that other people may need to be involved and consulted for more complicated compensation events (e.g. programmers to agree delay and disruption), and that each Party's representative needs to be reasonable and must be able to build up costs from first principles. But beyond that, there is nothing too hard about it, is there?

3.8. Resolving issues and disputes

3.8.1 General

The procedures of the NEC, if implemented, should decrease the likelihood of conflict developing and the need to refer individual disputes to the *Adjudicator* when compared with other conditions of contract. Within the original research sample of 29 contracts, only two matters were referred to adjudication. Certainly, in the early use of the ECC, adjudication was a rare occurrence. However, disputes will inevitably occur on some contracts.

My original research corroborated findings from the USA (Rowley, 1996).

- Adjudication, or a similar alternative dispute resolution (ADR) process, is rarely used, as the Parties perceive that they have failed if they have to use it. It is seen as the last resort.
- If a dispute is referred to the *Adjudicator*, it is rare for it to proceed beyond this stage. In the years since, with the introduction of statutory adjudication in the UK, wider practice would seem to bear this point out.

Other observations regarding factors which give a greater incentive for the Parties to resolve the matter themselves, without reference to the *Adjudicator*, are as follows.

- Once a dispute is referred to the *Adjudicator*, the Parties have lost the ability to manage and influence the outcome.
- An *Adjudicator's* decision has to be fairly cut and dried, allocating blame and therefore financial liability to one Party. In practice, disputes are often more complicated, with both Parties being at fault to varying degrees.
- The referred matter was the tip of the iceberg. In the two disputes in the original research sample, the *Contractor* was not happy generally with the *Project Manager's* decisions and was almost looking for an excuse to make a point. The decision therefore only addressed the symptoms of the dispute, and not the underlying cause. As someone who supports adjudications and acts as an independent witness, what drives the dispute is the commercials (i.e. someone is losing money), and the battleground is the contract. If the commercials were OK, there would be no dispute.

My consultancy experience would confirm this and, having been involved in helping organisations on both sides of the contractual divide, I would also point out – against my short-term financial interests as a consultant – that the costs of getting all your ducks in a row in order to win an adjudication can be quite substantial.

For these reasons, some participants suggested that the *Adjudicator* should be encouraged to act in a more conciliatory role, trying to make the Parties accommodate and understand each other's views, and only give a decision when all else has failed. This role is not facilitated by the words in the main contract or by the requirements of the two UK Construction Acts. Indeed, the *Adjudicator* would be acting as a conciliator or mediator, and not as an adjudicator, if they were to do this!

There are other options available before going to adjudication. The rest of this section explores these options, before briefly discussing tactics if you do decide to go down the adjudication route.

3.8.2 The DIY solution: dispute-avoidance ladders

Dispute ladders came about as a result of the introduction of (first-generation) partnering in the late 1990s. Essentially, it is attempted to resolve issues or disagreements at the lowest relevant level within a set timescale. If the issue is not resolved at that level, it is escalated up a level, to be resolved within a normally slightly longer timescale, and so on until the issue is resolved.

For each level, a person or named role in one party's organisation is given a 'buddy' of equivalent authority in the opposite organisation. For instance, for technical issues

1. the first level could be the *Contractor's* sectional engineer and the *Supervisor's* assistant or the relevant *Client's* designer, who are to resolve issues within 3 working days, after which the issues are referred to
2. the second level, which could be the chief engineer on site and the *Supervisor* or head designer, who are to resolve issues within 1 working week, before referring them to
3. the third level, which could be the *Contractor's* non-site-based 'expert' and a design expert, who are to resolve issues within, say, 3 weeks from referral.

There could be an equivalent ladder for commercial and programming issues. At some point, these ladders merge, say at *Project Manager*/site agent or regional director level. Indeed, NEC4 options W1 and W2 now have the '*Senior Representative*' provisions, which provide a formal route for trying to reach agreement at director level before starting any adjudication.

The advantage of this method is that it does not generally require external services to be called in, but the following points need to be considered.

- People need to be empowered to make decisions.
- There needs to be a clear starting point for each stage so that people know when the time to resolve a dispute starts and when it has to be concluded by. The inclusion of 'Dispute' in the title implies that the Parties are about to go to war, when in reality it might just be a disagreement. This can put people off from escalating unresolved issues and disagreements.
- The process should not – and does not legally – override the contractual mechanisms and timescales.

These three points can be placed within an NEC context as follows.

- People need to be properly delegated their authorities in accordance with clause 14.2.
- I suggest that the timescales generally start, for the first level, when informal face-to-face agreement cannot be reached, as there is no point in formally submitting something for acceptance knowing it will not be accepted. However, I would suggest that the timescale ends a day or two before the formal submission is required, whereby they escalate it and their seniors agree to extend the contractual timescale – in accordance with clause 13.5 – by the time stated in the agreed dispute-resolution ladder. This is then repeated until the issue is resolved.

  In the case of the technical example above, the starting point could be when the *Supervisor's* assistant refuses to sign something off, saying that, in its current state, it is a Defect.
- You might wish to call it something other than a 'dispute' resolution ladder. On one framework I am heavily involved in, we call it the 'Issue Resolution and Dispute Avoidance' (IRaDA) process, which we wrote into the Framework Information.

3.8.3 External help

This section considers one-sided expert opinions, independent and impartial expert opinions, mediation and conciliation, and/or Dispute Avoidance Boards, the last of which has been introduced into NEC4 through option W3. The first three of these are generally used when the DIY option described in the previous section has failed. Dispute Avoidance Boards involve a bit more proactiveness and upfront cost, hence they have typically only been used on larger projects or programmes.

3.8.3.1 One-sided expert opinion

Probably every 2–3 months, I am asked by an individual organisation – whether a *Client*, *Contractor* or Subcontractor – to come and give an independent view on whether their case has merits and, if it has, to help support and develop the case.

Sometimes the case has no merit, in which case I stop there! On the other hand, sometimes it does, and 'developing the case' may, at its simplest, be writing a short letter pointing out the relevant clauses (with a bit of explanation) which can be shown to their opposite number. For simple cases, I would say this resolves the issue around 75% of the time (and one of my friendly reviewers confirmed this to be their experience also).

However, it may be a bit more substantial and involve following the communications trail to establish who wrote what and when and whether these communications are of contractual significance, including at times consulting with friendly lawyers on points of contract law. Sometimes this helps and sometimes it does not, for two reasons.

- With the best intentions of my client, I am only given one perspective, and therefore in a letter rebuffing mine the other party points out other factors that I am not aware of. This is despite asking my client for all relevant information! I then normally have to write another letter refining the original communication.
- The other party perceives me as being hired to find the argument that reaches the conclusion my client wants, and not as being an independent.

Nevertheless, I would say this is fully successful about 50% of the time. It is also partially successful a lot of the time, in that it normally narrows down what is in dispute.

3.8.3.2 Independent and impartial expert opinion

A few times – I can count them on the fingers of two hands – I have been asked in by both Parties to either give an independent and impartial view on contractual wording or help prompt the team into a more constructive way of working, which has included providing some clarity on unresolved contractual issues.

This method overcomes the two obstacles mentioned in the previous section – that is, I gain both Parties' perspectives and, crucially, I am seen as independent. In addition, part of resolving the issue has been to give a simple plain English explanation of why I have arrived at that view, which both parties can understand and, hopefully, accept. If I miss a factor out in my explanation, they can just come back to me and I either give further explanation or modify my independent view to take account of it.

In all but one case where I have done this, it has worked. Where it did not was because the relationship in terms of trust and respect at site level, and a lesser extent senior level, was gone. This approach is far cheaper than adjudication.

The only contract in the NEC4 family which actually has provisions for independent expert opinion is the Alliance Contract, in which the Alliance Board can refer the dispute 'to an independent expert for an opinion'. However, in the other members of the family there is nothing to stop an independent expert giving an opinion on an issue or disagreement, provided it is by agreement between the Parties.

3.8.3.3 Mediation or conciliation

Mediation is a process where an independent person, normally with a mediation qualification, comes in and mediates between the two parties. Often, they initially talk to one party and then the other, then scuttle between the two. They help the parties to isolate and take out of the equation the matters on which they actually agree, enable each party to see the other's perspective, and generally build consensus and agreement until the parties are sufficiently close that they can reach a face-to-face agreement. At this meeting, the mediator chairs.

The advantage of this approach is that the parties have ownership of the solution, assuming a solution is found. There is no external expert 'telling' the parties how to sort out their differences, or which of them was right and which was wrong, although the Parties might ask the mediator to give a non-binding opinion on the likely legal outcome. However, as the mediator cannot impose a solution, both parties need to enter into the arrangement willingly and without intransigence. It can also be quite time-consuming, and therefore expensive, to do, both in terms of the cost of the mediator and of senior management's time in having meetings with the mediator.

As I understand it, the difference between mediation and conciliation is that the conciliator can have a more assertive role in making recommendations and proposing agreements for the parties to accept.

Under both of these arrangements, the end result is ideally a signed and legally enforceable agreement. Statistics suggest that these routes have a success rate greater than 75%. (CEDR – the biggest accredited mediation/conciliation body – regularly reports that approximately 75% of disagreements are settled during a mediation or conciliation, with another 15% settled shortly afterwards. Anecdotal evidence further supports this.)

The last point to note is that the courts are increasingly expecting parties to try and resolve their dispute in some other way before going to court. In some instances, expenses have been awarded against a winning party because of its intransigence in trying to resolve the dispute. Therefore, if you choose litigation as your form of *tribunal* after adjudication, you almost certainly want to have tried some form of mediation or conciliation before you go to court, to avoid expenses being awarded against you – even if you win.

3.8.3.4 Dispute Avoidance Boards and option W3

This section gives a brief history and overview of the Dispute Avoidance Board (DAB), before focusing on the NEC4 W3 clauses.

DABs were first used in the USA, where there is a greater tendency for litigation, and where damages can be punitive and therefore payouts much larger. These characteristics can also be seen in the Arab states. Therefore, the economic case for using DABs is particularly strong in the USA and the Middle East. (As a result of FIDIC conditions of contract being the dominant form in the Middle East, DABs have, apparently, been in the FIDIC conditions of contract for years.)

In the US model, a DAB is appointed at the start of the contract, and normally consists of at least three experienced and independent people, who will have a range of experience relevant to the contract. Often, the DAB members visit the site on a regular basis (not more than monthly), in order to keep up to date on the contract and emerging issues. The idea is that, because they are familiar with the issues, they can nip potential disputes in the bud by informally resolving disagreements quickly, making constructive suggestions and mediating between the parties in a more relaxed way than a mediator or conciliator could.

In addition, if a dispute or disagreement is referred to them, they have a much more up-to-date understanding of the context, if not the specifics, of the matter. Lastly, having at least three independent people is seen as advantage, as if all three decide against one party on a matter, then that party really does have to be intransigent to take a matter onto the next stage.

While traditional DABs are probably the best form of dispute resolution – or even avoidance – they are also undoubtedly the most expensive, and therefore are only warranted on larger projects or programmes.

The words above are substantially the same as those in the previous edition of this book (on NEC3). So, how does the new NEC4 option W3 compare with this traditional DAB model?

For a full answer, and to avoid repetition of text, go back and read Section 1.2.3.3. However, for a short answer, the key differences are as follows.

- While option W3 can be used as described above, it can also be used in a reactive way only, so that the DAB is only brought in when there is an active disagreement or emerging dispute.
- The DAB can be one or three people. Logically, the provision to appoint one person would divide the normal cost by a third, making its use on smaller contracts more viable. However, the appointed person would potentially have to act with extreme diplomacy to avoid the perception of bias, which would undermine their role.
- Although it would stretch the meaning of the words in option W3, the DAB, especially if it is only one person, could take on much more proactive role by coaching the key players in a contract. I have played this role a number of times – although not under a DAB appointment – and can verify that it aids the smooth running of the contract and is effective in heading off disputes. On large contracts, a three-person coaching team might be warranted.

My last comment relates to the opening words in option W3, which state: 'Used when . . . the United Kingdom Housing Grants, Construction and Regeneration Act 1996 does not apply'. In other words, option W3 as written does not comply with UK legislation regarding the use of adjudication. This may also be the case in other jurisdictions with dispute resolution legislation. As I pointed out in Section 1.2.3.3, this is a shame because, with a little bit of tweaking of the clauses, I think DABs could form part of a structured issue-resolution ladder, where it is at a level before adjudication. The NEC Board seems to have recognised this and has since brought out NEC4 Practice Note 5, which provides the amendments required to replace the *Senior Representatives* provisions in option W2 with the DAB provisions from option W3 (NEC, 2019b).

3.8.4 Adjudication as a last resort

Adjudication should be, and usually is, seen as the last resort, for the reasons outlined in Section 3.8.1. It is normally commenced when the relationship has broken down and, as a result, you typically play to win, without any regard to maintaining the relationship. However, once this is the case, there is a positive in having a timely adjudication, in that a matter is contractually resolved (until taken to arbitration or litigation), removing the contractual uncertainty which can negatively affect many other aspects of the contract administration.

Almost always, it is the Party that thinks it is owed monies that refers a dispute to adjudication, and usually it is the *Client* holding onto monies that the *Contractor* thinks are owed to it. Under the Construction Acts, and therefore under option W2, with or without amendments, you cannot be time barred from instigating an adjudication, but once one is referred, timescales and time bars do apply. Typically, therefore, as a *Contractor* you want the following.

- To have all your ducks in a row before you instigate an adjudication. By this I mean a mature, punchy and succinct case, with back-up evidence, to be submitted to the *Adjudicator*, but with supporting arguments and justifications lined up to rebuke any of the *Client's* potential counter-arguments.

Figure 3.5 Timescales for adjudication under option W2

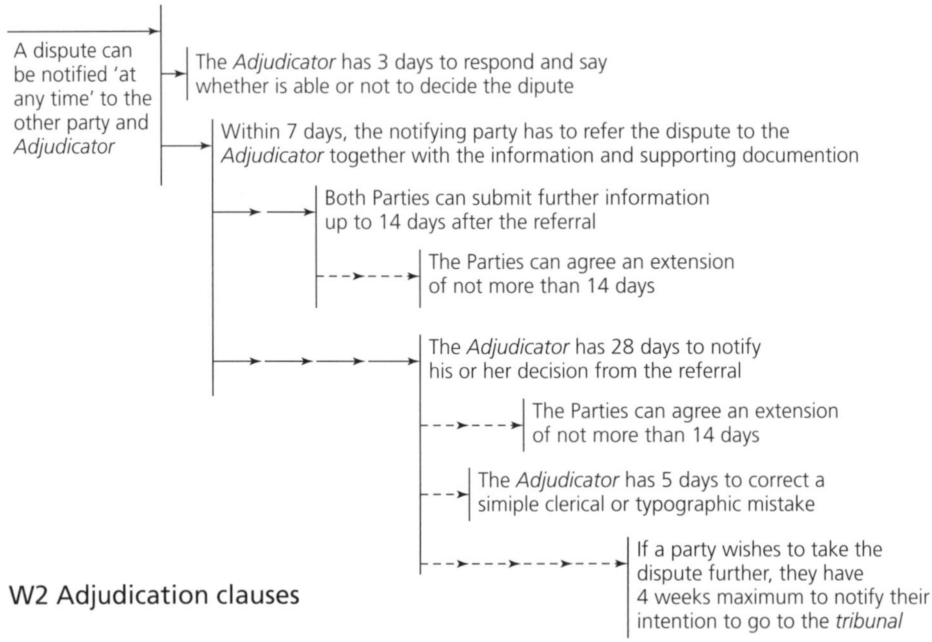

- To launch the adjudication at a time that suits you and does not suit the *Client*. The worst (or best) example of this was the *Contractor* launching an adjudication on the morning of the *Project Manager's* wedding, knowing that he was booked to go on a 2-week honeymoon!

Typically, as the *Client*, if you know or suspect an adjudication is coming, you too want to have as many of your ducks in a row as possible, in order to be able to respond to whatever the *Contractor* throws at you. These counter-arguments will need to be refined in the light of the specific arguments put forward by the *Contractor* to the *Adjudicator*, which you will be copied in on.

The timescales for an adjudication under option W2 are illustrated in Figure 3.5. If your contract is subject to the UK Construction Acts, from the time that a Party gives notice of a dispute, that Party has 7 days to refer the dispute and submit details of their case to the *Adjudicator*. This includes 3 days in which an adjudicator is appointed: most professional bodies in construction now have lists of adjudicators who (a) have relevant background and experience, (b) have gone on a training course and passed some sort of exam to get on the list and (c) periodically have an adjudication peer reviewed and assessed to stay on the list. The Parties can decide to choose jointly from such a list which adjudicator they want, assuming the chosen adjudicator is available and the Parties are still talking enough to reach agreement. Alternatively, the person or institution identified in Contract Data part one as the *Adjudicator nominating body* chooses an adjudicator.

Both Parties then have 14 days to submit any further information. If you are playing to win, it therefore makes senses, as the defending party, not to submit your defence until just before the 14 days are up, as this gives the other Party little or no time to tailor or even submit their pre-prepared rebukes to your counter-arguments. As a result, the referrer may just decide to submit their counter-arguments (together with additional evidence, justification etc.) on the deadline, just in case the defending party sneaks their defence in! It should be noted, however, that the *Adjudicator* can – and almost always does – instruct either Party to provide further information etc. after this 14-day deadline has passed.

As shown in Figure 3.5, the Parties can agree an extension of up to 14 days to submit further information and, if this is the case, the time that the *Adjudicator* has to make their decision will also be extended. If you are the referring Party and you have all your ducks in a row (and you know the other Party does not), it logically does not make sense to extend the timescale for further submissions. Having said this, the appointed *Adjudicator* often sets out the timetable and, in order not to get on the wrong side of them, the Parties agree to it.

The *Adjudicator* has 5 days from issuing their report in which to correct simple mistakes, whether in the mathematics or in the words, but not to fundamentally change their decision in any way. This is called the 'slip rule'.

Lastly, under clause W2.4(2), if a Party is not satisfied with the *Adjudicator's* decision and intends to take the dispute on to the tribunal, they must notify the other Party of this intention within 4 weeks of the *Adjudicator's* decision. I am led to believe that this contract requirement is legally debateable, but you do not want another dispute over this. So, if you are the aggrieved Party, unless you really are throwing in the towel, you should automatically notify the other Party that you intend take the dispute to the tribunal, even if you subsequently decide not to.

From this overview, I hope that you can see how adjudication is generally not a constructive process, instead relying on a series of ambushes and counter-ambushes to make your point and avoid the other Party effectively making theirs. It also tends to lead to all-or-nothing decisions, when my experience of resolving disputes is that rarely is either Party completely free of blame: in other words, it is a bit of a lottery. Adjudication is also expensive in terms of internal staff costs and, potentially, external consultant costs, if help is needed to prepare cases – and adjudicators are not cheap!

So, I finish with words I started with: adjudication is a 'last resort' when all other methods have failed.

3.8.5 A reference and a final thought

First, the reference: Rob Gerrard, the long-standing secretary of the NEC Users' Group, and Patrick Waterhouse have written an informative book on resolving disputes under NEC. It very much describes how the procedures should work under the contract, and it has a whole chapter on using the NEC4 Dispute Resolution Service Contract. The title of the book is *NEC4 Resolving and Avoiding Disputes* (Gerrard and Waterhouse, 2019).

My final thought is this: if you are going to pay for someone to come in and give independent input and advice, why not do it proactively? By this I mean, help to set the contract up properly, so that participants can focus on avoiding or reducing the effect of events and rapidly resolve those that do occur, as opposed to devoting their time and energy (and hence cost) into reactively containing issues and resolving disputes? (One of my reviewers made the observation that nearly all of the disputes he gets involved in revolve around poorly set up contracts, giving a figure of 95%. I would not disagree with that.)

To me, this would seem to be a far more cost-effective strategy. While I specifically refer you here to Section 3.2 on training and common understanding, I am actually referring you to all the other chapters in this book.

REFERENCES

APM C&P SIG (2020) *Procuring for Agile Projects.* Opinion Piece. Association for Project Management, Princes Risborough, UK. Available at: https://www.apm.org.uk/media/42324/procuring-for-agile-projects-white-paper.pdf (accessed March 2020).

Association for Project Management (2019) *APM Body of Knowledge*, 7th edition. APM, Princes Risborough, UK.

Broome J (2008) How to stop your team 'storming' and get them 'performing'. *Project Manager Today*, November. Available at: www.jonbroome.com/resources/articles (accessed March 2021).

Broome J (2009a) ECC programme – the six levels of use. *NEC Users' Group Newsletter* **48**.

Broome J (2009b) Critical Chain project management: a response to Glenn Hide's article on Contractor's float in programmes. *NEC Users' Group Newsletter* **43**. Available at: www.jonbroome.com/resources/articles (accessed March 2021). Video presentation available at http://www.thenbs.com/topics/ContractsLaw/videos/NEC3_six-levels-of-use.asp or www.jonbroome.com/resources/videos.

Broome J (2016) How to rapidly build a high performing NEC project team. Blog. Available at: www.jonbroome.com/blog/december-2016/how-to-rapidly-build-a-high-performing-nec-project (accessed March 2021).

Faniran OO, Oluwoye JO and Lenard D (1994) Effective construction planning. *Construction Management and Economics* **12**(6): 485–499.

France G (2006) *Tricks and Tips in Planning and Scheduling.* The Planning Engineers Organisation, London, UK

Freeman M (2011) NEC3 taking off in NZ development. *NEC Users' Group Newsletter* **53**: 4.

Gerrard RA and Waterhouse P (2019) *NEC4 Resolving and Avoiding Disputes.* ICE Publishing, London, UK.

Goodwin S (2015) NEC early warnings – the need for a more positive perception. *NEC Users' Group Newsletter* **27**. Available at: http://www.gvecs.co.uk/articles.htm (accessed April 2021).

Hide G (2010a) Producing a programme under the NEC3 form of contract. *Proceedings of the Institution of Civil Engineers – Management, Procurement and Law* **163**(MP3): 111–119. Available at: http://www.gmhplanning.co.uk/publications (accessed March 2021).

Hide G (2010b) Managing a programme under the NEC3 ECC contract. *Proceedings of the Institution of Civil Engineers – Management, Procurement and Law* **163**(MP2): 59–64. Available at: http://www.gmhplanning.co.uk/publications (accessed March 2021).

Laufer A, Tucker RL, Shapira A and Shenhar AJ (1993) The multiplicity concept in construction planning. *Construction Management and Economics* **11**(1): 53–65.

Ming Sun and Tejas Oza (2008) *Improving NEC Contract Change Management using CCM.* University of the West of England, Bristol.

Neilson GL, Martin KL and Powers E (2011) *The secrets to successful strategy execution.* In *HBR's 10 Must Reads: On Strategy.* Harvard Business Review, Boston, MA, USA.

NEC (2019a) *NEC4 ECC – Practice Note 1.1: Assessing delays due to compensation events.* NEC, London. Available at: https://www.neccontract.com/NEC4-Products/NEC4-Contracts/NEC4-Free-resources (accessed April 2021).

NEC (2019b) *NEC4 ECC – Practice Note 5: Using a Dispute Avoidance Board for contracts covered by the Housing Grants, Construction and Regeneration Act 1996.* NEC, London.

Available at https://www.neccontract.com/NEC4-Products/NEC4-Contracts/NEC4-Free-resources (accessed March 2021).

Patterson R (2010) Six key links in ECC. *NEC Users' Group Newsletter* **51**: 6–7.

Patterson R (2020) Helping NEC clients decide the best way to get a first programme in place. *NEC Users' Group Newsletter* **108**: 10. Available at: https://www.neccontract.com/About-NEC/News-and-Media/Helping-NEC-clients-decide-the-best-way-to-get-a-first-programme-in-place (accessed March 2021).

Rowley F (1996) Engineer demise, adjudicator rise. *New Civil Engineer*, 14 November.

Thompson P and Perry J (1979) *Engineering Construction Risks – A Guide to Project Risk Analysis and Risk Management.* Thomas Telford, London, UK.

ice Publishing

Broome, Jon
ISBN 978-0-7277-6205-4
https://doi.org/10.1680/necaug.62054.299

Chapter 4
Summary and last thoughts

The ECC is as much a tool for project managing a contract as a conditions of contract but, like most tools, its effectiveness depends on how it is used by people. So here is my recipe for success.

As the *Client*.

A. Choose the right contract strategy in terms of member of NEC4 contract family, main payment option and secondary options to create alignment of commercial interest while protecting the downside (see Section 1.2).

B. Intelligently fill in the Contract Data and do not undermine your chosen contract strategy by writing excessive and legalistic option Z (*additional conditions of contract*) clauses, which can both undermine the stimulus to good project management and create ambiguity and inconsistency at the *Client*'s expense (see Section 2.2.2).

C. Do not skimp on writing good 'well-structured, concise and sufficiently precise' Scope and Site Information, the former of which needs to be in line with your chosen contract strategy (see Section 2.2.3).

D. For the priced payment options A and B, as well as choosing a Contractor with realistic Prices, include a realistic evaluation of the amounts, rates and percentages submitted by the Contractor as part of its tender to give adjusted tendered Prices. If the Prices, additional rates and percentages are too low, consider not awarding the contract to that Contractor. For a cost-based payment option, do this with the target Prices and *fee percentage*, but verify any other submitted rates and percentages for realism prior to entering the contract or after the contract is signed (i.e. do not have them as part of the tenderer's offer) (see Section 2.4.1).

E. In your selection, take account of the tendering Contractors' abilities to effectively project manage **and** work 'in a spirit of mutual and co-operation', in line with NEC principles and practices, as demonstrated for your specific contract (e.g. quality of tender programme, pre-contract Early Warning Register and contract-specific proposals etc.). Include these contract specifics as part of the contractual obligations by reference from Contract Data part two (see Section 2.4.2).

As the potential *Contractor*.

F. Bid realistically given the risks which are allocated to you under the contract (see Section 2.3.1).

G. Programme out the first months in detail so that both Parties can 'hit the ground with their feet running' (see Sections 2.3.2 and 2.4.2).

H. Understand the rates and percentages you have to tender and that are applied in the relevant Schedule of Cost Components, and do not excessively manipulate them (see Section 2.3.3).

I. Make sure your Subcontractors' subcontracts are on compatible terms.

Obviously, realism in the Prices, rates, percentages and what else you offer has to be tempered by the need to win work to stay in business, yet make a profit.

For both the *Client* and the *Contractor*, on award of the contract.

J. Staff the contract with a sufficient quantity of proactive and project-management capable, collaborative and empowered people, who can make the right decisions in the right way at the right time (i.e. within the timescales of the contract) (see Section 2.2.2.3).

K. Ensure key participants from all organisations are on the same playing field when it comes to understanding what the NEC4 contract says, its implications and how they wish the contract to be run, as well as the more general context of the project (see Section 3.2).

L. Discuss and agree pretty much all contractual communications prior to submission. Acceptance should then become a rubber-stamping exercise. Have regular 'wrap-up' meetings for early warnings, updating the programme and the assessment of compensation events. This saves time, cost and hassle for everyone involved. Put in place a system for communications, which could include using a cloud-based NEC4 software-as-a-service system (see Section 3.3).

M. Operate the payment process properly and, if under one of the cost-based options C, D or E, agree a sensible level of proactive inspection and auditing that gives value for money in terms of assurance (see Section 3.4).

N. Having decided at which of the six levels you wish to operate the programme (and what information you do, or do not, want to be in it), keep the Accepted Programme up to date, closing out otherwise unresolved issues as the contract progresses. As a minimum, make it easy to assess the contractual consequences of compensation events; even better, use it to minimise the consequences of early warnings; ideally, use it as a proactive tool to knock barriers out of the way and speed up delivery, which should also have positive effects on cost. Regardless, make sure the programme is a living document that is used to deliver the contract efficiently (see Section 3.5).

O. View early warnings as a positive, as they give the team an opportunity to avoid, minimise or mitigate the consequences of emerging negative risks and issues, regardless of contractual ownership. Set up a system to record, resolve and track the status of both the early warning matter and, most importantly, the agreed actions. Hold people accountable for doing these actions (see Section 3.6).

P. Follow a structured framework/process for quickly and amicably agreeing the effect of compensation events on the Prices and Completion Date in accordance with the contract. Prioritise the prospective assessment of compensation events where the physical work has not yet been done, as this encourages proactive action and gives greater Price certainty to both Parties. If you can crack the assessment of compensation events, then you have overcome a major hurdle to effective and collaborative project management (see Section 3.7).

Q. Nip disputes in the bud by resolving them while they are still disagreements (see Section 3.8).

R. Lastly, do not be afraid to use the contractual sanctions (such as withholding a quarter of the Price for Work Done to Date for a first programme that does not show the information required (clause 50.5)) in a constructive way. Be clear about why it is being used and give clear direction on what needs to be done to avoid it being used in the future. Use of contractual sanctions highlights to senior management that something is amiss, rather than it being swept under the carpet only to re-emerge later as a bigger problem with unconstructive emotion attached.

Going back to my original research, I found that the more both Parties followed the NEC contract conditions in spirit and clause (but perhaps not to the exact letter), the greater the degree of satisfaction with the outcome. For Clients, given external factors, this means a contract delivered on or better than your time, cost and quality objectives. For Contractors, this means better cash flow and greater security over your profit margin (if not an improved margin). For the individual participants, it means less hassle and stress, and greater job satisfaction.

A phrase one of my mentors uses is that 'Just having a good theory is the booby prize in life'. Theory has to be implementable to be worth anything, so I hope you agree that most of what I have said is common sense pointed out. This is the feedback I have received when giving training and presentations, and from previous editions of this book. However, I cannot implement it: that is down to you!

Lastly, if there is something you do not understand or want clarifying, my view is that it is because I have not expressed it clearly. Therefore, if you want a clarification, or want to give me feedback, good or constructive (which I welcome), or just you want 5 minutes consultancy, feel free to contact me, ideally by phone on +44(0)7970 428 929 or at jon@leadingedgeprojects.co.uk. In addition, a range of resources is available on my website, at www.jonbroome.com/resources.

Best wishes and good fortune (a combination of luck and doing the right things consistently, consistently well).

Dr Jon Broome
Managing Consultant, leading edge projects consulting ltd

ICE Publishing

Broome, Jon
ISBN 978-0-7277-6205-4
https://doi.org/10.1680/necaug.62054.303

Appendix 1
Who pays for COVID-19? Contract risk allocation under the NEC Engineering and Construction Contracts

Article published on LinkedIn on 6 April 2020

CONTEXT: It may seem a little vulgar to be discussing 'who pays' for the financial effects of the Coronavirus pandemic when people are dying from it, so I apologise for anybody offended. Yet, at the same time, the survival of companies and the people who depend on them for their livelihood are also at risk due to lack of finance.

This article attempts to give some contractual clarity on who is most likely to bear the financial costs when contracting under the NEC Engineering and Construction Contracts. It covers both the third and fourth editions (NEC3 and NEC4) as there are nuances between the two. Much of what is said will also be relevant to other members of the family.

Before we get into the contractual situation, what often sits above the legal situation is the commercial situation. Indeed, my experience is that most legal disputes are because the Parties are not happy commercially. So while the contract provides a starting point for 'who pays', if you are a *Client* and your *Contractor* is going to be late, then it's your project that is going to be late. If your *Contractor* is losing money on your contract through no reason of its own making – i.e. the Coronavirus pandemic – and you are not helping out, then it will have both immediate and longer term affects. If you are a government client – or are in contract with a government client – this is very much the thrust of Procurement Policy Note (PPN) 02/20: *Supplier Relief due to COVID-19*. Massively shortening the guidance, it says, be practical, be prepared to fast-track payment to keep the supply chain going etc., and in return expect the supplier to open their books so you can see where the money is going etc., all of which might mean agreeing a Supplementary Agreement/ Deed of Variation. Indeed, in PPN 02/20: *Guidance Notes for Construction Contracts*, the government gives some hastily drafted model terms for a Deed of Variation to NEC.

The rest of this article assumes that there are no relevant modifications to the standard NEC Engineering and Construction Contracts.

FIRST THINGS FIRST: FOLLOW THE EARLY WARNING PROVISIONS. If you haven't done already, notify an early warning. This kickstarts the collaborative process of dealing with the effects of coronavirus and will be beneficial on whichever Party the financial effects ultimately fall. In reality, to varying degrees, it will be both, so it is worth collaborating and actively project managing events so as to minimise their impact.

Giving an early warning is particularly important if you are the *Contractor* as the assessment of any subsequent compensation event could be reduced if you don't do it as soon as you become aware of an event. At a big picture level, this is the coronavirus pandemic itself. At a smaller picture level, it will be as events unfold. For example, I know some contractors are struggling to get hold of soap for their sites, which might ultimately mean it cannot operate them safely, particularly as cleanliness is one of the key ways to reduce transmission of the virus. If the Client is a supermarket or soap manufacturer, they might be able to help!

REIMBURSEMENT UNDER THE COST-BASED OPTIONS, by which I mean both the target cost contract options C and D, the cost-reimbursable option E and the management contract option F.

During the contract, options C to E work pretty much the same in terms of how the *Contractor* is reimbursed: it is reimbursed Defined Cost incurred within the Working Areas plus the *fee percentage(s)*[1] applied to this Defined Cost. Defined Cost is both:

- the cost that the *Contractor* incurs directly – for instance, on its own People, what it pays directly for hire of construction Equipment, purchase of Plant and Materials etc., and
- what it pays to its Subcontractors in accordance with these subcontracts – so, for instance, if the *Contractor* made an additional payment to a Subcontractor to 'keep them going' which was not in accordance with its subcontract conditions, then this cost could be disallowed.

Option F is a lit bit different in that the management contractor does not get paid directly for its own costs as these are presumed to be covered by its *fee percentage(s)*.

With two exceptions, components of Defined Cost have to be incurred within the Working Areas, which comprise:

- the Site, which is specified by the *Employer* (under NEC3) or *Client* (under NEC4),
- any areas identified by the *Contractor* in its tender, and
- any additional areas proposed by the *Contractor* post-contract.

So *Contractor's* staff whose normal place of working is not now on Site, but at home, no longer qualify as a Defined Cost. However, note the last bullet above. The *Contractor* can propose that home addresses are added to the Working Areas and the *Project Manager* has only two reasons to not accept this addition: it's either not necessary to Provide the Works or it won't be used for work in the contract.

Defined Cost can be reduced by disallowing costs for the reasons stated in the contract. These are Disallowed Costs, which effectively fall completely on the *Contractor* even under the target cost options. There are eight main bullets in the definition of Disallowed Cost. The ones that maybe are, but hopefully are not, relevant to the coronavirus pandemic are:

- additional costs incurred through not giving an early warning, which reinforces the importance to a *Contractor* of giving early warnings, and
- 'resources not used to Provide the Works (after allowing for reasonable availability and utilisation) or not taken away from the Working Areas when the *Project Manager* requested'. So standing people or Equipment, particularly hired Equipment, need to be off-hired, especially if the *Project Manager* requests. This may well involve some grown-up

discussions around what resources it is sensible to keep on Site. The ideal forum for this discussion would, for example, be a risk reduction meeting (under NEC3) or early warning meeting (under NEC4).

The rest of this article looks at the four likely reasons for a compensation event related to the coronavirus pandemic. When a compensation event occurs, this is the point at which the risk transfers to the *Employer* (under NEC3) or *Client* (under NEC4). But note the wording I used here: it 'is the point at which the risk transfers', by which I mean up to this point the risk is within the Prices.

PROJECT MANAGER'S **INSTRUCTION:** Perhaps the most clear-cut of the four reasons is when the *Project Manager* gives an instruction which falls into one of the compensation event definitions listed in clause 60.1 (where all the core compensation events are listed).

This could be:

- an instruction changing the Works Information (NEC3) or Scope (NEC4) – see clause 60.1(1) – which modifies what the *Contractor* has to achieve or constrains how they achieve it,
- an instruction to stop or not to start any work – see clause 60.1(4).

The *EMPLOYER* (NEC3)/***CLIENT*** (NEC4) **OR OTHERS DON'T DO SOMETHING THEY SHOULD DO.** An Other is any person or organisation who is not a 'player' in the *conditions of contract*, i.e. not the *Employer*/*Client*, *Project Manager*, *Supervisor*, *Adjudicator*, Subcontractor or supplier. So this would be where the *Employer*/*Client* or Other can't or doesn't do what they were meant to do under the contract due to the effects of the virus.

This could be:

- the *Employer*/*Client* not providing access to a part of the Site by the later of its *access date* or a date shown in the Accepted Programme – see clause 60.1(2),
- the *Employer*/*Client* not providing 'something' by the date shown on the Accepted Programme, or
- the *Employer*/*Client* or Others not working within the times on the Accepted Programme or within the conditions stated in the Works Information (NEC3)/Scope (NEC4) – see clause 60.1(5).

A CHANGE IN THE LAW is only a compensation event if option X2 is specified, although a change in the law may become relevant for reasons explored later.

There have been changes in the law as a result of the coronavirus pandemic.

- On 21 March 2020, the Health Protection (Coronavirus, Business Closure) (England) Regulations 2020, were enacted. As well as giving power to quarantine individuals, it required cafes, restaurants and canteens etc. to close down, but had minimal direct effect on the construction industry.
- On 25 March 2020, the Coronavirus Act 2020 was enacted. This subsumed the above Regulations and gave the government (for the time being) certain powers to deal with the pandemic. Most of the 150 pages is not directly relevant to construction, but Schedule 22

gives the Secretary of State power to restrict gatherings – which could in extremis be a meeting of two people – and to close premises, both individual ones and, which is more likely, premises 'of a specified description'. In construction, this would be construction sites, and also other premises which need to be open for the construction industry to operate, e.g. builders' merchants, storage yards etc. However, these – at the time of publishing on 6 April 2020 – are *not* listed as specified premises.

So at the time of publishing (6 April 2020), the shutting down of Sites or other construction related premises is not due to changes in the law and hence not a compensation event.

In addition, the Health Protection (Coronavirus, Restrictions) (England) Regulations 2020 came into being on 26 March 2020. These sit below the Act, and each of the UK's four countries have their own separate regulations, albeit using substantially the same wording. I believe Regulations can be modified by the Secretary of State without the need to go before Parliament. These Regulations are the ones that impose the restrictions on movement – to only go outside for food, health or work (but only if you cannot work from home) – and on gatherings of more than two people in public places.

However, so far as I can tell, the guidance to stay 2 metres (6 feet) away from other people at all times and to wash hands frequently with soap for 20 seconds is only guidance. I can't find it in the Regulations or Act. So are the Site Operating Procedures[2] published by the Construction Leadership Council. Significantly, version 1 of the Site Operating Procedures says that workers should do their best to observe the 2-metre rule, not that they should stop work if they cannot.[3]

A Site not following this guidance could, in theory, be closed down as a category of premises, but this would be by order of a Secretary of State so is very unlikely. Far more applicable are the provisions of the Health and Safety at Work etc. Act 1974, which lays down wide-ranging duties on employers to protect the 'health, safety and welfare' at work of all their employees, as well as others on their premises. The Act allows the government to issue regulations, Approved Codes of Practice and guidance for employers. Because we only have 'guidance' at the moment, it is not a change in the law which you have to follow.

And under the Act, employers' duties are qualified with the words 'so far as is reasonably practicable'. This means that employers can argue that the costs of a particular safety measure are not justified by the reduction in risk that the measure would produce. But this comes with a sting in that the Health and Safety Executive, if they do not agree with an employer's view of what is reasonably practicable, can enforce the Act with penalties for non-compliance and by shutting down the site.

In addition, in construction we have the Construction (Design and Management) Regulations 2015. Quite clearly, safe ways of working are affected by the coronavirus, with the Principal Contractor being responsible for safe working on its site and having to submit method statements to the Principal Consultant.

So to **summarise** this section, at the moment, this compensation event – which has to be specified by the *Employer/Client* to apply – has not occurred. But a site which cannot comply with the 'guidance' may have to shut down and hence be a compensation event under clause 60.1(19), which is what we will look at next.

Stop press 1: On 7 April 2020, the Welsh Government added Regulation 6A to the Health Protection (Coronavirus Restrictions) (Wales) Regulations 2020. It states that 'A person responsible for work being carried out at premises' must 'take all reasonable measures to ensure that a distance of 2 metres is maintained between any persons on the premises' – i.e. it is now law to 'take all reasonable measures', not just guidance, and hence is a change in the law.

Stop press 2: Apparently, the Scottish Government ordered all non-essential sites to shut down late on 6 April 2020.[4] However, on the morning of 7 April, I could find no amendment to the Health Protection (Coronavirus) (Restrictions) (Scotland) Regulations 2020 to state this is the case.

THE FORCE MAJEURE CLAUSE OF NEC4: Clause 60.1(19) of both NEC3 and NEC4 is often referred to as the 'force majeure' clause, as well as the 'prevention clause'. It was added into NEC3 contracts to cover the occurrence of high-impact, low-probability events which would otherwise frustrate the contract, i.e. make the contract impossible to perform. Before we start, it is worthwhile pointing out that the words discussed below – in both NEC3 and NEC4 – are replicated in section 9 as a reason for terminating the contract. More on this later.

We will consider the wording in NEC4 first because it has cleared up an area of uncertainty from NEC3.

The wording of NEC4 clause 60.1(19) is:

> 60.1 The following events are compensation events.
>
> (19) An event which
>
> - stops the *Contractor* completing the whole of the *works* or
> - stops the *Contractor* completing the whole of the *works* by the date for planned Completion shown on the Accepted Programme,
>
> and which
>
> - neither Party could prevent,
> - an experienced contractor would have judged at the Contract Date to have such a small chance of occurring that it would have been unreasonable to have allowed for it and
> - is not one of the other compensation events stated in this contract.

So, taking the last three bullets first, if the Contract Date was back in January and certainly December, and probably the first half of February, it would be hard to argue that it would have been reasonable for the *Contractor* to have allowed for its effects in their Prices. Reversing this, if you are a contractor, I would currently absolutely not enter into a contract without an additional compensation event under clause 60.1(21) to cover the effects of coronavirus.

Going back to the first bullet point, unless there are unique circumstances, I also think it is hard to argue that the virus stops – as in makes impossible – the *Contractor* completing the whole of the *works*.

In the second bullet, 'Completion' is the equivalent of Practical Completion under the JCT and Substantial Completion under other civils-based forms. So once it becomes impossible to achieve the date in the *Contractor's* Accepted Programme for planned Completion, providing the last three bullets are satisfied, it becomes a compensation event. This would be after the *Contractor* has exhausted all options to achieve this date. This could include bringing in extra resource (if available), working longer hours and so on. I suggested in a LinkedIn discussion forum that this was equivalent to a 'best endeavours' requirement – i.e. if, despite the *Contractor's* proposed best endeavours, it becomes impossible to meet the *Contractor's* planned Completion as shown in their Accepted Programme, then it becomes a compensation event.

However, and very importantly, this comes into play not when planned Completion as shown in the Accepted Programme is actually exceeded, but when it is projected to become impossible to achieve.

So, for the compensation event to occur and hence become, from that point onwards, the *Client's* risk, the *Contractor* wants the planned Completion to be as early as possible and not drift backwards due to delays, whether caused by the effects of Coronavirus or other matters. Consequently, the wording of the second bullet encourages the *Contractor* either to not submit a revised programme for acceptance or, if they do, to make sure that it is unacceptable by the *Project Manager* – a perverse disincentive to actually project manage events which goes against its stated objective of 'stimulus to good management'.

As previously mentioned, sites are still allowed to work. But if a total lockdown was ordered, as has happened in various countries, then it would fairly rapidly become impossible to meet planned Completion and this compensation event would be triggered (if it hadn't been already).

One thing to note is the wording of clause 19 of the NEC4 Engineering and Construction Contract, which is, in my view, wrongly titled 'Prevention'.[5] It basically states that if the same circumstances occur as per clause 60.1(19) – it uses the same words as the five bullet points – then the *Project Manager* gives an instruction stating how the event is to be dealt with. This potentially takes us back to the first category of compensation events. Peter Higgins, chair of the NEC Steering Group, explains in an article and subsequent video[6] that this wording is used almost to force the *Project Manager* to start managing the contract once these circumstances occur. Or, to put in another way, once a clause 60.1(19) event occurs, the *Client* is going to be paying so they might as well provide leadership and direction to the *Contractor*, via its *Project Manager*, on how to deal with the consequences.

THE FORCE MAJEURE CLAUSE OF NEC3: In NEC3, clause 60.1(19) is substantially the same, but with some differences to the first two bullet points. I have illustrated this by crossing out the new NEC4 wording:

(19) An event which

- stops the *Contractor* completing ~~the whole of~~ the *works* or
- stops the *Contractor* completing ~~the whole of~~ the *works* by the date ~~for planned Completion~~ shown on the Accepted Programme,

and which …

The addition of the 'the whole of' the *works* to NEC4 probably only comes into play if option X7 for sectional Completion has been specified. Under NEC3, the second bullet point therefore applies for each section, whereas under NEC4 it is 'the whole of' the *works*. The NEC3 wording is therefore to the advantage of the *Contractor*.

In contrast, the addition of 'planned Completion' to the second bullet in NEC4, whilst aiding clarity, changes the meaning of the clause. Let me explain.

Under NEC3, there was sometimes – because clause 60.1(19) is a rare occurrence – a bit of an argument over whether 'by the date' meant planned Completion (as per NEC4) or the contractual Completion Date, i.e. the date at which liquidated damages for delay come into play. This is likely to be later than planned Completion and hence the *Employer's* preferred interpretation. Conversely, 'planned Completion' is going to be the *Contractor's* preferred interpretation.

Below are the arguments for the differing interpretations, some of which have been put forward on various discussion forums.

Arguments for the contractual 'Completion Date' as shown on the Accepted Programme:

1. The whole clause was added into NEC3 to avoid frustration of the contract (i.e. it becomes impossible to perform). The first bullet covers the impossibility of physically completing the *works*, while the second covers the impossibility of meeting the time obligations, i.e. by the contractual Completion Date. This, to me, has significant weight as an argument.

2. Planned Completion will naturally move around a bit from programme to programme due to better or worse productivities etc. It seems illogical that the date on which a force majeure event cuts in is due to *Contractor's* risks. This argument has some force.
A counter-argument to this point is that planned Completion will also move around, in part due to compensation events which are the *Employer's* risk. This partly neutralises the original argument.
However, the counter-counter-argument is that the contractual Completion Date will move around less and will only go one way, i.e. there is more certainty over the Completion Date.
So overall, while not a strong argument for the 'Completion Date', this point does have some sway.

3. The Guidance Notes published by the then NEC Panel gives insight into what they intended and would be persuasive in a court of law. The following relevant statements are picked out (with underlining for emphasis):

> A delay to planned Completion which can be recovered [...] does not stop the *Contractor* completing on time.
> The *Contractor* must demonstrate that that there is no reasonable means by which he can complete on time [...].

Contractually, 'on time' must mean the contractual date when the contract has to be delivered by, i.e. the Completion Date.

4. The NEC3 Engineering and Construction **Short** Contract, published at the same time as the full contract, also had a very similar compensation event added in to cover force majeure, but it does *not* have an Accepted Programme to be referenced. Instead it references the Completion Date, not planned Completion (or any other equivalent word for 'planned'). This also implies that the Completion Date is what was intended.

Argument for 'planned Completion' as shown on the Accepted Programme:

1. They clarified in NEC4 what 'they' meant as planned Completion, so it must be that. On the face of it, this has some weight. However, the counter-arguments are:
   a. NEC3 came out in June 2005 and NEC4 came out in June 2017. That's a gap of 12 years, with a significant turnover of people in the meantime. To claim 'what they really meant to say 12 years ago was . . .' is not provable without a detailed examination of the documents leading up to the publication of NEC3 or a very strong memory – because memories are fallible – from someone who was there. And for the latter, you would have to ask, if that was the stated intention, why they did not put the wording in the contract.
   b. If it was the intention, the Panel could have corrected it in the June 2006 and April 2013 editions of NEC3, but they did not.
   c. Arguments 3 and 4 for the 'Completion Date' further negate this point.

   Taken together, in my view, these counter-arguments negate this argument.
2. Assessment of a compensation event is based on reasonable timescales and this somehow equates to planned Completion because that is reasonable. I personally cannot make this argument stack up logically and, in any case, it seems to confuse assessment of the quantum of a compensation event – see clauses 63.3, 63.6 and 63.7 – with whether it is or is not a compensation event under clause 60.1.
3. 'If it meant the Completion Date, surely it would have said so', to which the obvious response is 'If it meant planned Completion (as shown in the Accepted Programme), surely it would have said so'. There is equal or no validity to both arguments, so they cancel each other out.
4. It would not apply if it was already impossible to achieve planned Completion by the contractual Completion Date, i.e. the *Contractor* was already in significant delay. Consequently, in this circumstance, all the risk for a force majeure type event would be on the *Contractor* and this cannot be what the NEC Panel intended. This is the most powerful argument for the planned Completion view, but it still has a counter-argument, which is: in the NEC3 Engineering and Construction **Short** Contract it states the Completion Date, so it was what the Panel intended under these contracts. And it is still the case in the NEC4 Engineering and Construction **Short** Contract.

   Further, in a recent court case[7] – not under NEC – the Court of Appeal held that if, but for the unexpected event, the relevant party would not have performed its obligations regardless, then that party's failure to perform cannot be said to have resulted from the unexpected event, and the force majeure clause did not provide relief. In other words, the Court of Appeal does not seem to endorse this argument.
5. The intention of the compensation event clauses is to ensure the *Contractor* is in no better or worse position than existed prior to the compensation event occurring. As someone involved in NEC's development from 1993, I can confirm this is true. But it is from the point at which the compensation event occurs, not from the position of the risk occurring and then all risk transferring to the *Employer*. For example, under both the physical conditions and weather compensation event clauses 60.1(12) and (13), the *Contractor* has to price for some risk and not assume perfect ground and perfect weather. But once the compensation event is triggered, the <u>additional</u> consequences are taken by the *Employer*.[8]

To **summarise**, under NEC3 there is no killer argument for it being the 'Completion Date' when the compensation event occurs, but all four reasons have some strength. In contrast, all four reasons for 'planned Completion' have reasonable counter-arguments, which reduce or eliminate their persuasiveness.

TERMINATION: Both in NEC3 and NEC4, the respective criteria stated in clause 60.1(19) and clause 19 are replicated as a reason R21 for termination in clause 91.7, but this can only be done by the *Employer/Client*. In this circumstance, the *Contractor* effectively gets paid for the work done (in accordance with the respective main payment options) and costs committed.

In addition, either Party may terminate if the *Project Manager* instructed the *Contractor* to stop or not start work and not restart within 13 weeks. As the pandemic is not either Party's fault, this would be for reason R20 in clause 91.6.

Whatever the reason, the golden rule is to follow the contract, as rather than ending up in a 'neutral' financial position, the *Employer/Client* may effectively be liable for loss of profit to the *Contractor* or the *Contractor* may be liable for the costs of the *Employer/Client* getting someone else in to finish the *works*.

ARTICLE CONCLUSION: It seems to me that we are close to the point at which compensation events will occur as a result of the coronavirus pandemic. We may even have passed the point on many sites. This could be because:

- the *Employer/Client*, via their *Project Manager*, wishes to proactively take control of the contract by changing the Works Information/Scope in some way or stopping the work,
- the *Employer/Client* or Other does not do something they should do under the contract,
- the government changes the law in some way – it could be a new regulation, a modification to the existing Coronavirus Act or orders for the closure of specified type of premises, i.e. construction or other related sites, or
- clause 60.1(19) cuts in, whether that is, under NEC4, at planned Completion as shown in the Accepted Programme or, under NEC3, most likely the Completion Date.

At this point, the financial and time consequences effectively transfer to the *Employer/Client*. In addition, without any compensation event, under the target and cost-reimbursable options, the *Employer/Client* is paying Defined Cost plus Fee. Under the target cost options C and D, this means the risk is currently shared under the pain/gain mechanism.

So given this, cooperation and good project management – responding to change as it occurs – seem to me paramount. And that starts with using the early warning mechanisms to notify matters and manage them.

Postscript: Steven Evans, another NEC expert in my associate group, has also written an article on this, titled 'NEC4 ECC and COVID-19'.[9] It focuses more on NEC4 and the contract processes, rather than the legal nuances of the words. The exception to this is his discussion on the fourth bullet of clause 60.1(19). It is certainly worth a read.

NOTES

1 In NEC3, there are two fee percentages, the *direct fee percentage* (applied to the *Contractor's* directly incurred costs) and the *subcontract fee percentage*; while in NEC4 there is only one, the *fee percentage*, which is applied to both.

2 Available at: http://www.constructionleadershipcouncil.co.uk/wp-content/uploads/2020/04/Site-Operating-Procedures-23-March-2020-v1.pdf.

3 Note: a version 2 was briefly published, which said the 2-metre rule should be strictly observed, but this was rapidly withdrawn due to feedback saying it was inoperable.

4 See *Construction Enquirer*, 6 April 2020. Available at: https://www.constructionenquirer.com/2020/04/06/scottish-sites-ordered-to-close-unless-on-essential-list/.

5 This is because the 'prevention principle' in English law is a general principle that a party cannot, in the absence of clear terms, take advantage of his own wrong. The specific words of this clause and 60.1(19) do not imply that the *Client* has done any wrong.

6 Article available at: https://neccmsmediaprod.azureedge.net/mediacontainer/nec/media/nec/document%20downloads/covid-19-and-nec4_1.pdf. Video available at: https://www.neccontract.com/About-NEC/News-and-Media/Covid-19-Guidance-Video.

7 *Classic Maritime Inc* v. *(1) Limbungan Makmur Sdn BHD and (2) Lion Diversified Holding BHD* (2019).

8 This point was added on 7 April 2020 following a comment from Steve Honey.

9 Available at: www.linkedin.com/pulse/nec4-ecc-covid-19-steven-c-evans/?trackingId=2Nl3%2BX39TvGL9y9ippOhjg%3D%3D.

ICE Publishing

Broome, Jon

ISBN 978-0-7277-6205-4
https://doi.org/10.1680/necaug.62054.313

Appendix 2
Write-up of workshop on collaborative working under NEC, held at the 2019 NEC People Conference

This is the work of approximately 40 people over an hour and 20 minutes and was co-facilitated with Neil Earnshaw.

During the workshop, the participants were split into seven mixed-up groups and were asked the following questions. There was a lot of overlap in their responses. This write-up has categorised the responses and attempted to take out obvious duplication while keeping the nuances.

Question 1

- What features of NEC help create collaboration?
- Briefly, how does the identified feature help create collaboration?

Flexibility

- Suite of contracts that can be adapted to suit the project circumstances.
- Flexibility of contracting options within the long forms: risk allocation can be appropriate for the circumstances of the contract.
- Draws out early upfront discussions on risks: what they are, how and when they can be managed and who is best placed to manage them.
- Flexibility forces thinking about best options to use etc.
- Joint objectives and outcomes – Parties' drivers, improved understanding of individuals.
- NEC Alliance Contract means joint stakeholder management and working towards common objectives.
- Option C pain/gain share helps align motivations.
- Options – flexibility to incentivise right/appropriate behaviours and focus.
- Secondary option X12 for multiparty collaboration – setting agreed joint targets.
- Option Z clauses – allow flexibility but need care, should be known before tender.
- Design extent varies in design and build – multi-disciplinary collaboration.

Clarity

- Prescriptive processes give clarity on status of any communication.
- Greater objectivity versus subjectivity – everyone is clearer where they stand.
- Stated reasons for non-acceptance reduce arguments.
- Clearer than JCT – easier for all participants to understand and use.
- Simplicity of language – aids collaboration by allowing more people to access the contract.
- Clear and plain English – much less need for lawyers to resolve contractual ambiguities.

Clause 10.1

- Sets the tone for the contract.

- Mutual trust and cooperation – encourages respect and good behaviour.
- Cooperation – giving the responsibility to work with others.
- (Mis)trust – mutual balance of understanding of positions.

Defined roles
- Project Manager – single point of contact: competent professionals = great; procrastinators = burden.
- Much clearer responsibilities.
- Clarity of authority and duties.
- Gives understanding of who/when somebody (Project Manager/Contractor) must take the lead in a given scenario – drives leadership in collaborative meetings
- Clarity over what everyone is expected to do and how the overall system operates.
- Ability to delegate – people closest to the issue are able to decide the course of action.
- Related to this: *key people* – leadership and culture, known quality of people.

Communications generally
- Clear to and from accountability, clear roles, formalises requirements, clears ambiguities, provides instruction.
- Transparency – each party able to see the other's actions.
- Key to collaboration (TALTEO!), notifications/information between parties, in agreed form = no verbal instruction.
- Set response times – prevents issues building up, applies to all.
- Timely management – keeping to defined timescales allows decisions to be made in a timely manner.
- Response times – ensures communication channels are opened, no procrastination.
- Timeframes – encourages people to act and therefore parties are clear about the status of issues.
- Time barring – promotes prompt responses, early warning notifying, as long as it is used to solve issues.
- Period for reply – sets timescales to respond promptly and effectively.
- Communication – formal (processes) and informal (meetings), promotes dialogue, regular meetings help understanding each other's drivers/issues.
- Regular meetings – openness and building trust, establishing common goal.
- Open communication – challenges all issues collectively, facing up to issues head-on.
- Records – note it, record it, move on in 'agreement', trading of decisions/actions.
- Stated criteria for acceptance – aids collaboration by parties accepting (not approving) how something will be done.

Quality
- Defect management – remedy or accept it.
- Defects procedure – promotes proactive cooperation between the *Supervisor*/*Contractor* and *Contractor*/Subcontractor.

Payment
- Default payment period – quicker cash flow to help promote relationships.
- Forecasting – certainty/understanding of costs.
- Open-book accounting – once the money is transparent there's not much more to hide, but does this help avoid Contractor losses?

- Project Bank Account – certainty of payments in a reduced timescale, protection against insolvency, how else do Contractors protect against loss of retention and defects in the supply chain?

Programme

- Helps in understanding risks and their reduction, provides a transparent and indisputable plan.
- The process of acceptance gets everyone on the same page, the heart of the contract, used as a tool to proactively manage the contract.
- Advises Client of its commitments early on
- Regular submissions drive collaborative reviews, programme detail explained as part of delay analysis.
- Regular programme agreement – no surprises or conflict at the end of the project.

Early warning (procedure)

- Gets issues on table that affect quality/time/cost, communication.
- Leads to early resolution of issues giving cost savings, culture, joint problem-solving, creates goodwill.
- They are conversation starters, can request risk reduction meetings to talk about the risk/issue and can request who attends meetings, allocate actions to mitigate/reduce risks.
- Stimulates a conversation about things that could go wrong, rather than hoping for the best.
- Promotes proactive rather than reactive risk management.
- Anyone can raise an early warning, bring attention to issues/opportunities when they arise to enable all to resolve/capitalise on, drives/encourages open dialogue/meetings to resolve issues.
- Risk reduction meetings mean both Parties have to share and understand how to close out risks.
- Risk reduction meetings actions are assigned with deadlines for completion, forces dialogue.

Compensation events

- Clear definitions reduce arguments.
- (N) compensation event – visible notification of change/cost/programme etc.
- Time for responses – structured process for change control, everyone knows what to expect.
- Early/prospective agreement, agree and move on, actual/forecast basis.
- Compensation event timeline and process – promotes good practice for managing change.
- Time barred response to compensation events means issues get dealt with and promotes actions by Client to reduce cost.
- Compensation event actual cost – Contractor cannot exploit Client.
- Contractor's margin stated and known upfront.
- Understanding and agreeing assumptions means event can be dealt with now and assessment corrected later if needed for mutual benefit, agreed way forward to overcome future uncertainty.
- Dividing date – clarity for basis of assessment and time focus.
- Prospective compensation events – on programme, aids understanding of direction of project.

Dispute resolution

- Senior representatives referral in NEC4 means disputes can be dealt with sensibly and promptly.
- Escalation procedure means there is clear avenue to resolve issue at a higher level.

Miscellaneous

- Subcontractor acceptances by project manager give feedback.
- Scope/Works Information – clarity on what in/what out.

Question 2

- What specific behaviours would you 'see' and 'hear' people doing when working collaboratively in accordance with an NEC contract?

Pre-contract

- Getting stakeholder input.
- Identifying and sharing objectives with the supply chain.
- Sharing risks/problems/challenges with the supply chain.
- Getting feedback early.
- Getting feedback on similar procurements.
- Using 360° feedback from similar projects.
- Listening and taking account of the feedback.

Pre-contract or during any Early Contractor Involvement phase (whether contractual or otherwise)

- Share project goals, including commercial ones.
- Share major constraints.
- Discussions about risks, assumptions and other areas of uncertainty.
- Shared working spaces, both real and virtual.
- Develop a behavioural charter together.
- Build in sufficient time for innovation and opportunity.
- A joint Site visit and 'walkthrough'.

Contract start-up

- People share insights into their character.
- Integrated team – everyone sharing.
- Generous listening to each other.
- Developing shared objectives.
- Social aspects: develop a programme of events.
- Everyone in one room.
- Read the contract together.
- Joint training.
- Joint understanding of programme, cost plan, risk etc.

Project organisation and systems

- Agreeing clear responsibilities.
- Integrated seating plan and organisational chart.
- Visual communications for all, e.g. information, goals, values.
- Single set of systems for accounts.
- Transparency of information.
- Joint IT systems give 'one version of the truth'.

Technical resolution of early warnings
- People empowered to make decisions.
- Be transparent and honest about what the problem or issue is.
- Don't take it personally, e.g. if your design is not right.
- Be open and approachable.
- Spend time to understand the problem from all perspectives.
- Feel free to ask if you don't understand.
- Be innovative and willing to challenge.
- Decide and agree next steps.

At an early warning meeting
- No absentees.
- Have the right people there, e.g. specialist Subcontractors.
- Understand personalities.
- No accusations to avoid blame/finger-pointing.
- Be factual.
- Be open and frank.
- Give people time to speak.
- Listen.
- Establish gaps in understanding.
- Focus on resolving the matter, not contract liability.
- Record actions and outcomes.
- Take responsibility for the actions.

Design meetings
- Have them regularly.
- Listen to others and be respectful of their expertise/experience.
- It wouldn't be obvious who works for whom.
- Be open to innovation.
- Encourage all to contribute.
- Positive body language.

Programme
- Have regular joint reviews.
- At the reviews, agree the Accepted Programme.
- Make sure the team understand what it is saying that and it is meaningful to them, i.e. they understand what they have to do to deliver it.
- It's a project programme for all parties, i.e. it is integrated.
- The team use it for constant communication.
- It helps the team to understand and discuss issues and find a way to overcome them.

Quality
- Strive for 'right first time'.
- The *Supervisor* contributes constructively and proactively to how work is done, but does not tell the *Contractor* how to do it.
- The *Supervisor* tells the *Contractor* if they have concerns as work is being done, rather than notifying a Defect when it has been done incorrectly.
- The *Contractor* does not hide Defects, but notifies and corrects promptly.

- An acceptance that mistakes are made.
- The *Supervisor* notifies Defects promptly.

Payment

- There is a pre-application meeting.
- A reliable forecast.
- The process is defined and transparent, as is the application.
- No 'red pen' items in application as it is already pre-agreed.
- Amount due is certified before it is due.
- Payment is made before it is due.

Compensation events

- Ongoing dialogue as a quotation is developed.
- No surprises.
- Risk is discussed and considered rationally.
- Open with challenges.
- Timely responses by both the *Contractor* and the *Project Manager*/QS.

Completion

- If not stated in the Scope, agree specifics early on.
- Involve the end user.
- Do the above for sectional Completions.
- Differentiate, if necessary, between take over and Completion.
- Agree actions.
- Honour commitments.
- Learn from and share what worked both for any remaining (sectional) Completions and future contracts.
- Celebrate!

Disputes

- Don't hide differences of opinion: recognise and resolve promptly.
- Resolve at the lowest, but right, level.
- Have a disputes ladder.
- Have respect for the other Party's and people's positions, knowledge and perspectives.
- Listen to understand.
- Be prepared to compromise – give and take.
- Agree a way forward.
- Address root causes so the dispute does not happen again, which includes being honest about your part in it.
- Having resolved it, move on: don't hold grudges.

ice Publishing

Broome, Jon
ISBN 978-0-7277-6205-4
https://doi.org/10.1680/necaug.62054.319

Index

Tables are indicated by an italic *t* following the page number; figures are indicated by an italic *f*, footnotes are indicated by an italic *n*.